FLORA OF BHUTAN

INCLUDING A RECORD OF PLANTS FROM SIKKIM

VOLUME 1 PART 2

A.J.C. GRIERSON & D.G. LONG

འབྲུག་གི་རྩ་ཤིང་སྔོ་ཚོགས།

ROYAL BOTANIC GARDEN, EDINBURGH.
1984.

Published by the Royal Botanic Garden, Edinburgh, for the Overseas Development Administration, Eland House, Stag Place, London, and the Royal Government of Bhutan.

ISBN 0 9504270 2 0

Typeset by PGA Studios (Edinburgh) Ltd. and printed by Econoprint, 42a Albany Street, Edinburgh.
Marketing Agents: Wheldon & Wesley Ltd., Lytton Lodge, Codicote, Hitchin, Herts. SG4 8TE, England.

CONTENTS

ADDITIONAL BIBLIOGRPAHY 130–139

(For bibliography 1–129 see Vol. I Part 1 pages 37–43)

130. Ayensu, E.S., et al. 1980. *Firewood Crops*. Washington.
131. Bennet, S.S.R. 1983. Ethnobotanical studies in Sikkim. *Indian Forester* 109: 477–481.
132. Das, S. & Ghosh, R.B. 1982. A preliminary census and systematic survey of climbing taxa of West Bengal with special reference to their importance. *J. Econ. Tax. Bot.* 3: 565–574.
133. Ghosh, R.B. & Ghosh, A. 1978. Some additions to the flora of Buxa division of Jalpaiguri district of West Bengal. *Bull. Bot. Soc. Bengal* 31: 78–83.
134. Grierson, A.J.C. & Long, D.G. 1983. *Flora of Bhutan* Vol. I. Part 1. Edinburgh.
135. Hara, H., Chater, A.O. & Williams, L.H.J. 1982. *An Enumeration of the Flowering Plants of Nepal* Vol. 3. London.
136. Malla, S.B, (ed.) 1976. Flora of Langtang. *Bull. Dept. Med. Plants Nepal* No. 6. Kathmandu.
137. Mathew, B. 1983. Plant hunting in Sikkim. *Greenhouse & Garden* Dec. 1983: 30–33.
138. Tamang, K.K. & Yonzone, G.S. 1982. A brief note on the vegetation from Lebong to Tiger Hill in Darjeeling. *J. Bengal Nat. Hist. Soc.* N.S. 1(1): 93–95.
139. Yonzone, R. & Mandal, S. 1982. A preliminary report on the ethnobotanical survey of Kalimpong in Darjeeling district, West Bengal. *J. Bengal Nat. Hist. Soc.* N.S. 1(1): 96–99.

Family 25. PHYTOLACCACEAE

by A.J.C. Grierson & D.G. Long

Fleshy perennial herbs. Leaves alternate, pinnately veined, entire, exstipulate. Flowers in leaf-opposed or terminal bracteate racemes, bisexual, actinomorphic. Perianth of 4–5 segments, connate at base. Stamens 8–12, inserted at base of perianth, filaments persistent. Ovary superior, subglobose, of 6–8 carpels connate only at base, ovule 1 per carpel, style filiform, terminal on each carpel. Fruit berry-like, subglobose, of 6–8 fleshy 1-seeded carpels.

1. PHYTOLACCA L.

Description as for Phytolaccaceae.

1. P. acinosa Roxb. Dz: *Kashakani*. Fig. 17m,n.

Stems erect, often purplish, 45–130cm. Leaves elliptic, 8–30 × 2.5–12cm, acute, base attenuate, entire, glabrous but with numerous minute linear cystoliths on lower surface visible when dry; petioles 1–4cm. Racemes dense, 8–20cm, pinkish; pedicels 6–10mm, bracts and bracteoles linear. Perianth segments elliptic, c 5 × 3mm, obtuse, green at first becoming purplish. Fruit depressed-globose, 8–10mm across; carpels c 4mm, dark purple.

Bhutan: S—Chukka district, **C**—Thimphu, Punakha, Tongsa, Bumthang and Tashigang districts; **Sikkim**. Clearings and roadsides in Warm broad-leaved forests, around cultivation and habitation, 900–2500m. April–September.

Family 26. NYCTAGINACEAE

by A.J.C. Grierson & D.G. Long

Perennial herbs or climbing shrubs. Leaves opposite or alternate, simple, entire, pinnately veined, exstipulate, with minute linear cystoliths visible when dry. Flowers in bracteate cymes; bracts small, free, or large and obscuring flowers, or connate into a calyx-like involucre or cup; actinomorphic, bisexual. Perianth tubular, of 5 fused petal-like segments, base of tube constricted and persisting around ovary. Stamens 1–10, filaments united at base into a collar surrounding ovary. Ovary superior (sometimes appearing inferior by constriction of perianth tube), 1-celled, ovule 1, style simple. Fruit a 1-seeded achene surrounded by persistent base of perianth tube.

1. Woody climbers; stems spiny; leaves alternate; flowers concealed by large showy bracts .... **3. Bougainvillea**

\+ Unarmed herbs; leaves opposite; flowers not concealed by bracts 2

2. Flowers in corymbose cymes or panicles; bracts conspicuous, united into a calyx-like involucre; fruit ovoid, eglandular .... **1. Mirabilis**
\+ Flowers in small dense subumbellate cymes; bracts minute, free; fruit narrowly obovoid, glandular-pubescent .... **2. Boerhavia**

1. MIRABILIS L.

Erect herbs. Leaves opposite. Flowers in terminal corymbose cymes or panicles, bracts united into a 5-lobed persistent calyx-like involucre. Perianth funnel-shaped, trumpet-shaped or campanulate. Stamens 3–6, unequal, exserted. Ovary with slender exserted style, stigma capitate. Fruit ovoid, eglandular.

1. Flowers in corymbose cymes; perianth large, 45–50mm long...**1. M. jalapa**
\+ Flowers in panicles; perianth short, c 5.5mm long ....**2. M. himalaica**

1. M. jalapa L. Eng: *Four o'clock, Marvel of Peru*. Fig. 17e,f.

Robust herb, 60–150cm. Leaves triangular-ovate, 4.5–10 × 3–5cm, acuminate, base truncate, glabrous or sparsely puberulous; petioles 1.5–3cm. Involucre campanulate, 1–1.3cm, divided to middle into 5 acute teeth. Perianth red or white, trumpet-shaped or narrowly funnel-shaped, tube 4.5–5cm, limb 5-lobed, 2.5–3.5cm across. Stamens 5–6. Fruit 7–8 × 5mm.

Bhutan: **S**—Phuntsholing district (Phuntsholing), **C**—Tashigang district (Tashigang). Cultivated in gardens and naturalised, 300–900m. April–November.

Native to tropical America; cultivated as an ornamental.

2. M. himalaica (Edgeworth) Heimerl; *Oxybaphus himalaicus* Edgeworth

Similar to *M. jalapa* but more pubescent; flowers in lax leafy panicles; involucre c 4mm, glandular-pubescent; perianth smaller, campanulate, c 5.5mm, pink; stamens 3–5; fruit compressed-ovoid, c 5.5 × 3mm.

Bhutan: **C**—Thimphu district (Paro). 2500m. July–September.

FIG. 17. **Nyctaginaceae, Basellaceae, Chenopodiaceae** and **Phytolaccaceae. Nyctaginaceae.** a–c, *Boerhavia coccinea*: a, portion of shoot with leaves and inflorescences; b, dissected flower; c, fruit. d, *Bougainvillea spectabilis*: inflorescence. e & f, *Mirabilis jalapa*: e, leaf and flower; f, dissection of base of flower. **Basellaceae.** g & h, *Basella alba*: g, portion of shoot with leaves and inflorescences; h, dissected flower. **Chenopodiaceae**. i & j, *Axyris prostrata*: i, dissected male flower; j, dissected female flower. k & l, *Chenopodium ambrosioides*: k, portion of shoot with leaves and inflorescences; l, flower. **Phytolaccaceae**. m & n, *Phytolacca acinosa*: m, flower; n, fruit. Scale: a, d × ⅔; g, k × 4/5; e × 1; f × 2; m, n × 2½; h × 7; c × 8; b × 12; l × 17; i, j × 20. Drawn by J. Chisholm.

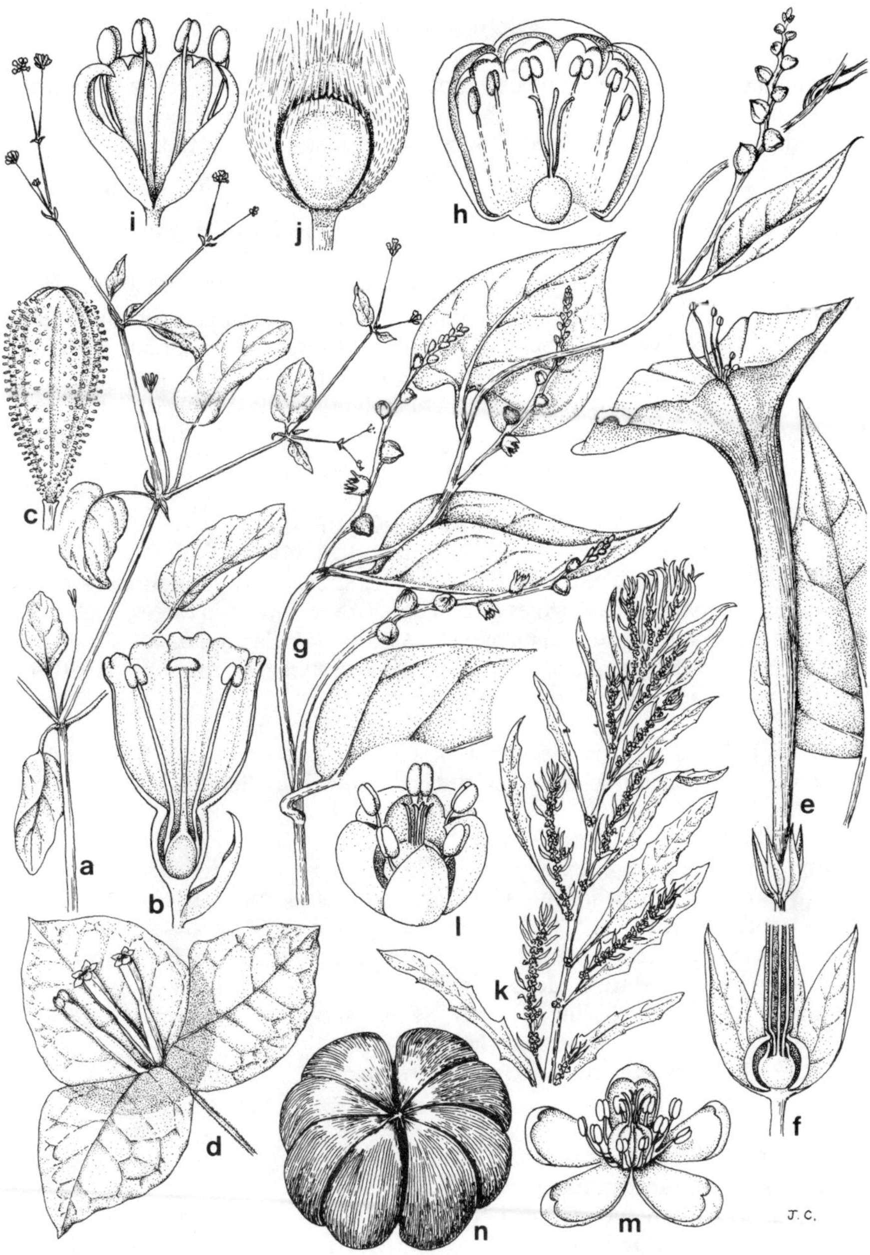
i
j
h
c
g
a
b
e
l
k
f
d
n
m
J. C.

2. BOERHAVIA L.

Perennial herbs with stout woody rootstock. Leaves opposite. Flowers in small dense axillary subumbellate cymes; bracts minute. Perianth campanulate, shallowly 5-lobed. Stamens 1–3, exserted. Ovary ellipsoid, style slender, exserted; stigma capitate. Fruit narrowly obovoid, glandular-pubescent.

1. B. coccinea Miller; *B. diffusa* auct. non L., *B. repens* L. var. *diffusa* sensu F.B.I. non (L.) Hook. f. Fig. 17a–c.

Stems 30–60cm, diffuse. Leaves ovate-elliptic to broadly ovate, 2.5–5 × 2–4cm, obtuse or subacute, base rounded or cordate, margins shallowly sinuate, subglabrous; petioles 1–2.5cm. Cymes 3–5-flowered; peduncles slender 1.8–3.5cm; bracts c 1mm. Perianth purple, c 3mm. Fruit 3–3.5mm.

Bhutan: S—Phuntsholing district (Phuntsholing), **C**—Tashigang district (Gamri Chu and Dangme Chu); **Sikkim**. Roadsides and disturbed ground, 300–1050m. April–August.

3. BOUGAINVILLEA Jussieu

Climbing shrubs with recurved supra-axillary spines. Leaves alternate. Flowers in terminal, leafy, bracteate cymes or panicles. Bracts showy, in threes, each bearing and concealing a flower on its midrib. Perianth tubular, angular, limb shortly 5-lobed. Stamens 8–10, not exserted. Ovary linear, style filiform, stigma feathery. Fruit ellipsoid, angular, often not developing.

1. Leaves pubescent; perianth tube pubescent ....**1. B. spectabilis**
\+ Leaves sparsely pubescent; perianth tube minutely puberulous ..**2. B. glabra**

1. B. spectabilis Willdenow. Fig. 17d.

Large climber. Leaves ovate or broadly elliptic, 3–5 × 2–3.5cm, abruptly acute, base rounded or somewhat attenuate, crispate pubescent; petioles 0.5–1.5cm. Bracts papery, ovate, 3–4.5 × 2.5–3cm, obtuse or subacute, base cordate, red, purple or orange, sessile, bearing a single flower below middle. Perianth c 2cm, pubescent. Fruit 11–14mm, densely hairy.

Bhutan: S—Phuntsholing district (Phuntsholing); **Sikkim:** Darjeeling. Cultivated in gardens, 200–250m. February–May.

Native of S America cultivated as an ornamental.

2. B. glabra Choisy

Very closely allied to *B. spectabilis* but leaves sparsely pubescent, more attenuate at base; perianth tube minutely puberulous.

Bhutan: S—Phuntsholing district (Phuntsholing) and Gaylegphug district (Gaylegphug); **Sikkim:** Kurseong. Cultivated in gardens, 200–300m. February–May.

Family 27. AIZOACEAE

by A.J.C. Grierson & D.G. Long

Annual herbs, sometimes stellately pubescent. Leaves simple, entire, basal or on stems, opposite or apparently whorled, exstipulate. Flowers in axillary clusters or in much-branched cymes, actinomorphic, hermaphrodite. Perianth segments 5, free, persistent. Stamens 3–20, free or connate at base. Ovary superior, 3–5-celled, placentation axile. Styles 3–5. Capsule 10–many-seeded, enclosed by perianth and falling off with it.

1. Prostrate, greyish stellate-pubescent herbs; flowers in axillary clusters **1. Glinus**
\+ Erect, glabrous herbs; flowers in much-branched cymes ....**2. Mollugo**

1. GLINUS L.

Prostrate stellately pubescent annuals. Flowers few in axillary clusters, shortly pedicellate. Stamens (3–)5–10(–20), free or, when numerous, fascicled. Styles small, 3–5-lobed. Seeds numerous, hilum with a long slender appendage almost encircling the seed.

1. G. lotoides L.; *Mollugo hirta* Thunberg. Fig. 18g.
Stems 15–90cm, greyish pubescent. Leaves obovate, 1–2.5 × 0.5–2cm, acute, base cuneate, pubescent on both surfaces or the upper often glabrescent; petioles 5–10mm. Flowers in fascicles of 3–5(–8). Perianth segments ovate, c 6mm, acute, greenish, scarious-margined. Stamens whitish, c 4mm. Ovary ovoid, c 5mm.
Bhutan: S—Phuntsholing district (Torsa River). On dry shingle, 200m. May.

2. MOLLUGO L.

Erect, slender, glabrous annuals. Leaves basal or on stems. Flowers in terminal much-branched cymes, minutely bracteate. Stamens 3–5. Ovary ovoid; styles 3, minute, filiform. Capsule membranous, 3-celled; seeds without appendage.

1. Leaves all basal, oblanceolate, obtuse ....**1. M. nudicaulis**
\+ Stems leafy throughout; leaves elliptic, acute......................**2. M. stricta**

1. M. nudicaulis Lamarck. Fig. 18h,i.
Leaves all basal, oblanceolate, 2–4 × 0.7–1.5cm, obtuse, base attenuate, subsessile. Flowering stems 10–25cm, leafless, cymosely branched from middle. Perianth segments elliptic, c 3mm, pinkish. Capsule ellipsoid, ± as long as perianth.
Bhutan: C—Tashigang district (Dangme Chu). Dry slopes, 900m. August.

2. M. stricta L.; *M. pentaphylla* auct. p.p. non L.

Stems 10–20(–30)cm, much branched. Leaves elliptic, 1.5–3.5 × 0.2–1cm, acute, base cuneate, subsessile. Perianth segments elliptic, c 2mm, whitish. Capsule subglobose.

Bhutan: S—Gaylegphug district (Gaylegphug), **C**—Punakha district (Nahi), Mongar district (Lingmethang) and Tashigang district (Shali and Gamri Chu valleys); **Sikkim.** Riverbanks and cultivated ground, 300–1600m. May–July.

Records of *M. pentaphylla* L. (a mostly S Indian plant with obovate leaves) from the E Himalaya refer to this species.

Family 28. PORTULACACEAE

by A.J.C. Grierson & D.G. Long

Decumbent or prostrate fleshy annuals. Leaves simple, entire, subopposite or spirally arranged, shortly petiolate, exstipulate but with a tuft of hair in the axils, otherwise glabrous. Flowers actinomorphic, bisexual, in axillary or terminal clusters surrounded by bracteoles and hairs. Sepals 2, keeled above, connate at base. Petals 4–6, obovate, connate at base. Stamens (7–)10–numerous, adnate to petals. Ovary semi-inferior, 1-celled, placentation free central; style 4–9-branched. Capsule surrounded by persistent calyx base, dehiscing transversely by means of a cap. Seeds laterally compressed, kidney-shaped, minutely tuberculate.

1. PORTULACA L.

Description as for Portulacaceae.

1. P. oleracea L. Sha: *Phagpa Jakpo*; Eng: *Purslane*. Fig. 18j–m.

Stems 15–30cm, decumbent or prostrate. Leaves obovate or spathulate, fleshy, 1–2.5 × 0.5–2cm, rounded or slightly emarginate; base cuneate; axillary hairs c 1mm; petioles 2–8mm. Flowers mostly 5–10 in clusters. Sepals c 6mm, strongly keeled in upper half. Petals 5–7mm, yellow. Stamens 7–10(–15). Style c 3mm, 4–5-branched from middle. Capsule ovoid, c 4 × 3mm.

Bhutan: S—Phuntsholing district (Phuntsholing), **C**—Punakha district (Punakha) and Tongsa district (Mangde Chu); **Sikkim.** Waste and cultivated ground in foothills, 200–1370m. January–December.

Two subspecies are known from Bhutan: subsp. **oleracea** with seeds longer than 0.85mm and subsp. **granulato-stellulata** (Poellnitz) Danin & H.G. Baker with seeds smaller than 0.85mm.

2. P. pilosa L.

Creeping herb rooting at nodes. Leaves linear-elliptic, 1–2.5 × 0.2cm, acuminate, base attenuate; petioles c 2mm; axillary hairs 3–4mm. Flowers 3–5 in terminal clusters. Sepals c 8mm, weakly keeled at apex, margins scarious. Petals c 2.5 × 2cm, orange, red or purplish. Style branches 5–9. Capsule globose, c 5mm.

Bhutan: C—Tashigang district (Tashigang). Cultivated in garden, 1400m.

Native of Tropical America grown as an ornamental. The above description refers to the large-flowered subsp. **grandiflora** (Hooker) Geesink, to which the Bhutan plant belongs.

Family 29. BASELLACEAE

by A.J.C. Grierson & D.G. Long

Fleshy, glabrous, perennial twining herbs. Leaves alternate, entire, pinnately veined, exstipulate. Flowers in axillary spikes, actinomorphic, bisexual, each subtended by a minute deciduous bract and 2 bracteoles connate with perianth tube. Perianth segments 5, connate and tubular at base, persistent. Stamens 5, borne on perianth segments. Ovary superior, 1-celled; ovule 1; styles 3, linear. Fruit indehiscent, 1-seeded, surrounded by persistent fleshy perianth.

1. BASELLA L.

Description as for Basellaceae.

1. B. alba L.; *B. rubra* L. Fig. 17g,h.

Leaves broadly ovate, sometimes elliptic above, 1.5–9(–12) × 1–6(–9)cm, acute, base rounded or cordate; petioles 1–3cm. Spikes 5–15cm, bearing flowers in upper half. Bracteoles and perianth 3–4mm, segments obtuse, pinkish, tube green. Fruit subglobose, 4–10mm diameter, black.

Bhutan: S—Samchi district (Dhoan Khola); **Sikkim**: Pankabari. Cultivated ground, 300–400m. September–February.

Cultivated as a vegetable and used medicinally.

Family 30. CARYOPHYLLACEAE

by A.J.C. Grierson

Annual, biennial or perennial herbs. Leaves opposite, simple, entire, rarely serrulate, often narrowly connate at base; stipules scarious or absent. Flowers actinomorphic, usually bisexual, solitary or in cymes. Sepals 4–5, free or united into a tube, often with scarious margins. Petals usually as many as sepals, sometimes with a basal claw concealed within calyx, sometimes small or absent. Stamens up to 10, free, filaments of those opposite sepals sometimes with a glandular thickening at base. Ovary superior, 1-celled or rarely incompletely 3-celled at base, ovules 2–many, placentation free-central, styles free or connate. Receptacle sometimes elongated bearing petals, stamens and ovary. Fruit a capsule opening by apical tooth-like valves, rarely fleshy and dehiscing irregularly.

1. Sepals united 2
\+ Sepals free to base 6

2. Epicalyx of 4 bracts present ....**1. Dianthus**
\+ Epicalyx absent 3

3. Petals borne on calyx tube; plants forming hard cushions; leaves densely crowded ....**11. Thylacospermum**
\+ Petals not borne on calyx tube; plants not cushion-forming; leaves not densely crowded 4

4. Petals without scales ....**2. Gypsophila**
\+ Petals bearing 2 scales at base of limb 5

5. Fruit a dry capsule dehiscing by 3–6 valves ....**3. Silene**
\+ Fruit fleshy, dehiscing irregularly ....**4. Cucubalus**

6. Petals present 7
\+ Petals absent 17

7. Petals distinctly notched or bifid 8
\+ Petals entire or subentire, rarely toothed or lacerate (sometimes notched in *Pseudostellaria* but then rhizomes bearing small tubers) 10

8. Petals notched at apex, not divided to middle ....**5. Cerastium**
\+ Petals bifid at least to middle 9

9. Stipules absent ....**6. Stellaria**
\+ Stipules present, entire or lacerate..................................**13. Drymaria**

10. Plants forming hard cushions; stems and leaves of previous years persisting and densely crowded ....**9. Arenaria**
\+ Plants not cushion-forming; stems and leaves of previous years not persisting and crowded 11

11. Leaves ovate, elliptic or oblanceolate 12
\+ Leaves linear or linear-lanceolate 15

12. Stipules present, scarious ....**14. Polycarpon**
\+ Stipules absent 13

13. Climbing herbs; petioles becoming recurved and hardened; sepals entirely scarious ....**8. Brachystemma**
\+ Erect or prostrate herbs; petioles where present not becoming recurved and hardened; sepals herbaceous in middle 14

14. Rhizomes bearing small tubers; styles filiform, minutely capitate **7. Pseudostellaria**
\+ Rhizomes absent or without tubers; styles filiform not minutely capitate **9. Arenaria**

15. Leaves long, up to 30mm, in dense axillary fascicles ....**12. Spergula**
\+ Leaves short, less than 15mm, not in dense axillary fascicles 16

16. Sepals acute or acuminate ....**9. Arenaria**
\+ Sepals obtuse ....**10. Sagina**

17. Sepals obtuse ....**10. Sagina**
\+ Sepals acute or acuminate 18

18. Suberect or prostrate herbs; stems more than 10cm ....**6. Stellaria**
\+ Erect or suberect herbs; stems less than 10cm ....**9. Arenaria**

1. DIANTHUS L.

Biennial or perennial herbs. Leaves elliptic or lanceolate, sometimes strongly connate at base. Flowers in diffuse or corymbose cymes. Calyx tubular, 5-toothed, finely striate, base surrounded by an epicalyx of 4 imbricate bracts. Petals with a narrow basal claw broadened at apex into broadly obovate or fan-shaped limb, toothed or fimbriate at the margin. Stamens 10. Receptacle elongated, c 3–4mm, bearing petals, stamens and ovary at apex. Styles 2, filiform. Capsules oblong, 4-valved.

1. D. chinensis L. Eng: *Chinese* or *Indian Pink*.

Stems 15–45cm. Leaves elliptic, 5–8 × 0.5–1cm, acute, subsessile, margins smooth, entire. Flowers in loose cymes. Calyx 1.7–2cm, basal bracts linear 1.3–1.5cm. Petal limb broadly obovate, 1.5–2cm, reddish-purple, white or streaked, margin dentate, sometimes deeply so.

Bhutan: C–Punakha district (Wangdu Phodrang) and Tashigang district (Tashigang). Cultivated, 1350–1450m. April–June.

Native of China, cultivated in gardens as an ornamental.

2. D. barbatus L. Eng: *Sweet William*.

Similar to *D. chinensis* but leaves lanceolate, 6–10 × 1–2.5cm, deeply (c 5–7mm) connate at base, with margins finely ciliate-serrulate; flowers in dense corymbs; calyx 1.5–1.7cm, basal bracts ± as long, finely ciliate-serrulate; petal-limb fan-shaped, 5–10mm, white or crimson, margin dentate.

Bhutan: C–Thimphu district (cultivated around Thimphu). August.

Native of Europe; cultivated as an ornamental.

2. GYPSOPHILA L.

Perennial herbs. Leaves oblanceolate or obovate. Flowers in lax leafy cymes. Calyx campanulate, 5-lobed. Petals 5, limb entire, narrowed to a basal claw ± as long as calyx. Stamens 10. Receptacle not elongated. Ovary ellipsoid; styles 2, filiform. Capsule 4-valved in upper half.

1. G. cerastioides D. Don

Prostrate herb; stems 7–30cm, often reddish, sparsely pubescent. Basal leaves oblanceolate, 1.5–5 × 0.4–1.3cm, acute, base attenuate, sparsely pubescent on both surfaces especially at margins; stem leaves obovate, 0.7–1.3 × 0.3–0.7cm, acute or obtuse, subsessile. Calyx c 5mm, divided to middle into 5 obtuse lobes, purplish at apex. Petal-limb c 5 × 3.5mm, white or mauve or white with purple streaks. Capsule c 6mm.

Bhutan: C – Ha to Mongar districts, **N** – Upper Mo Chu to Upper Kulong Chu districts; **Sikkim**. Common by streams and on grassy hillsides, 2750–4700m. April–July.

3. SILENE L.

Annual, biennial or perennial herbs. Leaves ovate, elliptic or linear-lanceolate. Flowers solitary or in cymes, sometimes corymbose. Calyx tubular or campanulate, sometimes inflated, 5-toothed, 10- or rarely many-nerved. Petals 5, limb emarginate or bifid, usually with 2 scales at the base; claw bearing an auricle on each side at apex. Stamens 10, 5 usually adnate to petals. Receptacle short or elongated, bearing petals, stamens and ovary at apex. Ovary 1-celled above but incompletely 3-celled at base; styles 3–5, filiform. Capsule 6–10-valved.

1. Flowers numerous in compact corymbose cymes .... **12. S. armeria**
\+ Flowers solitary or several in loose elongate cymes 2
2. Lower leaves ovate-oblong or oblanceolate, 2–4(–8) × as long as broad; mostly 1–2cm broad (but in *S. julaensis* 2–3mm broad) 3
\+ Lower leaves narrowly elliptic, linear-lanceolate or narrowly oblanceolate, more than 10 × as long as broad, 3–7mm broad 5

FIG. 18. **Caryophyllaceae, Aizoaceae** and **Portulacaceae. Caryophyllaceae.** a, *Arenaria glanduligera*: portion of shoot with leaves and flowers. b & c, *Cerastium glomeratum*: b, dehisced capsule with persistent calyx and corolla; c, petal. d, *Stellaria vestita*: petal. e & f, *Silene nigrescens*: e, plant in flower; f, petal. **Aizoaceae**. g, *Glinus lotoides*: dissected flower. h & i, *Mollugo nudicaulis*: h, plant in flower; i, dissected flower. **Portulacaceae**. j–m, *Portulaca oleracea*: j, flower; k, infructescence with dehisced capsules; l, cap-like upper part of capsule surrounded by persistent calyx; m, seed. Scale: h × ⅔; e × 1; f,i × 2; a × 2½; b × 3; j × 3½; k, l × 4; g × 4½; c, d × 7; m × 8. Drawn by M. Bates.

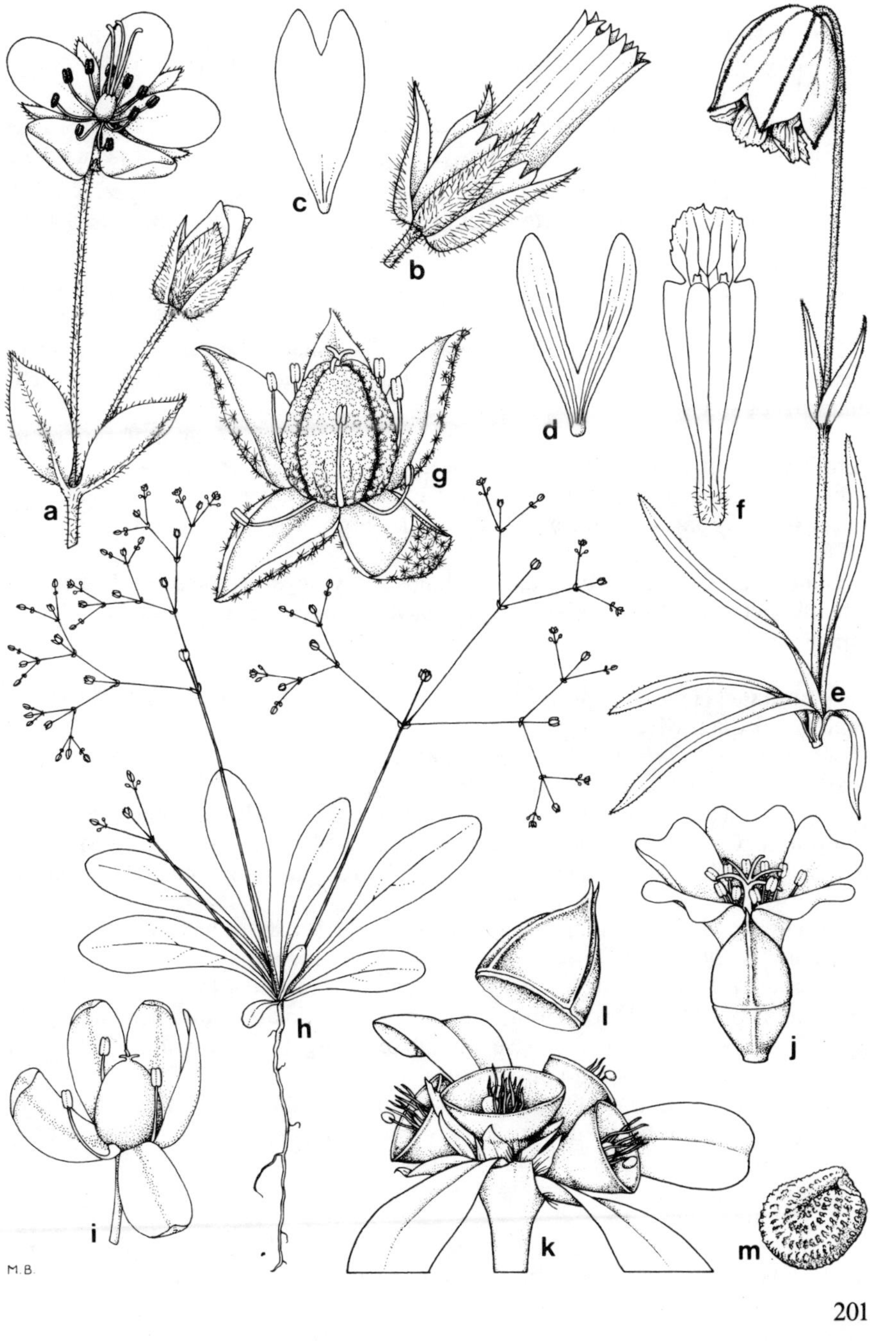
a
b
c
d
e
f
g
h
i
j
k
l
m
M.B.

3. Calyces cylindrical, 3 × as long as broad; scrambling herb ..**11. S. stracheyi**
+ Calyces campanulate to subglobose, 1–2 × as long as broad; erect or ascending herbs 4
4. Flowers several in loose corymbs ....**Species 1–3**
+ Flower solitary ....**Species 4–6**
5. Petals longer than calyx, exserted ....**Species 7 & 8**
+ Petals not or scarcely exserted from calyx ....**Species 9 & 10**

1. S. indica Otth; *Lychnis indica* (Otth) Bentham, *L. nutans* Bentham, *S. thomsonii* Majumdar

Perennial herb. Stems 20–70cm, whitish glandular-pubescent. Leaves all on stems, ovate-lanceolate, 2–6 × 1–2.25cm, acute or acuminate, base rounded, glabrous or sparsely pubescent. Cymes (3–)7–20-flowered. Flowers nodding at first, becoming erect. Calyx green, herbaceous, 9–12 × 5–7mm in flower, up to 10mm broad in fruit, glandular-pubescent on the nerves, teeth ovate, 3–4mm, subobtuse, green or blackish. Petal limb exserted 3–4mm from calyx, green or purplish. Receptacle scarcely elongated, 1–2mm. Capsules ovoid, 10–12 × 7–8mm.

Bhutan: C–Thimphu, Tongsa, Bumthang and Mongar districts, **N**–Upper Mo Chu district (Shingke La); **Sikkim**. Streamsides in open Fir forests, 2440–4000m. June–August.

2. S. bhutanica (W. W. Smith) Majumdar; *S. indica* Otth var. *bhutanica* (W. W. Smith) Bocquet. Med: *Sukpa, Sugpa*.

Similar to *S. indica* but often blackish glandular-pubescent; calyx larger, 1.6–2cm, somewhat inflated, 1cm broad in flower and up to 1.5cm broad in fruit, pale green, membranous, ribs conspicuous, blackish, densely glandular-pubescent.

Bhutan: C–Thimphu district (Barshong and Shodug), **N**–Upper Mo Chu (Chebsi); **Sikkim**. Alpine meadows and cliff ledges, 3650–3960m. July–August.

3. S. nepalensis Majumdar; *Lychnis multicaulis* Royle, non *Silene multicaulis* Gussone, *L. brachypetala* sensu F.B.I. p.p., non Hornemann.

Similar to *S. indica* but usually more slender; stems 20–45(–85)cm, whitish eglandular-pubescent throughout; leaves linear, lanceolate or oblanceolate, 5–11 × 0.5–0.9(–1.5)cm, acute, base attenuate, margins slightly inrolled; calyx herbaceous, green, ribs not conspicuous.

Bhutan: C–Thimphu district (Chapcha, Paro, Taba), and Punakha district (Tinlegang); **Chumbi** ; **Sikkim**. Grassy banks in dry valleys, 2130–3040m. June–July.

4. S. birgittae Bocquet

Stems 7–25cm, pubescent. Leaves elliptic or oblanceolate, 3–6 × 0.7–1.5cm, acute, base attenuate, pubescent. Calyx ± inflated, 1.5–2.3 × 1.3–2.3cm, pale green with densely blackish-pubescent ribs. Petals brownish-purple, exserted 3–6mm from calyx mouth, shallowly bifid and bluntly toothed at apex.

Bhutan: C –Tongsa district (Omta Tso and Changsethang), **N** – Upper Kulong Chu district (Me La). Alpine cliffs and stony slopes, 3960–4880m. July–August.

5. S. linae Bocquet

Similar to *S. birgittae* but smaller; stems 3–6cm; leaves elliptic or oblanceolate, 2–3 × 0.7–1cm, acute or obtuse, ciliate; calyx less inflated, 2 × 1–1.5cm, ribs pale-pubescent; petals pink, deeply bifid and with narrow teeth at apex, exserted 7–10mm from calyx.

Bhutan: C – Punakha district (Upper Tang Chu valley) and Tongsa district (Rinchen Chu valley). Alpine turf, scree and cliffs, 4265–4730m. August.

6. S. julaensis Grierson

Slender herb 4–10cm, with densely purplish-pubescent stems. Leaves ovate-oblong, 5–7 × 2–3mm, acute, base attenuate, purplish-ciliate. Calyx campanulate, scarcely inflated, 7–8 × 4–5mm, purplish-tinged, pubescent. Petals yellow, scarcely exserted from calyx, truncate or sub-bilobed at apex.

Bhutan: N – Upper Mangde Chu district (Ju La). Open grassy slopes, 4265m. July.

7. S. nigrescens (Edgeworth) Majumdar; *Lychnis nigrescens* Edgeworth. Fig. 18e,f.

Stems 8–20cm, blackish pubescent above. Leaves linear-lanceolate, 3–8 × 0.3–0.6cm, acuminate, base attenuate, ciliate. Flowers solitary. Calyx inflated, 1.7–2 × 1.3–1.8cm, pale green with dark purplish-pubescent ribs. Petals pink, exserted 4–5mm from calyx, shortly toothed at apex.

Bhutan: C –Thimphu district (Wasa La, Paro), **N** – Upper Mande Chu to Upper Kuru Chu districts; **Sikkim**; **Chumbi**. Open sandy mountainsides, 4265–4570m. July–September.

8. S. purii Bocquet & Saxena

Similar to *S. nigrescens* but more slender; stems 5–15cm, pubescent; leaves linear-lanceolate or oblanceolate, 5–7 × 0.3–0.7cm, pubescent; flowers solitary or rarely in pairs; calyx scarcely inflated, 8–10 × 5–7mm, pale pubescent on ribs; petals exserted 6–7mm from calyx, slightly bilobed and toothed at apex.

Bhutan: C –Thimphu district (Phajudin); **Chumbi**. Rock crevices, 3650–4170m. August–September.

9. S. gonosperma (Ruprecht) Bocquet; *Lychnis macrorhiza* Bentham, *L. apetala* sensu F.B.I. non L., *L. himalayensis* (Rohrbach) Edgeworth, *S. himalayensis* (Rohrbach) Majumdar, *S. wahlbergella* sensu Majumdar non Chowdhuri

Perennial herb up to 30cm. Basal leaves usually present, oblanceolate, 2–5.5 × 0.3–0.6cm, acuminate, base attenuate, finely pubescent. Flowers solitary or rarely 2–3, nodding at first. Calyx slightly inflated, campanulate, (0.75–)1–1.3

× 0.5–1.2cm, glandular-pubescent, green with broad dark purple ribs, rarely completely dark purple, teeth acute, with scarious margins. Petals not or scarcely exserted from calyx, reddish purple. Receptacle scarcely elongated, c 1mm. Capsules broadly ovoid, c 1.2 × 0.8cm.

Bhutan: C –Thimphu district (Sharna and Tremo La), **N** – Upper Mo Chu to Upper Kulong Chu districts; **Sikkim**. Screes and open grassy slopes, 3000–4880m. June–September.

The above description refers to subsp. **himalayensis** (Rohrbach) Bocquet, to which the Sikkim and Bhutan plants belong.

10. S. caespitella Williams; *S. maheshwari* Bocquet

Similar to *S. gonosperma* but more slender; stems (3–)10–45cm; leaves linear-lanceolate, 2–6 × 0.2–0.5cm, finely pubescent; flowers 3–9 in loose cymes; calyx narrowly campanulate, 7–8 × 3–4mm, with usually darkly puberulous ribs; petals reddish, bifid at apex, equalling or shorter than calyx.

Bhutan: N – Upper Mo Chu district (Laya) and Upper Bumthang Chu district (Weitsang); **Chumbi; Sikkim**. Among dwarf Rhododendron, 3650–4880m. June–July.

11. S. stracheyi Edgeworth

Slender scrambling herb. Leaves ovate, 2–4 × 0.9–2cm, acuminate, base rounded or cuneate, subsessile, sparsely pubescent. Flowers 1–3 in loose cymes at branch ends. Calyx narrowly cylindrical, 1.5–1.8 × 0.5mm, green, sparsely glandular-pubescent, ribs indistinct. Petals white, limb exserted c 5mm from calyx. Receptacle elongated, c 1cm.

Bhutan: C – Ha district (Tegong La) and Tongsa district (Rinchen Chu); **Sikkim**. Clearings in conifer forests, 3200–3800m. August.

12. S. armeria L.

Glaucous annual or biennial herb. Stems erect, 20–45cm, little-branched, viscid below upper nodes. Leaves ovate, 2–5 × 0.8–2.5cm, acute, base cordate, glabrous. Flowes in dense corymbose cymes. Calyx narrowly tubular, 1.3–1.7 × 0.2–0.3cm, teeth c 1.5mm, ovate-obtuse. Petal limb obcordate, c 6 × 3.5mm, pink or white, with two lanceolate scales at base. Receptacle elongate, c 10mm. Ovary cylindrical, c 5mm, styles 3. Capsule c 8mm, 6-valved.

Bhutan: C –Tongsa district (cultivated at Tongsa Dzong). May.

Native of Europe, cultivated as an ornamental.

4. CUCUBALUS L.

Slender, often scandent, herbs. Leaves ovate-lanceolate. Flowers few in leafy cymes at branch ends, nodding. Calyx divided to below middle into 5 teeth, ribs obscure. Petals 5, claw narrow ± as long as calyx, limb bifid with two scales at base. Receptacle slightly elongated (c 2 – 3mm) bearing petals, stamens and ovary at apex. Ovary 1-celled above but partially 3-celled at base; styles 3, filiform. Fruit globose, fleshy, bursting irregularly.

1. C. bacciferus L.

Stems up to 75cm. Leaves 3 – 7 × 1 – 2.5cm, acuminate, base rounded, sparsely pubescent; petioles up to 1cm. Calyx 1 – 1.3cm. Petals white, limb c 3 × 2mm. Fruit black when ripe, 1 – 1.5cm diameter.

Bhutan: S – Chukka and Deothang districts, **C** – Thimphu to Tashigang districts; **Sikkim**. Common on grassy banks and open Blue Pine forests, 1370 – 2133m. July – August.

5. CERASTIUM L.

Slender herbs with small leaves. Flowers in terminal cymes. Sepals 5, free, margins scarious. Petals 5, notched at apex. Stamens 10. Receptacle not elongated. Ovary 1-celled, styles 5, filiform. Capsule cylindrical, 10-valved.

1. C. glomeratum Thuillier agg.; *C. vulgatum* sensu F.B.I. p.p. non L. Med: *Tirham*. Fig. 18b,c.

Annual or (?) perennial herb, stems suberect up to 30cm. Leaves ovate or oblanceolate, 0.5 – 2 × 0.3 – 0.8cm, acute or obtuse, rounded or narrowed at base, glandular- or eglandular-pubescent. Sepals lanceolate, 5 – 6 × 1 – 1.5mm, glandular- or eglandular-pubescent (but without tufts of hair exceeding apex). Petals ± as long as sepals, white. Capsule 9 – 12mm, scarious, straw-coloured.

Bhutan: C – Thimphu, Punakha, Tongsa, Bumthang and Tashigang districts, **N** – Upper Mo Chu, Upper Pho Chu and Upper Bumthang Chu districts; **Sikkim**. Streamsides and grassy slopes, 2400 – 3960m. May – September.

A variable species that has been subdivided and names such as *C. fontanum* Baumgarten subsp. *triviale* (Link) Jalas (*C. holosteoides* Fries) and *C. glomeratum* Thuillier, in the strict sense, have been applied here. These taxa were originally recognised on the basis of European material and are clearly different from the plants of E Himalaya. Subdivision of this aggregate must await further research.

6. STELLARIA L.

Diffuse or mat-forming herbs, rarely scrambling. Leaves ovate-lanceolate or linear. Flowers in cymes, rarely solitary. Sepals 5 or 4, ovate-lanceolate, often scarious-margined. Petals 5 or 4, bifid to middle or to base, white, never longer than sepals, sometimes absent. Stamens 10 or 8, rarely as few as 3, those opposite sepals gland-thickened at base. Receptacle not or scarcely elongated. Ovary 1-celled, styles 3 – 5. Capsule short, opening by 6 valves. Seeds numerous or rarely few.

1. Stems 60–130cm; leaves 3.5–14 × 1–3cm ....**1. S. monosperma**
+ Stems not more than 60cm; leaves not more than 2.5 × 1.5cm 2
2. Leaves greyish stellately-pubescent or tomentose at least beneath **Species 2 & 3**
+ Leaves glabrous, or if pubescent then hairs simple and never tomentose ...3
3. Leaves pilose or pubescent ....**Species 4 & 5**
+ Leaves ciliate at margins but otherwise glabrous, or completely glabrous ..4
4. Leaves ovate or elliptic, usually more than 5mm broad ....**Species 6–8**
+ Leaves ovate-lanceolate or linear-lanceolate, always less than 2mm broad ..5
5. Leaves ± herbaceous, margins not thickened ....**Species 9 & 10**
+ Leaves stiff, ± hardened, margins thickened ....**Species 11 & 12**

1. S. monosperma D. Don; *S. crispata* Edgeworth & Hook. f., *S. paniculata* Edgeworth

Scrambling herb, stems with a line of fine hairs, otherwise glabrous. Leaves elliptic or lanceolate, acuminate, base rounded, sessile or with petioles up to 1cm, margins smooth or minutely crisped, fine intramarginal vein present, glabrous. Cymes elongate, glandular-pubescent. Sepals lanceolate, 4–6mm, acuminate. Petals divided to middle, as long or half as long as sepals. Capsule subglobose, c 2.5mm, 1–2-seeded.

Bhutan: C–Thimphu district (Simtoka) and Tashigang district (Gomchu); **Sikkim**. Open forests, 2270–2745m. August–October.

Two varieties are recognised: var. **monosperma** with sepals and petals 5–6mm and var. **paniculata** (Edgeworth) Majumdar, with sepals c 4mm and petals only half as long. A specimen from Sikkim (Pankabari) has flowers similar to those of var. *paniculata* but lacking petals and with only five stamens.

2. S. vestita Kurz; *S. saxatilis* D. Don non Scopoli, *S. hamiltoniana* Majumdar. Fig. 18d.

Weak, greyish, stellately-pubescent herb; stems prostrate up to 50cm. Leaves ovate, 0.6–3 × 0.3–1cm, acute or acuminate, base rounded, sessile. Flowers few, in loose cymes. Sepals lanceolate, c 6mm. Petals c 4mm, bifid almost to base. Capsule ovoid, ± as long as sepals. Seeds c 10.

Bhutan: S – Chukka district, **C** – Ha, Thimphu, Punakha and Tongsa districts, **N** – Upper Kulong Chu district (Lao); **Sikkim**. Amongst shrubs and on margins of cultivated ground, 1400–2750m. April–July.

3. S. lanata Edgeworth & Hook. f.

Similar to *S. vestita* but thinly lanate at least at first; leaves narrowly ovate or lanceolate, 0.7–3 × 0.2–0.4cm, glabrescent above, white tomentose beneath; flowers in terminal cymes; sepals c 3mm; petals absent or reduced to narrow bifid threads; capsules 4–5mm.

Bhutan: N – Upper Mangde Chu (Jiutang), Upper Mo Chu (between Kohina and Yabu Thang) and Upper Kulong Chu (Me La) districts; **Sikkim**. Marshy ground on mountains, 3050–3960m. February–September.

4. S. sikkimensis Edgeworth & Hook. f.; *S. hamiltoniana* Majumdar var. *sikkimensis* (Edgeworth & Hook. f.) Majumdar

Similar to *S. vestita* but sparsely yellowish pilose throughout; leaves ovate-lanceolate, 0.7–2 × 0.3–0.8cm; flowers in loose terminal cymes; sepals 4–5mm; petals 3–4mm; capsules 5–6mm.

Bhutan: S – Phuntsholing and Gaylegphug districts, **C** –Thimphu, Punakha, Mongar and Tashigang districts; **Sikkim**. Moist roadside banks in Warm broad-leaved forests, 1200–2000m. April–July.

5. S. patens D. Don; *S. longissima* Edgeworth & Hook. f.

Similar to *S. vestita* and *S. sikkimensis* but with whitish pilose hairs on stems and leaf-undersides; leaves 1–2.5 × 0.1–0.2cm; cymes few-flowered; sepals c 5mm; petals somewhat shorter.

Bhutan: C –Thimphu district (between Thimphu and Chimakhothi); **Sikkim**: Darjeeling. Roadsides and margins of cultivation, 2000–2450m. June.

6. S. media (L.) Villars. Eng: *Chickweed.*

Suberect or decumbent herb, stems 5–40cm, 4-angular, with a line of pubescence along one side, rooting at nodes. Leaves ovate, 1–2.5 × 0.5–1.5cm, acute, base rounded or cordate, glabrous or with a few ciliate hairs at base, sessile or the lower ones with petioles up to 1.5cm, venation indistinct. Flowers few or numerous in terminal leafy cymes, pedicels up to 1.5cm. Sepals ovate, 3–4mm, sparsely glandular-pubescent. Petals somewhat shorter than sepals, deeply bifid. Stamens 3–8. Capsule ovoid, 4–7mm.

Bhutan: S – Samchi district (Dorokha), Phuntsholing district (Phuntsholing) and Chukka district (Bunakha), **C** –Thimphu district (Dotena) and Punakha district (Menchunang); **Sikkim**. Weed of cultivation, possibly introduced, 610–2500m. February–November.

A variant with more densely hairy stems and ciliate leaves has been collected from Riang in Sikkim; it has much reduced flowers with three to four sepals and no petals.

7. S. reticulivena Hayata

Similar to *S. media* but leaves always sessile, acuminate, midrib impressed above, prominent beneath, minor veins reticulate, distinct in transmitted light; flowers in slender leafless branched cymes; bracts ovate, obtuse, 1–1.5mm, scarious; sepals c 2.5mm, glabrous; stamens 10; capsules c 3mm.

Bhutan: C –Thimphu, Punakha, Tongsa and Mongar districts, **N** – Upper Mo Chu district; **Sikkim**. By streams and in Hemlock forests, 1750–2750m. April–May.

8. S. uliginosa L.; *S. alsine* Grimm, *nom. inval.*

Similar to *S. media* but leaves elliptic, 0.5–2 × 0.15–0.5(–0.6)cm, acute or acuminate, sessile, glabrous; flowers few in terminal cymes; bracts ovate, acute, c 2mm, scarious; sepals 3–4mm, glabrous; stamens 10 or sometimes 3–5.

Bhutan: **S**–Samchi, Chukka and Gaylegphug districts, **C**—Tongsa and Bumthang districts. Streamsides, 610–2400m. February–May.

9. S. subumbellata Edgeworth & Hook. f.

Decumbent glabrous herb, stems 10–20cm. Leaves linear-lanceolate, 5–7 × 0.75–1.5mm, acuminate. Cymes 2–4, terminal, umbellately arranged, pedicels slender; bracts c 2mm, ovate. Sepals c 2mm. Petals 5 or absent. Stamens 5. Capsule ovoid, c 4mm.

Sikkim: Chumighata. 4880m. September.

10. S. gyangtsensis Williams

Similar to *S. subumbellata* but more erect, stems c 10cm, minutely puberulous at first; leaves linear-lanceolate, 4–10 × 0.5–0.75mm, acuminate; cymes divaricately branched; bracts ovate, c 2mm; sepals c 4mm, acuminate; capsule ± as long as calyx.

Sikkim: locality unknown.

11. S. decumbens Edgeworth; *S. cherleriae* Williams p.p.

Lax or densely tufted herb, stems pubescent, 5–15cm. Leaves linear- or ovate-lanceolate, 3–15 × 1–1.5mm, acuminate, glabrous or ciliate especially near base, sometimes clustered in axillary fascicles. Flowers solitary or in cymes. Sepals 4–5, lanceolate, 2.5–3.5mm. Petals deeply bifid, shorter than sepals. Stamens 8–10. Capsule ± as long as calyx.

Bhutan: **C**–Mongar district, **N**–Upper Mo Chu to Upper Kulong Chu districts; **Sikkim**. Damp rocks and scree on mountains, 3050–4100m. June–August.

A variable species of which several varieties have been recognised. The first two, with solitary flowers, are known from Bhutan and Sikkim: var. **decumbens** (including var. *minor* Edgeworth & Hook. f.) forming loose mats, leaves linear-lanceolate, usually from lower altitudes; and var. **pulvinata** Edgeworth & Hook. f. forming cushion-like clumps, leaves ovate-lanceolate, usually at higher elevations. A further two varieties have flowers in cymes and are known only from Sikkim: var. **polyantha** Edgeworth & Hook. f. forming rather dense clumps, leaves ovate-lanceolate, 4–7mm, cymes ± sessile; and var. **acicularis** Edgeworth & Hook. f. forming loose clumps, stems c 12cm, leaves linear-lanceolate, needle-like, up to 15mm, often in axillary fascicles, cymes pedunculate.

12. S congestiflora Hara

Similar to *S. decumbens* (especially to var. *polyantha*) but stems densely white lanate; leaves ovate-lanceolate, 7–9(–13) × 2–3.5mm, margins ciliate; flowers in terminal sessile cymes; sepals 5–6mm, acuminate; petals 2.5–3mm; stamens 10.

Bhutan: N – Upper Mo Chu district (Lingshi); **Sikkim**. Mountain cliff ledges and amongst boulders, 3960–4570m. July–September.

7. PSEUDOSTELLARIA Pax

Weak erect perennial herbs; rhizomes slender, bearing small tubers. Leaves elliptic, obovate or linear-lanceolate. Flowers solitary, axillary or few in terminal cymes (cleistogamous flowers sometimes present in lower leaf axils). Sepals 5, with scarious margins. Petals 5, white, entire or notched at apex. Stamens 10, those opposite sepals gland-thickened at base. Styles 2–3, filiform, minutely capitate.

1. P. heterantha (Maximowicz) Pax; *Stellaria bulbosa* sensu F.B.I. non Wulfen

Stems 3–15cm bearing a line of fine pubescence; rhizomes usually bearing a subglobose or ovoid 3–5mm tuber at each node. Leaves elliptic or obovate, 0.7–2.5 × 0.4–1.5cm, acute or acuminate, base rounded or attenuate into petiole up to 1cm, glabrous or sparsely ciliate at base. Pedicels up to 3.5cm. Sepals ovate-elliptic, c 5mm. Petals obovate, 7–8 × 3–5mm, entire.

Bhutan: C – Ha to Tashigang districts, **N** – Upper Mo Chu to Upper Kulong Chu districts; **Sikkim**. In moss or leaf-mould in moist mixed and coniferous forests, 2440–3800m. May–July.

There are two varieties, both widespread: var. **himalaica** Ohwi with elliptic, acuminate leaves and var. **nepalensis** (Majumdar) Hara with leaves, especially the upper ones, obovate and less pointed.

2. P. sylvatica (Maximowicz) Pax

Similar to *P. heterantha* but more slender, leaves linear-lanceolate, 1.5–2.5 × 0.15–0.2cm, acuminate, sessile, always glabrous; petals oblanceolate, c 7 × 3mm, entire or sometimes notched at apex.

Bhutan: N – Upper Kulong Chu district (Me La). In leaf-mould in Fir forest, 3650m. July.

8. BRACHYSTEMMA D. Don

Divaricate climbing herb. Leaves lanceolate. Flowers in axillary or terminal panicles. Sepals 5, somewhat scarious. Petals 5, narrow, entire, shorter than sepals. Stamens 5, sometimes also with 5 staminodes. Ovary 1-celled; styles 2. Capsule 4-valved, 1- or few-seeded.

1. B. calycinum D. Don

Stems 4-angular, glabrous, glossy. Leaves 2.5–7 × 0.6–2.5cm, acuminate, base rounded or cuneate, sparsely pubescent on upper surface; petioles up to 2cm becoming recurved, thickened and hardened. Panicle branches minutely pubescent, bracts linear-lanceolate up to 4mm, recurved. Sepals elliptic, 6–7mm, weakly 3-nerved. Petals lanceolate, 2.5–4mm, white. Stamens 1–2mm.

Bhutan: S – Samchi, Phuntsholing and Gaylegphug districts, **C** –Tongsa and Tashigang districts; **Sikkim**. In thickets, 300–1600m. December–April.

9. ARENARIA L.

Annual or perennial, erect, prostrate or cushion-forming herbs. Leaves ovate, oblong-linear or suborbicular. Flowers solitary or in cymes. Sepals 5, free, herbaceous throughout or with scarious margins. Petals 5, white or purplish, entire or lacerate, never bifid, rarely absent. Stamens 10, those opposite sepals glandular-thickened at base. Ovary 1-celled; styles 2–3, filiform. Capsules ± as long as calyx, 2–6-valved; seeds few or numerous.

1. Prostrate herb with trailing stems up to 25cm; leaves suborbicular or broadly obovate .... **1. A. rotundifolia**
\+ Erect, suberect or cushion-forming herbs; leaves linear, lanceolate, elliptic-ovate or oblanceolate 2
2. Plants forming dense cushions from persistent stems and leaves of previous years 3
\+ Plants not cushion-forming; stems and leaves of previous years not persistent 4
3. Leaves finely ciliate; sepals acuminate, hard, distinctly ribbed ..**Species 2–5**
\+ Leaves glabrous; sepals herbaceous, obtuse, indistinctly ribbed .**Species 6–8**
4. Sepals hardened, subcoriaceous at apex and margins ....**9. A. ciliolata**
\+ Sepals herbaceous, margins not thickened and hardened 5
5. Petals lacerate at apex ....**10. A. debilis**
\+ Petals entire or somewhat toothed, sometimes absent 6
6. Petals 6–10mm, pink or purplish ....**Species 11–13**
\+ Petals 1–5mm or absent, white (but deep rose-red in *A. ludlowii*) **Species 14–16**

1. A. rotundifolia Bieberstein; *A. orbiculata* Edgeworth & Hook. f.

Slender prostrate or decumbent perennial; stems 7–25cm, minutely recurved pubescent. Leaves suborbicular or broadly obovate, 2–7 × 2–7mm, obtuse, acute or mucronate, midrib visible, other veins obscure, subsessile or with petioles to 2mm, ciliate, otherwise glabrous. Flowers axillary or few in terminal cymes. Sepals ovate-lanceolate, c 4mm, pubescent along midrib, margins scarious. Petals oblanceolate, c 4 × 1.5mm, entire, white. Styles 3.

Bhutan: C – Ha, Thimphu, Punakha and Tongsa districts, **N** – Upper Kulong Chu district; **Sikkim**. On moist ground, 2500–3500m. May–June.

2. A. edgeworthiana Majumdar; *A. monticola* sensu F.B.I. non Buckley

Cushions up to 5cm tall, 15–30cm diameter, branches densely leafy. Leaves linear-lanceolate, 5–8 × 1mm, apex needle-like, rigid, recurved, base sheathing, ± scarious, margins minutely ciliate, uppermost leaves on flowering shoots bract-like, c 2mm broad, subscarious. Flowers solitary, sessile at branch ends. Sepals ovate-lanceolate, 8–9 × 2–3mm, 3-nerved, margins scarious. Petals obovate, 11–12 × 5–6mm, white. Styles 3.

Bhutan: C – Punakha district (Black Mountain), **N** – Upper Mo Chu, Upper Pho Chu, Upper Bumthang Chu and Upper Kuru Chu districts; **Sikkim**. Alpine mountainsides and screes, 4260–4880m. July–August.

3. A. bryophylla Fernald; *A. musciformis* sensu F.B.I. non Triana & Planchon

Similar to *A. edgeworthiana* but leaves 4–8mm; flowers smaller, sepals c 8mm, petals narrowly obovate, c 7 × 2mm.

Sikkim: Lhonak and Thango. Mountain slopes, 4265–5180m. August.

4. A. pulvinata Edgeworth & Hook. f.

Similar to *A. edgeworthiana* but leaves 3–5mm, strongly ciliate, upper ones not bract-like; flowers ± sessile; sepals 3–4.5mm; petals obovate 6–10mm.

Bhutan: N—Upper Mo Chu district (Lingshi La and Laya); **Sikkim**: Chakalung La. Mountain slopes, 4600–4880m. October.

5. A. ischnophylla Williams

Similar to *A. edgeworthiana* but leaves subulate, triquetrous above, up to 15mm, uppermost ones bract-like, c 4mm, subscarious, close below flowers; pedicels villous c 4mm; sepals 6–9 × 2.5mm; petals ± equalling sepals.

Bhutan: C – Ha district (Ya La); **Chumbi** : E of Phari. Mountain slopes, 3800–5000m. August.

6. A. polytrichoides Edgeworth & Hook. f.

Cushions 5–10cm tall. Leaves linear-lanceolate, 3.5–4 × 1–1.25mm, gradually narrowing to an aristate point, recurved, base sheathing, sessile, margins thickened. Flowers terminal, sessile at first, pedicels lengthening up to 4mm in fruit. Sepals oblong-elliptic, 2.5 × 1.5mm, obtuse, glabrous. Petals obovate, c 4 × 1.75mm, narrowed at base.

Bhutan: N – Upper Mo Chu district (Lingshi), Upper Bumthang Chu district (Hoopkye La) and Upper Kulong Chu district (Me La); **Sikkim**. On mountain rocks, 4570m. June–July.

7. A. densissima Edgeworth & Hook. f.

Similar to *A. polytrichoides* but leaves up to 2mm broad, ovate, abruptly acuminate, recurved; flowers axillary; petals linear-spathulate, c 4 × 1.25mm.

Bhutan: N – Upper Mo Chu district (between Laum Thang and Chawa Gassar); **Sikkim**: Goraphu Chu and Ningbil. On mountain rocks, 4570–4730m. August.

8. A. oreophila Hook. f.

Similar to *A. polytrichoides* but leaves up to 12mm; flowers on short (c 5mm) puberulous pedicels; sepals obovate c 4 × 1.5mm, minutely ciliate; petals c 5mm.

Sikkim: Momey. 5490m.

9. A. ciliolata Edgeworth & Hook. f.; *A. ramellata* Williams

Perennial tufted herb. Stems 2–6(–11)cm with a line of pubescence on either side. Leaves ovate-elliptic, 6–10(–15) × 2–4(–5)mm, sharply acuminate, base rounded or narrowed, sessile, margins pale, somewhat thickened, covered on both surfaces with coarse yellowish articulated hairs, sometimes only ciliate, rarely glabrous. Flowers solitary or few in leafy cymes, pedicels up to 1(–2)cm. Sepals lanceolate, 6–7 × 2–3mm, acuminate, margins and apex thickened and hardened, ciliate or scattered yellowish pilose, rarely glabrous. Petals oblanceolate, 7–10 × 3–3.5mm, white.

Bhutan: N – Upper Mo Chu to Upper Kulong Chu districts; **Sikkim**. Mountain cliffs and screes, 4265–4875m. July–September.

A. ramellata Williams appears to differ only in the absence of the coarse yellowish hair, leaves glabrous or with a few weak cilia at base. Intermediates occur between this condition and that of typical *A. ciliolata*.

10. A. debilis Edgeworth & Hook. f.; *A. benthamii* sensu F.B.I. non Fenzl, *A. glandulosa* (G. Don) Williams non Jacquin, *A. blinkworthii* McNeill, *Lepyrodiclis debilis* (Edgeworth & Hook. f.) Ohba, *L. glandulosa* (G. Don) Ohba

Slender, erect, branching herb 15–50cm, pubescent throughout with glandular and eglandular hairs intermixed. Leaves ovate, elliptic or oblanceolate, 1–3 × 0.3–1cm, acute, base narrowed, sessile or shortly (up to 5mm) petioled. Flowers few or many, nodding. Sepals ovate-lanceolate, 3.5–6mm, margins membranous. Petals white, spathulate or obovate, 4–10mm, lacerate at apex. Styles 2.

Bhutan: C – Ha to Mongar districts, **N** – Upper Mo Chu and Upper Kulong Chu districts; **Sikkim**. On river banks and clearings in Fir forests, 3200–4000m. July–August.

11. A. melandryoides Edgeworth & Hook. f.; *A. cerastiiformis* Williams

Slender rhizomatous perennial. Stems 3–7cm, finely pubescent, hairs often tinged with purple. Leaves elliptic, 6–12 × 3–5mm, acute, base rounded or narrowed, sessile, margins somewhat thickened, ± glabrous above, shortly and densely glandular-pubescent beneath especially at margins. Flowers solitary, ± sessile or nodding on pedicels 5–10mm. Sepals ovate-lanceolate, c 7 × 2–3mm, obtuse, finely glandular-pubescent, margins membranous, obscurely veined. Petals pink, oblanceolate or narrowly obovate, c 10 × 3–5mm, rounded at apex, narrowed and almost clawed at base. Styles 2.

Bhutan: C – Ha district (between Tremo La and Ha), **N** – Upper Mo Chu (Yale La), Upper Mangde Chu (Saga La) and Upper Bumthang Chu districts (Marlung); **Sikkim**. Mountain screes and rock crevices, 4730–4880m. July.

12. A. melandryiformis Williams

Similar to *A. melandryoides* but taller, 5–10cm, finely purplish-pubescent; leaves spathulate, 7–12 × 2–3mm; pedicels 1.5–4.5cm; sepals 5–6mm, dark purplish-pubescent; petals 6–8mm.

Bhutan: C–Thimphu and Tongsa districts, **N**–Upper Mo Chu, Upper Bumthang Chu and Upper Kulong Chu districts; **Sikkim**. Moist sandy or stony turf on mountains, 3950–4570m. June–August.

13. A. glanduligera Edgeworth & Hook. f. Fig. 18a.

Similar to *A. melandryoides* but leaves 3–7(–12) × 2–3(–4)mm, acuminate, finely glandular-pubescent; pedicels up to 2cm, pubescent; sepals c 5mm, often purplish; petals 6–7mm, white or purplish.

Bhutan: C–Ha, Thimphu and Punakha districts, **N**–Upper Mo Chu, Upper Mangde Chu and Upper Kuru Chu districts; **Sikkim**. On sandy slopes on mountains, 3650–5500m. June–August.

14. A. ludlowii Hara

Similar to *A. glanduligera* but leaves oblong-lanceolate, 3–5 × 1–2mm, pedicels 5–30mm; sepals 3–4mm, reflexed; petals ovate-cordate, limb 3–4 × 1.5–2mm, on basal claws c 1mm, deep rose red.

Bhutan: N–Upper Bumthang Chu district (Marlung). 4265m. July.

Endemic to Bhutan. Possibly not distinct from *A. glanduligera*.

15. A. depauperata (Edgeworth & Hook. f.) Hara; *Stellaria depauperata* Edgeworth & Hook. f., *A. bhutanica* Majumdar & Babu, *A. balfouriana* W. W. Smith

Slender suberect herb 5–10cm with 4-angular stems bearing a line of pubescence. Leaves linear-lanceolate, 5–10 × 1.5–2mm, acuminate, glabrous, surface minutely tuberculate. Flowers solitary or few, pedicels slender, 1–2.5(–3)cm, pubescent, recurving in fruit. Sepals lanceolate, 3.8–4 × 0.8–1(–1.2)mm, margins scarious. Petals white, 1.5–5mm, sometimes absent. Stamens 10, filaments blue, anthers blackish.

Bhutan: C–Thimphu district (Chelai La) and Mongar district (Rudong La), **N**–Upper Mangde Chu district (Ju La); **Sikkim**. Alpine rock crevices, 3800–4880m. August.

16. A. littledalei Hemsley; *A. thangoensis* W. W. Smith, *A. pharensis* McNeill & Majumdar

Minute, erect branching herb 2–4cm. Stems glabrous or with one or two lines of pubescence. Leaves ovate or linear-lanceolate, c 2 × 1mm, acute, base rounded, sessile or shortly petiolate, glabrous or ciliate. Cymes few-flowered, pedicels 3–16mm, glabrous or pubescent. Sepals 4–5, linear-lanceolate, 2–3mm, glabrous or pubescent. Petals white, 1–4 or absent, 1–1.5mm, ± toothed at apex. Stamens 2–5. Capsules c 2mm, few-seeded.

Sikkim: Chugya, Naku Chu and Thango; **Chumbi** : Phari plain. 4360–4570m. September–October.

10. SAGINA L.

Small mat-forming annual or perennial herbs. Leaves linear, connate at base. Flowers solitary, axillary and terminal. Sepals (4–)5, obtuse, margins scarious. Petals (4–)5, white, sometimes absent. Stamens 5–10. Ovary 1-celled; styles 5. Capsule slightly longer than sepals, 5-valved.

1. S. japonica (Swartz) Ohwi; *S. procumbens* sensu F.B.I. p.p. non L.

Annual or rarely perennial herb. Stems 5–15cm. Leaves 3–15 × 0.5–0.75mm, acuminate to a fine awn-like tip at least as long as half the breadth of the leaf, glabrous or the uppermost sometimes glandular-ciliate near base. Pedicels 5–15(–20)mm usually sparsely glandular-pubescent. Sepals usually 5, 2–2.5mm, rounded, not keeled on the back. Petals usually 5, broadly ovate, somewhat shorter than sepals. Seeds dark brown, rough with minute papillae.

Bhutan: **S**–Chukka district (Takhti Chu) and Gaylegphug district (Chabley Khola), **C**–Ha, Thimphu, Tongsa and Bumthang districts, **N**–Upper Mangde Chu district; **Sikkim**. Moist ground by roadsides, and on disturbed ground, 1750–4260m. May–June.

2. S. saginoides (L.) Karsten; *S. procumbens* sensu F.B.I. p.p. non L.

Similar to *S. japonica* but smaller, usually perennial, stems 1–4cm; leaves acuminate but scarcely awn-like at apex; pedicels always glabrous; sepals 1.5–2mm, slightly keeled; seeds pale brown, slightly tuberculate but not papillate, grooved dorsally.

Sikkim: Mome Samdong; **Chumbi** . On moist sand, 3960m. July–August.

Some records of this species from Bhutan may refer to *S. japonica*.

3. S. procumbens L.

Similar to *S. japonica* but completely glabrous; leaves not more than 10mm; sepals 4; petals minute or absent; stamens usually 4; seeds dark brown, smooth, weakly sculptured.

Sikkim: Darjeeling. On pavement in town, 2100m.

Possibly introduced from Europe.

11. THYLACOSPERMUM Fenzl

Densely tufted herb forming hard cushions. Leaves linear, densely crowded. Flowers solitary, terminal, sessile. Calyx-tube funnel-shaped, divided to middle into 4–5 lobes. Petals 4–5, oblong, inserted with 8–10 stamens on a glandular ring on calyx tube. Ovary 1-celled, shortly stalked; styles 2–3. Capsule 4–6-valved; seeds few, large with soft, loose seed-coats.

1. T. caespitosum (Cambessedes) Schischkin; *T. rupifragum* (Karelin & Kirilow) Schrenk

Cushions 30–100cm in diameter. Leaves c 3.5 × 0.75cm, acute, sessile, glabrous, unveined, keeled near apex, ± erect, scarcely recurved, tips sharp to touch. Calyx c 3mm, lobes acute. Petals c 1.5mm, white.

Sikkim: Kalaercc and Yak La. 4880–5180m.

12. SPERGULA L.

Suberect annual herbs. Leaves linear, opposite but apparently whorled due to presence of short dense leafy shoots in axils; stipules small, scarious. Flowers in terminal cymes. Sepals 5, margins scarious. Petals 5, entire, white. Stamens 10 or sometimes 5. Ovary 1-celled; styles 3–5. Capsule 3–5-valved.

1. S. arvensis L.

Stems 10–25cm, sparsely glandular-pubescent. Leaves 10–30 × 0.5mm, acute, glandular-pubescent. Sepals ovate, c 3.5mm, acuminate. Petals elliptic ± as long as sepals. Capsules ovoid, c 4mm. Seeds orbicular, slightly compressed, blackish, surrounded by a pale ridge and covered with whitish papillae.

Bhutan: C –Thimphu district (Thimphu) and Punakha district (Lometsawa); **Sikkim**. Weed of cultivated ground, 2000–2450m. May–August.

13. DRYMARIA Schultes

Prostrate or suberect branching herbs. Leaves rounded, stipules small, scarious. Flowers in axillary or terminal cymes. Sepals 5, margins scarious. Petals 5, white, bifid to below middle. Stamens 2–3(–5). Ovary 1-celled; styles 2–3 united nearly to middle. Capsule 3-valved, seeds 1 or many.

1. D. cordata (L.) Roemer & Schultes

Stems elongate, rooting at nodes. Leaves broadly ovate or suborbicular, 5–25 × 4–20mm, acute or obtuse, mucronate, base rounded, glabrous, usually 5-veined; petioles 2–5mm; stipules lacerate into 1–2mm filaments. Pedicels finely glandular-papillose. Flowers broadest above middle. Sepals elliptic-ovate, 3–3.5mm, keeled, 3-veined, inflexed, glandular-papillose on veins. Petals c 2.5mm. Seeds 1 or few, finely tuberculate.

Bhutan: S – Samchi district (Dorokha) and Gaylegphug district (Gaylegphug and Tama); **Sikkim**. Damp ground by streams, 300–1050m. May.

The above description refers to subsp. **diandra** (Blume) Duke, to which the Sikkim and Bhutan plants belong. Several medicinal properties are ascribed to this species (13).

2. D. villosa Chamisso & Schlechtendal

Similar to *D. cordata* but smaller, leaves 5–15 × 5–15mm, villous with long whitish hairs; stipules entire; pedicels and sepals sparsely pubescent; seeds numerous, rough with minute papillae.

Bhutan: S – Phuntsholing district (Phuntsholing); **Sikkim**: Darjeeling. Streamsides in Subtropical and Warm broad-leaved forests, 200–2000m. May–August.

14. POLYCARPON L.

Prostrate or suberect much-branched herb. Leaves opposite but appearing whorled due to presence of axillary fascicles of leaves; stipules scarious. Flowers numerous in dense cymes, bracts conspicuous, similar to stipules. Sepals 5, keeled, scarious-margined. Petals 5, entire, white. Stamens 3–5. Ovary 1-celled; style short 3-fid to middle. Capsule 3-valved.

1. P. prostratum (Forsskal) Ascherson; *P. loeflingiae* Edgeworth & Hook. f.

Stems 15–30cm, weakly pubescent. Leaves oblanceolate, 5–20 × 2–5mm, acute, base narrowed, subsessile, sparsely pubescent beneath; stipules lanceolate, 2–3mm. Sepals narrowly ovate, 2–2.5mm. Petals half as long as sepals. Capsule ovoid c 1.5mm.

Bhutan: S – Phuntsholing district (Phuntsholing); **Sikkim**: Darjeeling. River-banks in Subtropical forests, 200–250m. May.

Family 31. CHENOPODIACEAE

by D. G. Long

Annual herbs. Leaves alternate, simple, exstipulate, sometimes gland-dotted, mealy with white scales or stellate-pubescent. Flowers in axillary clusters, sometimes forming cymes or terminal panicles, unisexual or bisexual, actinomorphic. Perianth segments 3–5, free or connate, herbaceous or membranous, persistent (absent in female flowers in *Microgynoecium*). Stamens 2–5, opposite perianth segments, filaments free or connate at base. Ovary superior, 1-celled, ovule 1; styles usually 2, linear. Fruit an achene, often enclosed by persistent perianth.

1. Leaves stellate-pubescent; seeds obovoid .... **2. Axyris**
\+ Leaves with translucent yellowish gland-dots or mealy with minute whitish scales, never stellate; seeds depressed-globose or ellipsoid 2

2. Erect herbs 10–300cm; leaves 1–11cm; flower clusters forming conspicuous terminal panicles .... **1. Chenopodium**
+ Diffuse, prostrate alpine herb; leaves mostly 1–1.5cm; flowers concealed amongst leaves .... **3. Microgynoecium**

1. CHENOPODIUM L.

Erect herbs. Leaves entire, dentate, shallowly lobed or pinnatifid, mealy with whitish bladder-like scales, or yellowish gland-dotted. Flower clusters aggregated into terminal panicles; flowers bisexual or unisexual. Perianth segments 3–5, herbaceous, connate at base.

1. Leaves mealy with minute whitish bladder-like scales, eglandular, not aromatic 2
+ Leaves with translucent yellowish gland-dots, not mealy, aromatic 5
2. Robust herbs, 30–300cm, leaves usually large, mostly 3–11cm, ovate-deltoid, scarcely lobed but often dentate; upper part of leaf triangular, often acute 3
+ Weak herbs, 10–50cm, leaves often small, mostly 1–2.5cm, ovate-oblong, often 3-lobed but scarcely dentate; upper part of leaf triangular or oblong, obtuse 4
3. Plants green throughout; stems 30–100cm; leaves mostly 3–7cm, margins entire or shallowly dentate; petioles 0.5–4cm .... **1. C. album**
+ Plants often reddish tinged; stems 100–300cm; leaves mostly 7–11cm, margins coarsely dentate; petioles 4–10cm .... **2. C. giganteum**
4. Slender erect herb, 20–50cm; leaves usually lobed and bluntly toothed, upper part oblong, obtuse .... **3. C. ficifolium**
+ Decumbent herb, 5–20cm; leaves unlobed or shallowly lobed near base, otherwise entire, upper part triangular, obtuse .... **4. C. prostratum**
5. Petioles 0.2–1cm; leaves lanceolate, narrowly acute or acuminate, irregularly serrate-dentate to subentire; flower clusters subglabrous in slender spikes forming elongate panicles .... **5. C. ambrosioides**
+ Petioles 1–3cm; leaves oblong, pinnatifid with blunt teeth; flower clusters pubescent, cymosely arranged on branches of elongate panicles **6. C. botrys**

1. C. album L.; *C. album* sensu F.B.I. p.p.

Stems 30–100cm. Leaves ovate-deltoid, (1.5–)3–7 × 0.7–4cm, subacute, base cuneate, margins entire or shallowly dentate, sometimes weakly 3-lobed, mealy beneath; petioles 0.5–4cm. Flower clusters dense, sessile, forming slender panicles. Flowers bisexual or lower ones female, c 0.75mm diameter. Perianth segments 5.

Stamens 5. Seeds (with thin pericarp removed) black, 1.2–1.5mm, with faint radial striations.

Bhutan: **S** – Samchi and Gaylegphug districts, **C** – Ha, Thimphu and Bumthang districts; **Sikkim**. Weed of cultivated ground, 550–2800m. January–August.

Sometimes cultivated as a food-grain and pot-herb (126).

2. C. giganteum D. Don; *C. album* L. subsp. *amaranticolor* Coste & Reynier, *C. album* sensu F.B.I. p.p.

Similar to *C. album* but a larger, often reddish plant, 1–3m; leaves rhombic-ovate, 7–11 × 3–7cm, unlobed or shallowly 3-lobed, acute, base cuneate, margins coarsely irregular-dentate; petioles 4–10cm.

Sikkim: terai and foothills. Cultivated, 200–1500m. September.

Cultivated as a green vegetable.

3. C. ficifolium Smith; *C. album* sensu F.B.I. p.p. Dz: *Hethu*; Nep: *Bethu*.

Slender erect herb, 20–50cm. Leaves ovate-oblong, 1.5–2.5(–3) × 0.5–1.5cm, usually 3-lobed, margin usually bluntly toothed, mealy beneath, upper lobe oblong, obtuse; petioles 0.4–1.5cm. Panicles slender. Seeds c 1mm, minutely pitted.

Bhutan: **S** – Samchi and Phuntsholing districts, **C** – Punakha and Tongsa districts. Weed of cultivated ground, 500–1600m. January–June.

4. C. prostratum Herder; *C. karoi* (Murr) Aellen

Similar to *C. ficifolium* but a small decumbent herb 5–20cm; leaves ovate, 1–2.5 × 0.7–1.5cm, unlobed or shallowly lobed in lower part, obtuse, entire; petioles 0.5–1.2cm; seeds pitted.

Chumbi: between Phari and Tuna. Disturbed soil, 4260–4570m. August.

5. C. ambrosioides L. Fig. 17k,l.

Aromatic herb 60–125cm. Leaves lanceolate, 3–7 × 0.7–2cm, narrowly acute or acuminate, base attenuate, usually irregularly serrate-dentate, sometimes almost entire, densely yellowish gland-dotted beneath; petioles 0.2–1cm. Flower clusters subglabrous, rather distant on slender spikes forming elongate panicles. Flowers usually bisexual. Perianth segments 5. Stamens 4–5. Seeds c 0.7mm diameter, smooth.

Bhutan: **S** – Phuntsholing and Gaylegphug districts, **C** – Punakha and Tashigang districts; **Sikkim**. Weed of cultivation, 200–1650m. May–November.

Native of tropical America.

6. C. botrys L.

Similar to *C. ambrosioides* but a smaller herb, 10–55cm; leaves oblong, pinnatifid, 1–6 × 0.5–2cm, lobes bluntly toothed; petioles longer, 1–3cm; flower clusters pubescent, usually cymosely arranged on branches of elongate panicles.

Bhutan: S – Chukka district (Chasilakha), **C** – Ha district (Ha) and Thimphu district (Paro and Thimphu); **Sikkim**. Weed of cultivated fields, 1500–2500m. July–October.

2. AXYRIS L.

Prostrate or erect monoecious herbs. Leaves stellate-pubescent, entire. Flowers in small clusters in axils of uppermost leaves, unisexual, ebracteate; male flowers uppermost with 3–5 perianth segments and 3–5 stamens; female flowers beneath, with 3 perianth segments.

1. A. prostrata L. Fig. 17i,j.

Stems up to 20cm. Leaves ovate or obovate, 5–18 × 3–12mm, obtuse, base cuneate, brownish stellate especially beneath. Male flowers 4–5 in a short stalked cluster, perianth segments obovate, 0.5–0.7mm, membranous. Female flowers several, perianth segments c 0.5mm, increasing to c 1mm in fruit. Seeds obovoid c 1.5mm, enclosed by persistent perianth.

Sikkim: Temu La. Alpine screes, 4900m. September.

A second, possibly undescribed, species of *Axyris*, has been collected in the Lhonak valley, Sikkim, differing from *A. prostrata* as follows: habit erect, stems 2–6cm; leaves sometimes rosetted at base, ovate, 5–6 × 2–3mm, acute; seeds obovoid, c 1.8mm.

3. MICROGYNOECIUM Hook. f.

Similar to *Axyris* but leaves mealy with whitish scales, not stellate-pubescent; male flowers in small clusters, with 5 connate perianth segments and 2–3 stamens; female flowers below males, solitary or few, each with 2 bracteoles, perianth absent; seeds compressed-ovoid, warted.

1. M. tibeticum Hook. f.

Prostrate herb, stems up to 12cm. Leaves rhombic-ovate, (3–)10–15 × 1.5–9mm, acute or obtuse, base cuneate or rounded. Male perianth 1.5mm, yellowish; filaments 2–2.5mm. Female flowers with bracteoles c 0.7mm.

Sikkim: Tungu, Cholamo etc.; **Chumbi**: Phari. Alpine screes, 4200–5400m. August–September.

Family 32. AMARANTHACEAE

by D. G. Long

Herbs, shrubs or woody climbers. Leaves simple, alternate or opposite, entire, exstipulate. Flowers in heads, racemes, spikes or panicles, actinomorphic, bisexual or sometimes unisexual, each subtended by a papery or hardened bract and usually

2 bracteoles. Perianth segments 3–5, mostly free, papery or rigid. Stamens 3–5, opposite perianth segments, 1- or 2-celled, free, or filaments connate into a basal ring or tube, sometimes alternating with pseudostaminodes. Ovary superior, 1-celled, ovules 1 or several, styles 1–3. Fruit a capsule or utricle, sometimes circumscissile, rarely a berry.

1. Leaves alternate (some leaves also opposite in *Aerva*) 2
\+ Leaves opposite 6
2. Shrubs, sometimes climbing; flowers rather distant in spike-like racemes, often forming panicles; fruit a 3–5-seeded berry ....**1. Deeringia**
\+ Herbs, not climbing; flowers densely congested in heads, spikes or panicles; fruit a dry capsule 3
3. Flowers in elongate terminal panicles, unisexual, males above, females below; style short, thick, bearing 2–3 subulate stigmas; leaves ovate-elliptic; petioles up to 10cm ....**4. Amaranthus**
\+ Flowers in short, rarely elongate spikes, bisexual; style slender, stigmas 1–2, subcapitate; leaves elliptic, often narrowly; petioles up to 3cm 4
4. Some leaves usually opposite; flower spikes 6–9mm broad; bracts and perianth densely white pubescent ....**6. Aerva**
\+ All leaves alternate; flower spikes 10–16mm broad; bracts and perianth glabrous 5
5. Diffuse annuals; flowers in globose heads; capsules 1-seeded ...**3. Allmania**
\+ Erect annuals; flowers in elongate spikes; capsules 2–10-seeded ..**2. Celosia**
6. Flowers in diffuse panicles, unisexual; leaves reddish or purple with pink or yellowish bands ....**11. Iresine**
\+ Flowers in dense spikes or heads, bisexual; leaves usually green, sometimes purplish but never with paler bands 7
7. Large woody climber with pendulous spikes; perianth segments hairy only at apex, otherwise glabrous ....**7. Stilbanthus**
\+ Erect, decumbent or prostrate herbs or subshrubs; spikes or heads suberect or in axillary clusters; perianth segments glabrous or pubescent throughout...8
8. Bracts and perianth segments flat and papery, not needle-like or hooked (stiff and prickly in *Alternanthera pungens*); inflorescences mostly axillary (always terminal in *Gomphrena*, sometimes in *Aerva*) 9
\+ Bracts and perianth segments needle-like or hooked at apex; inflorescence of elongated or rounded terminal spikes 11
9. Inflorescences always terminal; each flower enclosed by two boat-like keeled bracteoles ....**10. Gomphrena**
\+ Inflorescences mostly axillary, also terminal in *Aerva*; flowers not enclosed by bracteoles 10

10. Some leaves usually alternate; spikes oblong; perianth segments densely silky hairy ....**6. Aerva**
\+ Leaves all opposite; spikes subglobose; perianth segments glabrous or sparsely pubescent ....**9. Alternanthera**

11. Flowers in threes forming spikes or dense heads; central flower of each group perfect, with unhooked perianth segments; lateral flowers imperfect, with perianth segments reduced to hooked spines ....**5. Cyathula**
\+ Flowers arranged singly in spikes, all perfect, subtended by needle-like bracts; perianth segments rigid, unhooked ....**8. Achyranthes**

1. DEERINGIA Brown

Scrambling or climbing shrubs. Leaves opposite. Flowers bisexual, rather distant in elongate spike-like racemes which form panicles. Perianth segments 5, concave, becoming reflexed in fruit. Stamens 5, anthers 2-celled, filaments united at base into a cup; pseudostaminodes absent. Ovary subglobose, stigmas 3, linear. Fruit a 3–5-seeded berry.

1. D. amaranthoides (Lamarck) Merrill; *D. celosioides* Brown. Nep: *Bakri Sag* (34), *Bangi Lahara* (117), *Benthuya* (117). Fig. 19a,b.

Stems up to 6m. Leaves ovate, 2–7 × 1.5–4cm, acute or acuminate, base rounded or truncate, often reddish-tinged, puberulous beneath; petioles 5–15mm. Spikes 10–25cm; pedicels 1–2mm. Perianth segments crimson, c 2.5mm. Berry subglobose, c 5mm, red.

Bhutan: S – Samchi, Phuntsholing, Chukka, Gaylegphug and Deothang districts, **C** – Punakha and Tashigang districts; **Sikkim.** Amongst shrubs at margins of Subtropical and Warm broad-leaved forests, 100–1800m. August–February.

2. CELOSIA L.

Erect glabrous annuals. Leaves alternate. Flowers in terminal spikes, bisexual. Bracts and 5 perianth segments whitish, papery. Stamens 5, anthers 2-celled, filaments connate into a tube at base, pseudostaminodes minute. Ovary ovoid; style 1, filiform, with capitate stigma. Fruit a 2–10-seeded circumscissile capsule.

1. C. argentea L.

Herb 30–60cm. Leaves narrowly elliptic, 4–15 × 0.5–4.5cm, tapering to acute base and apex; petioles 1–3cm. Spikes white or tinged pink, ovoid or oblong, 2–12 × 1.5cm. Perianth segments lanceolate, 5–6 × 1mm. Capsule obovoid, c 3.5mm; seeds black, glossy.

Bhutan: C –Tashigang district (Tashi Yangtsi); **Sikkim**: Darjeeling. Weed of cultivated and waste ground, 100–1200m. March–August.

The above description and records apply to the typical var. **argentea**, an introduced weed of unknown origin. The var. **cristata** (L.) Kuntze (*C. cristata* L.) is sometimes cultivated in Sikkim as an ornamental; it differs in having brightly coloured red, yellow or purple spikes in which the upper parts bear sterile flowers and are enlarged, crested, feathery or fasciated.

3. ALLMANIA Wight

Diffuse annuals. Leaves alternate. Flowers in leaf-opposed or terminal globose heads, bisexual. Bracts papery, long-pointed. Perianth segments 5, papery, free. Stamens 5, anthers 2-celled, filaments connate into tube at base, pseudostaminodes absent. Ovary ovoid, compressed; style slender with shortly bilobed stigma. Capsule ovoid, 1-seeded, circumscissile.

1. A. nodiflora (L.) Brown

Stems 30–60cm. Leaves narrowly elliptic or spathulate, 3–9 × 0.5–1cm, acute or obtuse, tapering at base, subsessile or with short petioles up to 10mm. Heads subglobose, 1–1.5cm diameter. Bracts and perianth segments narrowly lanceolate, 4–6mm, pale with green or purple midribs. Capsule c 3.5mm; seeds c 2mm.

Sikkim: Darjeeling, Siliguri etc. Weed of cultivation, 120m. May–July.

4. AMARANTHUS L.

Erect annuals, sometimes spiny. Leaves alternate. Flowers in clusters densely aggregated into spikes, unisexual, usually males above, females below. Bracts small, papery. Perianth segments 3 or 5, free, membranous with green or purplish midrib. Stamens 3 or 5, anthers 2-celled, filaments free to base, pseudostaminodes absent. Ovary ovoid, style very short, stigmas usually 3, filiform. Capsule 1-seeded, compressed, circumscissile or bursting irregularly.

FIG. 19. **Amaranthaceae** and **Cactaceae. Amaranthaceae**. a & b, *Deeringia amaranthoides*: a, portion of shoot with leaves and inflorescence; b, flower. c & d, *Aerva sanguinolenta*: c, flower surrounded by bracts and bracteoles; d, flower dissected to show stamens and ovary. e & f, *Achyranthes aspera*: e, upper part of inflorescence; f, portion of inflorescence with single maturing flower. g, *Cyathula tomentosa*: flower cluster. h–j, *Stilbanthus scandens*: h, portion of shoot with leaves and terminal inflorescence; i, flower with bract and bracteole; j, the same with perianth removed to show stamens, staminodes and ovary. k & l, *Amaranthus hybridus*: k, male flower; l, female flower. m & n, *Gomphrena celosioides*: m, flower dissected to show staminal tube and ovary; n, bracteole. o & p, *Alternanthera sessilis*: o, portion of shoot with leaves and flower clusters; p, flower with developing capsule. **Cactaceae**. q, *Opuntia vulgaris*: portion of branch segment with spines and flower. Scale: a, h × 2/5; o, q × ⅔; e × 1½; g × 5; m, n × 6; b, f, i, j, k × 7; l × 10; p × 12; c, d × 13. Drawn by M. Bates.

a
b
c
d
e
f
g
h
i
j
k
l
m
n
o
p
q
M.B.

1. Stems with axillary spines ....**4. A. spinosus**
\+ Stems unarmed 2
2. Robust herb 30–200cm; spikes 8–12mm broad; perianth segments 5; capsule circumscissile ....**1. A. hybridus**
\+ Weak herbs 10–50cm, spikes 3–7mm broad; perianth segments 3; capsule bursting irregularly or indehiscent 3
3. Erect herb; capsules not or slightly exceeding perianth; seeds somewhat glossy, minutely reticulate with scurfy warts ....**2. A. viridis**
\+ Erect or prostrate herb; capsules distinctly exceeding perianth; seeds strongly glossy, faintly striate without scurfy warts ....**3. A. lividus**

1. A. hybridus L.; *A. paniculatus* sensu F.B.I. non L. Sha: *Lasomo*. Fig. 19k,l.

Stems 30–200cm. Leaves ovate-elliptic, 5–18 × 2.5–9cm, acute, base attenuate, glabrous, green with large purplish triangular blotch beneath; petioles 1–10cm. Spikes robust, bright red, 4–15 × 0.8–1.2cm, male flowers uppermost, females closer to main rachis. Bracteoles c 5mm, with awn-like tips. Perianth segments 5, ovate, 2.5–3.5mm, keeled; male flowers with 5 free stamens and rudimentary ovary; females with oblong ovary. Capsule circumscissile, seed c 1.25mm.

Bhutan: C –Thimphu district (Paro, Thimphu and Dochong La) and Mongar district (Lhuntse), **N** –Upper Mo Chu district (Gasa); **Sikkim**. Cultivated, 2000–2850m. July–August.

Cultivated as a grain and leaf crop. The Bhutan plants belong to subsp. **hybridus** var. **erythrostachys** Moquin (*A. hypochondriacus* L., *A. hybridus* subsp. *hypochondriacus* (L.) Thellung) characterised by its bright red inflorescences (becoming green when dry).

2. A. viridis L.

Differs from *A. hybridus* in its smaller size, 10–50cm; leaves broadly ovate, 2–7 × 1–6cm, subacute or obtuse; spikes slender, 2–7 × 0.3–0.6cm; flowers mostly female with a few males above; perianth segments 3, c 1mm long; stamens 3; stigmas 2–3, minute; capsule warty, bursting irregularly or indehiscent, not or slightly exceeding perianth; seeds somewhat glossy, minutely reticulate with scurfy warts.

Bhutan: S –Phuntsholing and Gaylegphug districts, **C** –Tongsa and Tashigang districts. Weed of gardens and roadsides, 300–1300m. April–June.

3. A. lividus L.; *A. blitum* L., *A. oleraceus* L.

Very similar to *A. viridis* but erect or prostrate; capsules distinctly exceeding perianth; seeds strongly glossy, faintly striate without scurfy warts.

Sikkim: Terai and Darjeeling foothills. Weed of disturbed ground, 1050–1800m. April–August.

Two subspecies occur: subsp. **lividus** (*A. lividus* var. *ascendens* (Loiseleur) Hayward & Druce), an erect plant with larger leaves up to 12cm, closely resembling

A. viridis, and subsp. **polygonoides** (Moquin) Probst, a smaller prostrate plant with smaller leaves 1.5–2cm.

4. A. spinosus L.

Distinguished from other *Amaranthus* species by the presence of paired axillary spines 7–12mm; stems 20–100cm; leaves ovate-elliptic, 2–7 × 1–3.5cm, acute or obtuse; petioles up to 5cm; spikes 3–8cm; perianth segments 5; capsule circumscissile.

Bhutan: S–Samchi, Phuntsholing, Sarbhang and Gaylegphug districts, **C**–Punakha district; **Sikkim**. Weed of waste ground, roadsides and riverside shingle, 200–1500m. May–September.

5. CYATHULA Blume

Perennial herbs or subshrubs. Leaves opposite. Flower clusters forming terminal spikes or dense heads, each cluster composed of 1–2 central fertile flowers and an outer 1–2 sterile flowers with perianth segments reduced to hooked awns. Fertile flowers bisexual. Perianth segments 5, rigid or papery. Stamens 5, anthers 2-celled, filaments connate at base into a cup, alternating with short pseudostaminodes. Ovary obovoid or oblong, style filiform with capitate stigma. Capsule ellipsoid, 1-seeded, indehiscent; dispersed by hooks.

1. Creeping herb with slender erect stems 15–60cm; flower clusters forming slender spikes 4–6mm broad .... **3. C. prostrata**
+ Erect subshrubs 100–200cm; flower clusters aggregated into dense spikes or heads 13–20mm broad 2

2. Leaves tomentose; flower clusters forming spikes .... **1. C. tomentosa**
+ Leaves sparsely pubescent; flower clusters in oblong or subglobose heads **2. C. capitata**

1. C. tomentosa (Roth) Moquin, Fig. 19g.

Subshrub 1–2m. Leaves ovate-elliptic, 5–15 × 2–7cm, acute, base rounded, silky tomentose, paler beneath; petioles 5–15mm. Flowers in rounded white or yellowish heads 13–20mm diameter, densely or loosely aggregated into terminal spikes. Sterile flowers with hooked awns 6–7mm. Fertile flowers with perianth segments ovate-lanceolate, rigid. Stamens c 3mm, pseudostaminodes spathulate, fimbriate c 1mm. Seed c 1.75mm.

Bhutan: S–Deothang district (Tsalari Chu), **C**–Punakha district (Wangdu Phodrang, Lometsawa, Thinleygang); **Sikkim**. Margins of Warm broad-leaved forest, 900–2000m. August–September.

2. C. capitata Moquin

Similar to *C. tomentosa* but leaves sparsely pubescent, acuminate; flower clusters aggregated into short subglobose or oblong heads; sterile and fertile perianth segments 3–4mm, up to 7mm in fruit; pseudostaminodes minute, ciliate.

Bhutan: C – Tongsa district (Tashiling) and Tashigang district (Gomchu); **Sikkim**. Forest margins, 1800–2750m. July–September.

3. C. prostrata (L.) Blume

Creeping perennial herb with erect shoots 15–60cm. Leaves rhombic-elliptic, 2.5–7 × 1.5–4cm, acute, base cuneate, sparsely or densely hirsute, often reddish-tinged; petioles 2–8mm. Flower clusters forming slender spikes 4–6mm broad. Fertile flowers with perianth segments c 2mm, c 1mm in sterile flowers, enlarging to 2mm at maturity. Clusters becoming deflexed in fruit.

Bhutan: S – Sarbhang district (Lao Pani) and Gaylegphug district (Gaylegphug); **Sikkim**. Amongst leaf litter in terai forests, 150–600m. October–March.

6. AERVA Forsskal

Herbs, often woody at base, or subshrubs. Leaves alternate or opposite. Flowers in short terminal and axillary spikes, often becoming paniculate, bisexual or unisexual. Perianth segments 5, free, papery. Stamens 5, anthers 2-celled; filaments connate into a basal collar, alternating with 5 linear pseudostaminodes. Ovary compressed, style short, stigma minutely bilobed. Capsule compressed, 1-seeded, circumscissile.

1. A. sanguinolenta (L.) Blume; *A. scandens* (Roxb.) Moquin. Fig. 19c,d.

Slender erect herb or subshrub 60–200cm. Leaves elliptic or lanceolate, 3.5–7 × 1.5–2.5cm, acute, base cuneate, appressed whitish hairy especially beneath, often purplish tinged; petioles 3–10mm. Spikes ovoid, 10–20 × 6–8mm, white. Perianth segments ovate, c 2.5mm, densely white hairy. Capsule c 1mm.

Bhutan: S – Samchi, Phuntsholing, Gaylegphug and Deothang districts, **C** – Punakha, Tongsa and Tashigang districts; **Sikkim**. Margins of Subtropical and Warm broad-leaved forests, 200–1600m. November–May.

Sometimes cultivated as an ornamental border plant.

7. STILBANTHUS Hook. f.

Large woody climber, branches pendulous. Leaves opposite. Spikes terminal or axillary from uppermost nodes, often panicled, borne on pendulous peduncles. Perianth segments 5, rigid. Stamens 5, anthers 2-celled, filaments shortly connate at base, alternating with oblong, toothed pseudostaminodes. Ovary oblong, compressed, style slender, stigma minute. Fruit oblong, 1-seeded, indehiscent.

1. S. scandens Hook. f. Fig. 19h–j.

Leaves ovate-elliptic, 4–11 × 1.5–6cm, acuminate, base cuneate, appressed pilose; petioles 1–4cm. Peduncles 3–5cm. Perianth segments ovate-lanceolate, c 4.5mm, enlarging to 9mm in fruit, whitish woolly at apex. Fruit oblong c 3.5mm, hairy at apex.

Bhutan: S – Phuntsholing district (above Phuntsholing); **Sikkim**: Darjeeling, Rongbe etc. Warm broad-leaved forests, 900–2100m. June–August.

8. ACHYRANTHES L.

Perennial herbs. Leaves opposite. Flowers in elongate terminal or axillary spikes, becoming deflexed, bisexual. Bracts and bracteoles spine-tipped. Perianth segments 5, rigid. Stamens 5, anthers 2-celled, filaments connate at base into short cup, alternating with oblong, toothed pseudostaminodes. Ovary oblong, style filiform, stigma capitate. Fruit 1-seeded, indehiscent, shed with perianth and bracteoles.

1. A. aspera L.; *A. porphyristachya* Moquin. Fig. 19e,f.

Erect or sprawling herb with long branches, 25–100cm. Leaves ovate-elliptic, 4–11 × 1.5–6cm, acute, base cuneate, sparsely to densely appressed pubescent; petioles 5–10mm. Flowers at first erect-spreading and congested, becoming strongly deflexed and distant in long slender spikes up to 35cm. Bracts subulate, spinous, 3mm, basal part 1–1.5mm long, broadened and concave, adnate throughout to the spine. Perianth segments ovate-lanceolate, 3.5–4mm, greenish, rigid. Pseudostaminodes bearing a fimbriate scale, almost equalling stamens. Capsule oblong, 2mm.

Bhutan: S – Phuntsholing and Deothang districts, **C** – Punakha, Mongar and Tashigang districts; **Sikkim**. Weed of roadsides and margins of cultivation, 300–1450m. June–August.

Roots, leaves and seeds used medicinally (13).

2. A. bidentata Blume

Similar to *A. aspera* but texture softer; spikes shorter, mostly up to 10(–20)cm; basal part of bracts 0.75–1mm, free from spine except at base; pseudostaminodes very short or almost absent.

Bhutan: S – Chukka and Gaylegphug districts, **C** – Punakha district, **N** – Upper Mo Chu district; **Sikkim**. Margins of Subtropical and Broad-leaved forests, 300–2400m. August–November.

9. ALTERNANTHERA Forsskal

Prostrate or erect annual or perennial herbs. Leaves opposite. Flowers in sessile or pedunculate rounded heads, bisexual. Bracts and bracteoles papery or rigid, sometimes spine-tipped. Perianth segments 5, free. Stamens 2–5, sometimes some without anthers, anthers 1-celled, filaments united into basal cup, alternating with pseudostaminodes. Ovary with short style and capitate stigma. Fruit 1-seeded, indehiscent, falling with perianth.

1. Leaves sessile; stems with 2 lines of hairs .... **1. A. sessilis**
\+ Leaves petiolate; stems uniformly pubescent or glabrous 2

2. Leaves 5–10cm; flower clusters on short peduncles .... **4. A. brasiliana**
\+ Leaves 1–4cm; flower clusters sessile 3

3. Prostrate herb; leaves green; bracts and bracteoles glabrous .. **2. A. pungens**
\+ Erect herb; leaves purplish; bracts and bracteoles whitish pilose on back **3. A. bettzickiana**

1. A. sessilis (L.) DC. Fig. 19o,p.

Prostrate or decumbent herb rooting at nodes, often perennial, stems with 2 lines of hairs. Leaves elliptic or oblanceolate, 1–5 × 0.25–2cm, acute, attenuate at base, sessile, glabrous. Flower clusters sessile, globose, white, 5–8mm diameter. Perianth segments all similar, 2–3mm, papery, unarmed. Stamens 5, only 3 bearing anthers; basal cup very short, pseudostaminodes minute. Capsule rounded, c 2mm diameter, compressed, emarginate at apex; seed c 1mm surrounded by a thick wing c 0.5mm broad.

Bhutan: S – Samchi, Phuntsholing, Gaylegphug and Deothang districts; **C** – Punakha and Mongar districts; **Sikkim**. Weed of cultivated ground and damp roadsides, 200–1500m. March–August.

2. A. pungens Kunth

Similar to *A. sessilis* but a more compact prostrate herb; stems evenly pubescent; leaves broadly ovate, 1–2.5 × 0.75–2cm, apex rounded and apiculate, base tapering into distinct petiole up to 1cm; bracts and outer perianth segments spine-tipped, inner segments blunt, dentate, bearing barbed hairs at base; stamens 5, all bearing anthers; capsule orbicular, compressed.

Bhutan: C – Mongar district (Lingmethang, Shongar Chu) and Tashigang district (Shiri Chu). Weed of roadsides and footpaths, 950–1300m. June–July.

Native of tropical America.

3. A. bettzickiana (Regel) Voss

Erect herb 20–45cm, stems and leaves purplish. Leaves rhombic to oblanceolate, 1–2.5 × 0.5–1.5cm, acute or acuminate, often undulate. Flower clusters 5–6mm, white. Perianth segments acute with excurrent midrib, outer 3 distinctly 3-nerved and white-pilose on back, inner 2 shorter.

Bhutan: S – Samchi district (Samchi). Cultivated in garden, 500m. March.

Native of S America cultivated as an ornamental.

4. A. brasiliana (L.) Kuntze

Similar to *A. bettzickiana* but decumbent; leaves elliptic, 5–10 × 2–4cm, shortly acuminate, dark purple, base attenuate into petiole 1–2cm, flower clusters larger, 8–11mm diameter, borne on short peduncles; bracts pilose, outer perianth segments subglabrous.

Bhutan: S – Samchi district (Samchi). Cultivated in garden, 500m. March.

Native of S America cultivated for its purple foliage.

10. GOMPHRENA L.

Erect, prostrate or decumbent, annual or perennial herbs. Leaves opposite. Flowers in oblong or globose terminal heads, bisexual, each flower in axil of persistent bract and subtended by 2 bracteoles; bracteoles keeled, often crested on back, shed with fruit. Perianth segments 5, free, densely lanate on back. Stamens 5, anthers 1-celled, filaments united into tube; pseudostaminodes present or absent. Ovary with short or long style and 2 stigmas. Capsule 1-seeded, indehiscent, compressed.

1. Perennial; flower heads oblong or subglobose, white; bracteoles with short dorsal tooth near apex .... **1. G. celosioides**
\+ Annual; flower heads globose, pink to purple; bracteoles with dorsal crest from base to apex .... **2. G. globosa**

1. G. celosioides Martius. Fig. 19m,n.

Prostrate or decumbent perennial, to 60cm tall. Leaves elliptic-oblanceolate, subacute, base attenuate, pilose with long appressed hairs especially beneath; petioles 2–10mm. Flower heads subglobose or oblong, 1–4cm, closely subtended by uppermost leaves. Bracts ovate, c 3 × 2mm, bracteoles c 6 × 3mm, bearing a short dorsal tooth near apex. Perianth segments c 5.5mm, outer 3 almost flat, lanate near base, inner 2 with green midrib, lanate to apex. Capsule ovoid, compressed, c 1.75mm.

Bhutan: C –Tongsa district (near Shamgong) and Mongar district (Lingmethang, Shongar Chu). Weed of roadsides and cultivated ground, 850–1750m. June–July.

Native of S America.

2. G. globosa L.

Similar to *G. celosioides* but annual; flower heads globose, broader, 1.5–2.5cm, pink or purple; bracteoles c 8mm with a dorsal crest from base to apex.

Bhutan: S – Phuntsholing district (Torsa River); **Sikkim**. Weed on riverside shingle, 200m. May.

Possibly naturalised from cultivation; native of S America.

11. **IRESINE** Browne

Perennial herbs or subshrubs, dioecious. Leaves opposite. Flowers in spikes forming panicles, unisexual. Bracts and bracteoles membranous. Perianth segments 5, free. Stamens 5, filaments connate at base, pseudostaminodes minute. Ovary globose with very short style and 2–3 linear stigmas. Capsule 1–seeded, indehiscent.

1. I. herbstii Lindley

Erect perennial herb to 1.5m. Leaves succulent, broadly ovate or orbicular, 2.5–8 × 2.5–8cm, apex rounded or retuse, variegated purple to green with pink or yellow blotches along veins, sparsely yellowish pubescent beneath; petioles 1–2cm. Panicles 5–20cm. Bracts and bracteoles ovate, c 1mm, glabrous. Perianth segments oblong, c 1.25mm, greenish-white, with a few long hairs at base.

Sikkim. Cultivated.

Native of Brazil; only female plants are cultivated for their showy foliage.

Family 33. CACTACEAE

by D. G. Long

Large succulent shrubs; stems cylindric; branches segmented, deeply longitudinally ribbed or compressed, usually armed with spines and bristles borne on cushion-like areoles. Leaves usually absent or small and early deciduous. Flowers solitary, sessile, actinomorphic, bisexual, usually showy. Perianth of numerous free petal-like segments in several whorls. Stamens numerous, inserted at base of perianth. Ovary inferior, 1-locular; ovules numerous, parietal; style elongate with 5–7 stigmatic lobes. Fruit a berry bearing areoles, sometimes spiny, seeds numerous, enclosed in hard white aril.

1. Segments of branches columnar, deeply longitudinally ribbed, not compressed **3. Cereus**
\+ Segments of branches compressed, not ribbed 2
2. Segments usually with long spines; flowers yellow; stamens shorter than perianth **1. Opuntia**
\+ Segments without long spines; flowers red; stamens much longer than perianth **2. Nopalea**

1. **OPUNTIA** Miller

Much branched shrubs, segments of branches strongly compressed, obovate. Areoles woolly, bearing short bristles and 1–8 long spines. Flowers subterminal, perianth yellow, segments spreading. Stamens much shorter than perianth.

1. O. vulgaris Miller; *O. monacantha* (Willdenow) Haworth. Dz: *Jawairinga Tsang*; Eng: *Prickly Pear*. Fig. 19q.

Suberect or sprawling shrub 2–4m; trunk spiny or smooth; branch segments obovate, flattened, 10–30 × 5–12cm; areoles whitish woolly with a few brown barbed bristles; spines 1–2 per areole, 2–4(–6)cm, terete, pale greyish with brown tip. Leaves subulate, 2–3mm. Flowers 5–7cm diameter; perianth segments bright yellow, outer tinged with red. Fruit obovoid, 5–7 × 2–4cm, reddish, spineless but with numerous barbed bristles on areoles.

Bhutan: S – Phuntsholing district (Phuntsholing), **C** – Punakha district (Punakha and Wangdu Phodrang) and Tongsa district (Mangde Chu valley). Hillsides in dry valleys, 250–1500m. April–June.

Native of S America, naturalised in Bhutan. Formerly used as a host for the cochineal insect. Fruit edible, but barbed bristles highly irritant.

2. O. elatior Miller

Similar to *O. vulgaris* but branch segments often larger, to 40cm long; areoles brownish woolly; spines 3–8 per areole, 1.5–3cm long, slightly compressed at base, brown throughout, sometimes blotched.

Sikkim: introduced at Darjeeling (34).

2. NOPALEA Salm-Dyck

Similar to *Opuntia* but areoles without long spines; perianth segments red, erect, much shorter than long-exserted stamens.

1. N. cochenillifera (L.) Salm-Dyck; *Opuntia cochenillifera* (L.) Miller. Eng: *Cochineal Cactus*.

Shrub 2–4m. Branch segments obovate, 15–20 × 7–10cm, smooth. Areoles white, spines absent, bristles early caducous. Flowers numerous, subterminal, bright scarlet. Perianth segments erect, obovate, 6–13 × 4–6mm. Stamens exserted c 1.5cm beyond perianth. Stigma yellow, 6–7-lobed, shortly exserted beyond stamens. Fruit red, c 5cm long.

Bhutan: Phuntsholing district (Phuntsholing). Cultivated in garden, 250m. February.

Cultivated as an ornamental. Formerly cultivated in India as a host for the cochineal insect. Native of tropical America.

3. CEREUS Miller

Little-branched shrubs, segments of branches columnar, deeply longitudinally ribbed; areoles borne on ribs, woolly and spiny. Flowers large, nocturnal, showy. Perianth early deciduous. Stamens numerous, not exserted.

1. C. peruvianus (L.) Miller. Eng: *Hedge Cactus*.

Columnar shrub or tree up to 10m or more. Branch segments elongate, 10–12cm diameter, slightly glaucous-green. Ribs 6(–9), when young almost flat, becoming broadly triangular when old. Areoles whitish, bearing 9–12 yellowish-brown spines 1–2cm. Flowers white, red or brown, c 15cm long. Fruit orange-yellow, subglobose, c 3.5cm diameter.

Bhutan: S – Phuntsholing district (Phuntsholing). Cultivated in garden, 250m. Native of S America cultivated as a hedge plant.

Family 34. MAGNOLIACEAE

by A.J.C. Grierson

Trees or shrubs. Leaves mostly coriaceous, alternate, simple, entire, pinnately veined; stipules large enclosing young growth, later deciduous leaving annular scars around stem and often longitudinal scars on petioles. Flowers axillary or terminal, solitary, bisexual, enclosed at first by deciduous spathe-like bracts. Perianth parts 9–12(–15) in whorls of 3, petaloid or the outer whorl sometimes sepal-like. Stamens numerous, spirally arranged, anthers usually with an apical appendage. Carpels numerous, rarely few, free or united, spirally arranged on an elongated receptacle. Fruiting carpels dehiscing longitudinally or transversely; seeds 1 or more suspended at maturity by thread-like funicles.

1. Flowers axillary .... **2. Michelia**
\+ Flowers terminal 2

2. Stipules free from petiole; receptacle shorter than stamens ...**4. Alcimandra**
\+ Stipules adnate to petiole leaving scar 0.5–7cm; receptacle longer than stamens 3

3. Deciduous or evergreen; leaves ovate, elliptic or obovate, acute or obtuse; carpels free, dehiscing along dorsal suture .... **1. Magnolia**
\+ Evergreen; leaves oblanceolate, shortly acuminate; carpels united, dehiscing transversely around base .... **3. Talauma**

FIG. 20. **Magnoliaceae, Tetracentraceae** and **Eupteleaceae. Magnoliaceae**. a, *Michelia doltsopa*: fruiting receptacle. b & c, *Talauma hodgsonii*: b, fruiting receptacle with two carpels remaining attached, seeds suspended by thread-like funicles; c, dehisced carpel. d–g, *Magnolia campbellii*: d, apex of shoot bearing leaves and terminal bud enclosed by stipule; e, flower; f, fruiting receptacle; g, seed bearing part of thread-like funicle. **Tetracentraceae**. h–k, *Tetracentron sinense*: h, portion of branch bearing short-shoot with leaf and infructescence; i, portion of inflorescence; j, flower; k, dehisced fruit. **Eupteleaceae**. l–o, *Euptelea pleiosperma*: l, portion of shoot bearing leaves and infructescence; m, flower; n, carpel bearing stigmatic crest along upper margin; o, fruiting carpel surrounded by papery wing. Scale: d × 1/5; e, f, h, l × ½; b × 3/5; a, c, g × 1; i × 1½; o × 2; m × 2½; j, k × 7; n × 12. Drawn by M. Bates.

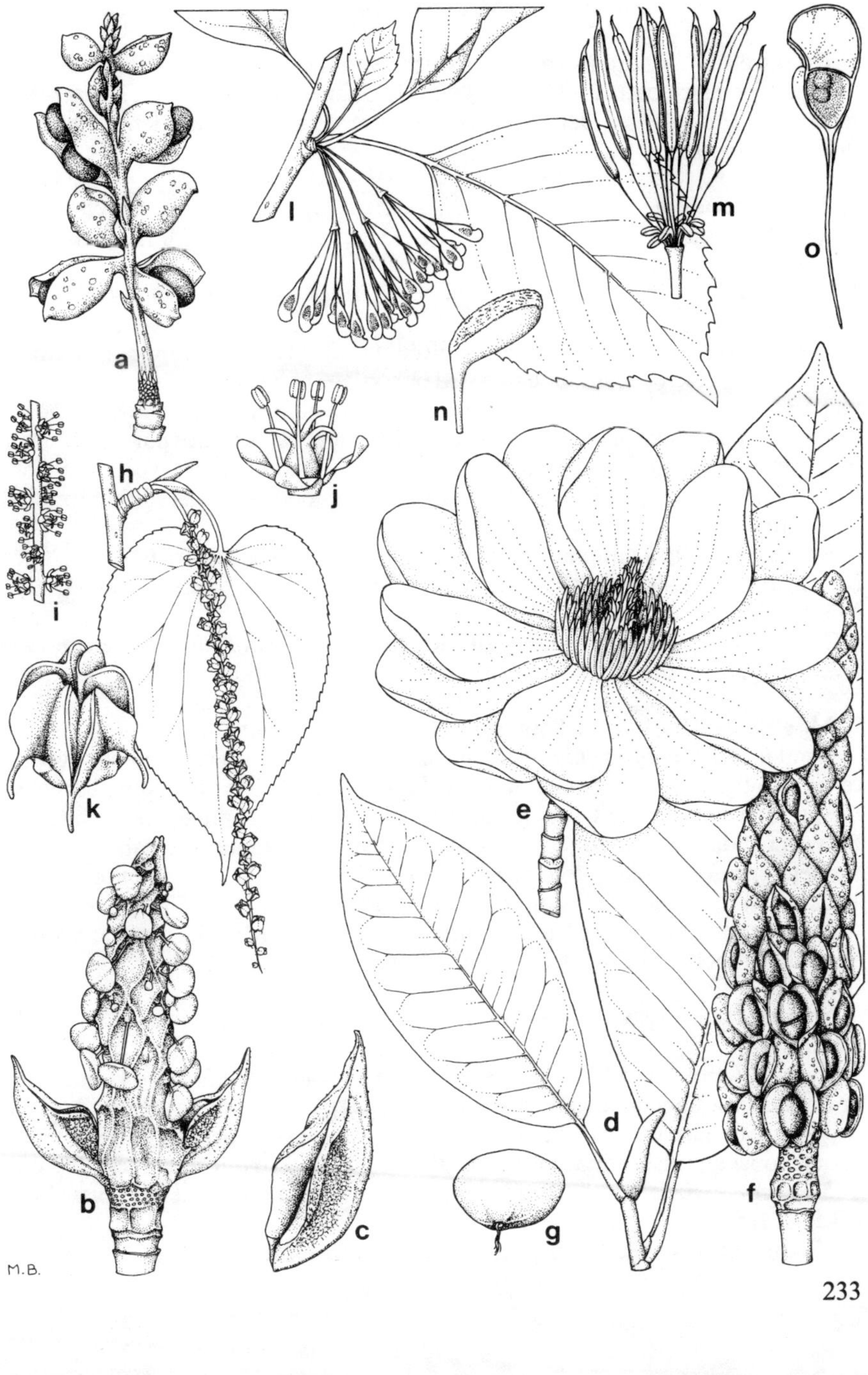
a
b
c
d
e
f
g
h
i
j
k
l
m
n
o
M.B.

1. MAGNOLIA L.

Evergreen or deciduous trees or shrubs. Stipules adnate to petioles. Flowers terminal, perianth parts 9–12 in 3–4 whorls. Stamens several-seriate, shorter than receptacle. Carpels free, densely crowded on receptacle, in fruit dehiscing along dorsal suture; seeds 1–2.

1. Leaves deciduous ....**Species 1 & 2**
\+ Leaves evergreen ....**3. M. pterocarpa**

1. M. campbellii Hook. f. & Thomson. Nep: *Ghoge Chanp, Lal Chanp* (34). Fig. 20d–g.

Deciduous tree, 12–15m. Leaves unfolding after flowers, elliptic to obovate, 25–33 × 10–14cm, acute, base rounded or cordate, pubescent at first, veins 10–15 pairs; petioles 3–4cm, stipular scar 5–7mm. Flowers erect c 15cm across, enclosed at first by spathe-like bract 4.5–6cm, densely sericeous; perianth parts c 12, outer ones yellowish, inner white, purple near base. Stamens and carpels purplish. Fruiting receptacle cylindrical, c 15 × 3cm; carpels obovoid, laterally compressed 1–1.5cm.

Bhutan: C – Ha, Thimphu, Punakha, Tongsa and Sakden districts, N – Upper Mo Chu district; **Sikkim**. Moist Hemlock/Rhododendron forests, 2000–3100m. March–May.

Two forms occur in Sikkim, pink-flowered (*Lal Chanp*) and white-flowered (*Ghoge Chanp*); the latter is the common form in Bhutan.

2. M. globosa Hook. f. & Thomson. Nep: *Kokre Chanp* (34)

Similar to *M. campbellii* but leaves unfolding before flowers, subcoriaceous, ovate-elliptic, 11–22 × 5–11cm, acute, base rounded, veins c 8 pairs, brownish hirsute beneath especially on veins; petioles 1.5–4cm, stipular scar ± as long; flowers pendulous c 15cm across; perianth parts c 9, white or cream-coloured, stamens crimson; fruiting receptacle 3–5.5 × 2cm, carpels 12–20, elliptic-ovoid, 1–1.5 × 0.7cm, shortly (c 2mm) beaked at apex.

Bhutan: C –Thimphu to Mongar districts, **N** – Upper Mo Chu district; **Sikkim**. Moist Hemlock and Fir forests, 2440–3200. May–July.

3. M. pterocarpa Roxb.; *M. sphenocarpa* Hook. f. & Thomson. Nep: *Patpate* (34), *Chanp* (34).

Evergreen tree up to 30m. Leaves elliptic to obovate, 18–35 × 9–18cm, acute or obtuse, base attenuate, veins prominent, 16–20 pairs, glabrous; petioles 2–4cm, stipular scar ± as long as petiole. Flowers fragrant, c 8cm across, perianth parts 9, outer ones greenish, coriaceous, inner thinner, white. Fruiting receptacle ellipsoid, c 15 × 7cm, carpels 2 × 1.2cm narrowed at apex into a flattened lanceolate beak c 3.5 × 1mm.

Sikkim: near Sivoke. Terai forests, 150m. April–June.

2. MICHELIA L.

Leaves and flowers similar to those of *Magnolia* but flowers axillary. Fruiting carpels forming a loose spike, dehiscing by dorsal suture.

1. Petioles unscarred; carpels 1–5 ....**6. M. kingii**
\+ Petioles bearing a stipular scar; carpels numerous 2

2. Stipular scar 3–7mm, less than half as long as petiole ....**Species 1 & 2**
\+ Stipular scar 0.7–2.5cm, more than half as long as petiole ....**Species 3–5**

1. M. doltsopa DC.; *M. excelsa* (Wall.) Wall. Nep: *Seto Chanp* (34), *Pulney Champ, Rani Champ*. Fig. 20a.

Evergreen tree 8–25m, buds greyish-brown sericeous. Leaves elliptic or oblanceolate, 15–25 × 5–7cm, acute or acuminate, base cuneate, densely pubescent beneath at first; petioles 2–3cm, stipular scar 5–7mm. Flowers white, 10–15cm across, perianth parts 9–12, oblanceolate. Stamens yellowish. Fruiting receptacle 7–12cm, carpels ovoid, 1–1.5 × 0.7–1cm, acute, slightly compressed.

Bhutan: S–Chukka and Gaylegphug districts, **C**–Punakha, Tongsa and Tashigang districts; **Sikkim**. Evergreen oak forests, 1830–2500m. February–March.

Often cultivated; a very valuable timber tree used for building and furniture-making (34).

2. M. velutina DC.; *M. lanuginosa* Wall. Nep: *Phusre, Guay Chanp* (34), *Gogey Champ*.

Similar to *M. doltsopa* but leaves narrower, 17–25 × 3–6cm, silky greyish pubescent beneath at maturity; petioles 1–2cm, stipular scar 3–5mm; flowers c 5cm across, perianth parts oblanceolate, 3.5×0.7cm; carpels densely pubescent at first, glabrous at maturity, 1.5–2 × 1–1.2cm, rounded.

Bhutan: S–Chukka and Gaylegphug districts, **C**–Punakha, Tongsa and Mongar districts; **Sikkim**. Warm evergreen broad-leaved forests, 1800–2100m. August–September.

Sometimes cultivated for its useful timber.

3. M. kisopa DC.

Tree 10–20m, glabrous except for grey sericeous buds. Leaves ovate-elliptic, 10–18 × 3–7cm, gradually acuminate, base rounded or cuneate; petioles 2–4cm, stipular scar 1–2.5cm. Flowers pale yellow, rarely white, 3–4cm across; perianth parts oblanceolate, 2–3 × 0.7cm. Fruiting receptacles c 5cm, carpels ellipsoid, 10 × 7mm.

Bhutan: C–Punakha district (Tinlegang to Gonchunang) and Tashigang district (Tashiyangsi Chu valley); **Sikkim**. Warm broad-leaved forests, 1600–1800m. August–October.

4. M. punduana Hook. f. & Thomson. Nep: *Chanp* (34).

Similar to *M. kisopa* but leaves smaller, elliptic-oblanceolate, 8–11 × 3–5.5cm, abruptly acuminate, base cuneate; petioles 1–1.5cm, stipular scar 0.7–1.2cm; flowers white.

Bhutan: S–Samchi district (Sangura); **Sikkim**. Warm broad-leaved forests, 1300m.

5. M. champaca L. Sha: *Chhampe Shing*; Nep: *Chanp, Aule Chanp* (34).

Similar to *M. kisopa* but leaves ovate-elliptic, 15–25 × 6–9cm, long acuminate, base cuneate, pubescent beneath at first; petioles 2–3cm, stipular scar 1–2cm; flowers c 6cm across, yellow or orange; fruiting receptacle c 12cm.

Sikkim: foothills and terai, rare. Subtropical forests, 200–600m. May–June.

A very valuable timber species, sometimes cultivated (34).

6. M. kingii Dandy; *M. montana* auct. non Blume

Tree. Leaves subcoriaceous, obovate, 15–19 × 7–10cm, acute, base cuneate, glabrous or greyish pubescent on veins; petioles 2–2.5cm. Flowers white, c 4cm across, perianth parts 9, oblanceolate, 2–2.5cm. Carpels (1–)3–4(–5), usually only one maturing, ellipsoid, 4–7cm.

Sikkim: near Jalpaiguri. Terai forests. September.

3. TALAUMA Jussieu

Evergreen trees. Leaves and flowers as in *Magnolia* but carpels united, dehiscing transversely around base, upper parts of pericarp falling away at maturity to leave seeds suspended within persistent lower portions; seeds solitary.

1. T. hodgsonii Hook. f. & Thomson. Dz: *Khem*; Sha: *Kadering Shing*; Nep: *Hare* (34), *Chiuri* (117), *Balu Kat (Kath)*. Fig. 20b,c.

Tree 8–12m. Leaves oblanceolate, 20–45 × 7–15cm, acuminate, base attenuate, glabrous, veins 18–22 pairs, prominent; petioles 2.5–7cm with stipular scar along entire length. Flowers 7–10cm across, fragrant, perianth parts 9, outer whorl thicker, white, inner whorls purplish. Stamens yellowish. Fruiting head ellipsoid, 8–10 × 6–7cm; carpels ovoid, 3–4.5 × 1cm, tapering to sharp beak.

Bhutan: S–Sarbhang, Gaylegphug and Deothang districts, **C**–Punakha, Mongar and Tashigang districts; **Sikkim**. Warm broad-leaved forests, 270–1850m. May.

Wood good for furniture-making (16).

A cultivated specimen from Rongbe, Sikkim, has been identified as *T. rabaniana* Hook. f. & Thomson, a native of Khasia. It is doubtfully distinct from *T. hodgsonii*, but reported as differing in its lanceolate leaves, and fruit with shortly beaked carpels borne in shallower pits.

4. ALCIMANDRA Dandy

Tree. Leaves and flowers similar to *Magnolia* but petioles without stipular scars and receptacle scarcely longer than stamens. Carpels free, dehiscing at maturity by dorsal suture; seeds 1–4.

1. A. cathcartii (Hook. f. & Thomson) Dandy; *Michelia cathcartii* Hook. f. & Thomson. Dz: *Khashi*; Nep: *Tite Champ*.

Leaves subcoriaceous, elliptic, 10–13 × 3–5cm, acuminate, base rounded or cuneate, ± densely pubescent on midrib; petioles 0.5–1cm. Flowers creamy-white often drying reddish, 7.5–10cm across, fragrant. Mature carpels ovoid, 1–1.5cm, compressed.

Bhutan: S – Chukka district (Chasilakha and Jumudag); **Sikkim**: Dumsong etc. Evergreen oak forests, 1500–2100m. April–July.

Timber valuable for house-building (34).

Family 35. ANNONACEAE

by A. J. C. Grierson

Trees, shrubs or climbers, indumentum sometimes stellate. Leaves alternate, entire, pinnately veined, exstipulate. Flowers usually bisexual, actinomorphic, terminal, axillary, extra-axillary or leaf-opposed, solitary or in cymes or panicles. Sepals 3, free, valvate. Petals 6, in 2 whorls, alike or unlike, valvate or imbricate, outer whorl sometimes sepaloid. Stamens usually numerous, minute, filaments short and thick, anthers opening by longitudinal slits, connective truncate or slightly elongated, sometimes laterally expanded hiding anther cells. Carpels minute, usually numerous, superior, free; styles truncate or shortly elongated. Fruiting carpels dry or fleshy, usually stalked, radiating umbellately from receptacle, rarely (*Annona*) becoming united into a many-celled fruit. Seeds 1 or several in each carpel, sometimes horizontal; endosperm ruminate.

1. Climbing shrubs with branches bearing conspicuous downwardly directed hooks **6. Artabotrys**
\+ Scrambling or erect shrubs or trees, branches without hooks 2

2. Sepals and usually leaf undersides stellately hairy ....**1. Uvaria**
\+ Sepals and leaf undersides with simple hairs or glabrous 3

3. Inner petals as long as or longer than outer ones 4
\+ Inner petals shorter than outer ones 6

4. Inner petals thick and triquetrous above ....**4. Fissistigma**
\+ Inner petals flat above 5

5. Sepals and outer petals equal and alike; inner petals sometimes pouched and coherent at base..**2. Miliusa**
\+ Sepals much shorter than and unlike petals; inner petals not pouched, free to base ....**8. Polyalthia**
6. Petals thick and fleshy, inner ones much reduced (1–2mm) or absent; carpels becoming fused to form ovoid-globose fruit..........................**9. Annona**
\+ Petals not thick and fleshy, inner ones half or more times the length of the outer; carpels remaining free at maturity..7
7. Inner petals 4–7cm, free; fruiting carpels elongated, constricted between seeds **3. Desmos**
\+ Inner petals not more than 2cm, lightly coherent above; fruiting carpels not constricted between seeds 8
8. Outer petals spreading, inner ones broadly triangular above, cohering to form a rounded dome over stamens and carpels........................**5. Mitrephora**
\+ Outer petals ± erect, inner ones narrowly triangular above, cohering to form a narrow cone ....**7. Goniothalamus**

1. UVARIA L.

Scrambling shrubs with stellate pubescence. Flowers solitary or 2–3, terminal or leaf-opposed. Petals imbricate, ± equal in size, spreading. Fruiting carpels ovoid or subglobose, fleshy, several-seeded, stalked.

1. U. hamiltonii Hook. f. & Thomson

Stems densely brownish pubescent. Leaves elliptic-obovate, 7–20 × 5–9cm, acute or acuminate, base rounded or shallowly cordate, brownish pubescent especially beneath; petioles c 5mm. Flowers 5–7cm across, pedicels 2–4cm. Sepals broadly ovate, 1–1.5 × 1cm, brown tomentose. Petals red, obovate-spathulate, 2–3 × 0.8–1.5cm, finely tomentose. Fruiting carpels obovoid-ellipsoid, 2–2.5 × 1.5–2cm, borne on stalks 1.5–3cm, reddish-brown tomentose.

Sikkim: Kalighara and Ryang. Terai forests, 300–600m. May–June.

FIG. 21. **Annonaceae**. a–e, *Artabotrys caudatus*: a, portion of shoot with leaf and inflorescence; b, inner petal; c, stamen; d, carpel; e, portion of shoot with persistent hook-like peduncle. f & g, *Polyalthia simiarum*: f, flower seen from above; g, fruiting carpels. h–j, *Desmos dumosus*: h, portion of shoot bearing leaves and flower; i, portion of stem with thickened loop; j, fruiting carpels. k & l, *Mitrephora harai*: k, portion of shoot with leaf and flowers; l, flower from above with one inner petal detached and folded back to expose stamens. m, *Annona reticulata*: fruit. Scale: i × 2/5; g × ½; e, h, k, m × ⅔; j × ¾; a × 4/5; l × 1 1/5; f × 1½; b × 5; c, d × 7. Drawn by M. Bates.

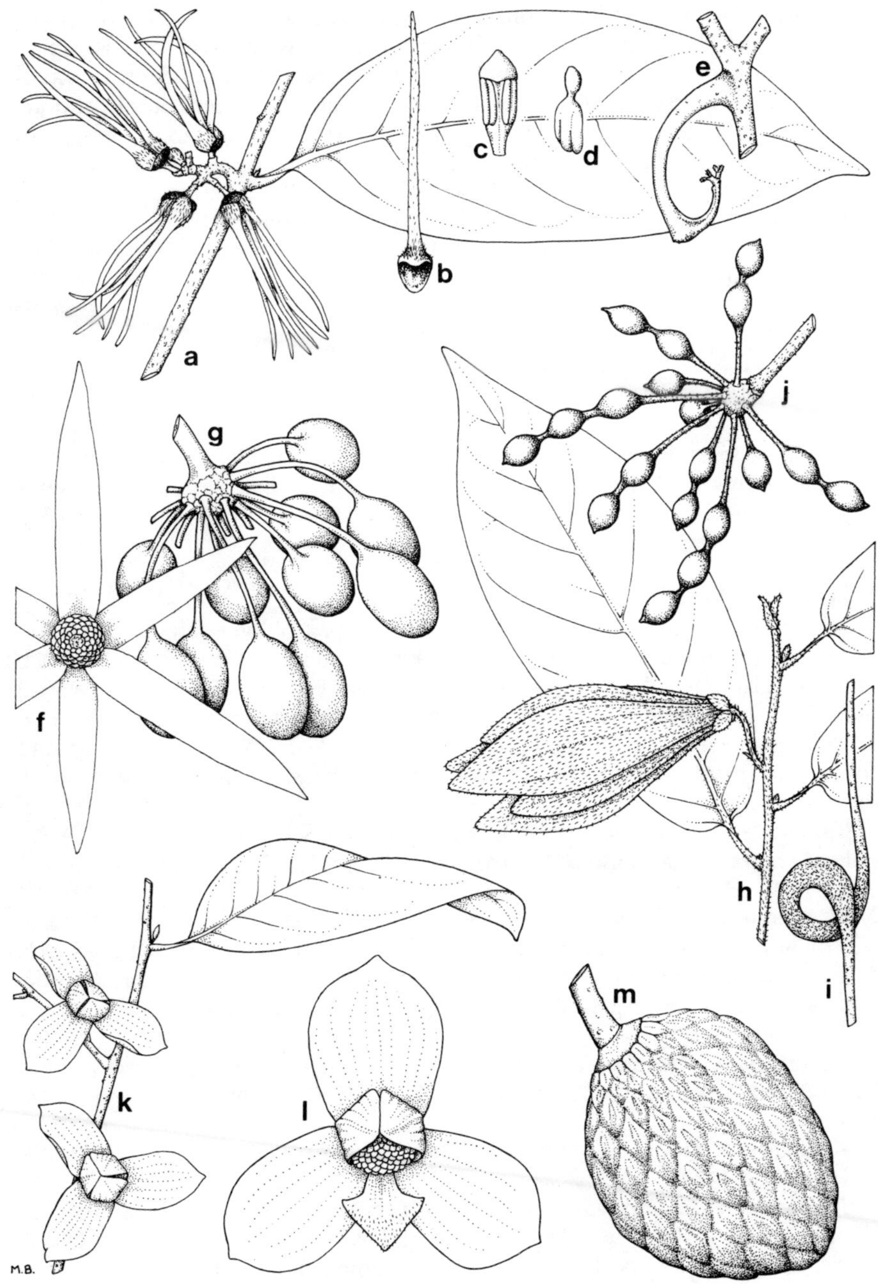
a
b
c
d
e
f
g
h
i
j
k
l
m
M.B.

2. U. lurida Hook. f. & Thomson. Nep: *Rabu Lahara* (34).

Similar to *U. hamiltonii* but stems becoming glabrous; leaves elliptic, 11–25 × 4–7cm, acuminate, base rounded, pubescent at first beneath, later glabrous; sepals and pedicels ± warted and densely covered with stellate hairs; petals purple, obovate 1.5–2 × 1–1.3cm, inflexed above, pubescent on both surfaces, densely so externally; fruiting carpels cylindric, 3–4 × 1.25–1.5cm, rounded and mucronate at apex, narrowed and 3-ridged at base, borne on stalks 10–15cm, glabrous.

Sikkim: Mungpo Peak. Subtropical forest, 450–900m. October.

Records from Sikkim of *U. macropoda* Hook. f. & Thomson, a Ceylonese plant, refer to this species.

2. MILIUSA A.DC.

Shrubs or trees. Flowers axillary or apparently extra-axillary, solitary or 2–3, bisexual or unisexual. Sepals and outer petals valvate, similar, ± equal; inner petals larger, thickened and ± saccate at base, cohering at edge at first, valvate. Fruiting carpels subglobose, thin-walled, stalked.

1. M. roxburghiana (Wall.) Hook. f. & Thomson

Dioecious shrub or small tree, 3–4m. Leaves oblong-elliptic, 5–15 × 2.5–4.5cm, acuminate, base rounded, appressed pubescent on midrib beneath, otherwise glabrous, subsessile or petioles up to 5mm. Pedicels 1.5–2cm with elliptic bract 3 × 1mm near middle. Sepals and outer petals narrowly ovate, 3–4 × 1.5mm, ± spreading. Inner petals crimson, ovate, 1–1.2 × 0.7cm, ± erect and coherent at first. Fruiting carpels obovoid or subglobose, 7–8mm diameter, borne on stalks 1.5–2cm.

Bhutan: S – Sankosh district (Pinkhua) and Gaylegphug district (Birti); **W Bengal Duars**: Buxa; **Sikkim**. Subtropical forests, 150–330m. March.

2. M. macrocarpa Hook. f. & Thomson. Nep: *Kalikat*.

Similar to *M. roxburghiana* but flowers bisexual, borne on ebracteate pedicels 4–5cm; sepals and outer petals ovate c 10 × 7mm, acute, glabrous outside, tomentose within; inner petals pink, oblong c 20 × 8mm; fruiting carpels oblong, c 15 × 7mm.

Bhutan: S – Gaylegphug district (Sham Khara), **C** – Tashigang district (Yonpu La); **Sikkim**. Warm broad-leaved forests, 1250–1650m. April–May.

3. M. longiflora (Hook. f. & Thomson) Sinclair; *Saccopetalum longiflorum* Hook. f. & Thomson

Similar to *M. roxburghiana* but flowers bisexual, precocious; leaves obovate-elliptic, 15–22 × 7–9cm, shortly acuminate, base rounded or cuneate, glabrous above, pubescent at first beneath; petioles c 5mm; inner petals erect, oblong-lanceolate, 3–3.5cm, acuminate from pouched base c 1cm broad, puberulous; fruiting carpels black, subglobose, 2.5–3cm diameter, on stalks c 2cm.

Sikkim: Ryang Valley and Mungpo. Subtropical forest, 300–1500m. March.

3. DESMOS Loureiro

Scrambling shrubs. Flowers extra-axillary, leaf-opposed or axillary. Petals valvate, elongate, inner shorter. Fruiting carpels elongate, constricted between seeds.

1. D. dumosus (Roxb.) Safford; *Unona dumosa* Roxb. Nep: *Male Lahara*. Fig. 21h–j.

Climbing shrub, stems developing thickened loops and coils. Leaves elliptic-oblanceolate, 7–16 × 3.5–6cm, acute or acuminate, base rounded or shallowly cordate, sparsely pubescent on midrib above, brownish pubescent beneath; petioles 2–5mm, pubescent. Pedicels 1.5–2cm with an ovate submedian bract 2–5mm, pubescent. Sepals ovate, 7–12 × 5–7mm, pubescent. Petals yellowish, oblanceolate, outer ones 6–7 × 1.5–2cm, narrowed at base, inner 4.5–5 × 0.8–1cm, densely fine-pubescent. Fruiting carpels dark red at maturity, up to 3cm, 1–4-seeded, basal stalks 0.7–1cm.

Bhutan: S–Samchi district (Gokti and Samchi) and Gaylegphug district (Gaylegphug); **Sikkim**. Subtropical forest, 270–700m. May.

2. D. chinensis Loureiro; *Unona discolor* Vahl

Similar to *D. dumosus* but more slender; leaves oblong-elliptic 7–11 × 2.5–4cm, glabrous above, greyish beneath, glabrous or minutely pubescent; pedicels slender, 3–5cm; sepals nearly glabrous; petals glabrous or minutely pubescent.

Bhutan: S–Samchi district (Chourini Hills) and Phuntsholing district (Phuntsholing); **Sikkim**. Subtropical forest, 300–600m. May–July.

3. D. praecox (Hook. f. & Thomson) Safford; *Unona praecox* Hook. f. & Thomson

Similar to *D. dumosus* but flowers precocious or appearing with young leaves; leaves narrowly ovate, 12–15 × 4–5cm, acuminate, base cuneate, glabrous; petioles c 5mm; flowers borne at base of leafy shoot, pedicels 1.5–3cm; sepals ovate-lanceolate, 10–15 × 3–4mm, sparsely pubescent, reflexed; petals narrowly elliptic, 5–8.5 × 0.3–0.6cm, glabrous.

Sikkim: Ryang. Subtropical forests, 600–1200m. April.

4. FISSISTIGMA Griffith

Scrambling shrubs. Flowers solitary or several in a cyme, terminal, leaf-opposed or axillary. Petals valvate, leathery, outer whorl ovate, ± flat, inner smaller, triquetrous above. Fruiting carpels ovoid or globose, stalked.

1. F. polyanthum (Hook. f. & Thomson) Merrill; *Melodorum polyanthum* Hook. f. & Thomson

Branches dark brown, glabrous. Leaves ± coriaceous, oblong-elliptic, 6–12 × 2.5–4cm, acute or acuminate, base rounded or cuneate, minutely appressed pubescent beneath; petioles up to 1cm. Flowers usually 3–5 in leaf-opposed subumbellate cymes, pedicels 8–10mm with an ovate bract c 2mm near middle. Sepals ovate, c 4 × 2.5mm, greyish tomentose. Petals orange-red, outer ovate, c 1cm, broadened and concave at base, tomentose outside, inner ones c 5mm, thicker, glabrous except near apex externally. Fruit unknown.

Bhutan: S – Gaylegphug district (Tama to Birti). Subtropical forests, 1000m. April.

2. F. bicolor (Roxb.) Merrill; *Melodorum bicolor* (Roxb.) Hook. f. & Thomson

Similar to *F. polyanthum* but branches brownish tomentose; leaves oblong-elliptic, 10–15 × 4–5cm, acute or obtuse, base rounded, appressed pubescent beneath; petals ovate, outer ones c 2 × 1cm densely brownish pubescent, inner ± as long but half as wide, glabrous; fruiting carpels globose, c 2.5cm diameter, pubescent, borne on stalks c 1cm.

Sikkim: foothills. Subtropical forests, 300m.

5. MITREPHORA Hook. f. & Thomson

Trees. Flowers solitary or 2–3, leaf-opposed. Petals valvate, outer whorl spreading, inner clawed at base with upper part broadened, incurved and cohering at first near apex. Fruiting carpels ovoid, stalked or sessile.

1. M. harai Ohashi. Nep: *Kalikat*. Fig. 21k,l.

Tree to 15m, branches brownish tomentose at first, later glabrous. Leaves ovate-elliptic, 9–18 × 4.5–6.5cm, acuminate, base rounded, glabrous above, appressed brownish pubescent beneath. Flowers 4–5cm across, pedicels 4–6mm. Sepals broadly ovate 6–7 × 6–7mm, densely brown-pubescent. Outer petals white, oblong-ovate 2–2.3 × 1.2–1.8cm, distinctly veined, inner petals purplish, basal claw 5–10 × 3–3.5mm, upper part triangular 7–8 × 10–14mm. Fruit unknown.

Bhutan: S – Gaylegphug district (near Gaylegphug), **C** – Tongsa district (Pertimi to Tintibi Bridge); **Sikkim**: Legship, Relli Chu and Tarku. Warm broad-leaved and Subtropical forests, 300–1200m. March–May.

6. ARTABOTRYS Brown

Shrubs climbing by means of persistent hook-like peduncles. Flowers solitary or several clustered on leaf-opposed peduncles. Petals valvate, concave at base, inner whorl connivent around stamens and carpels. Fruiting carpels ovoid, subsessile.

1. A. caudatus Hook. f. & Thomson. Nep: *Kali Lahara*. Fig. 21a–e.

Leaves oblong-elliptic, 10–15 × 4–5cm, acuminate, base cuneate, glabrous; petioles up to 1cm. Peduncles becoming flattened, angular, 2–2.5cm, pedicels c 5mm, bracteate at base. Sepals broadly ovate, 4–5 × 4–5mm, appressed pubescent. Petals yellow, linear, erect, 2.5–3cm, appressed pubescent, outer 2–3mm broad, inner c 1mm broad, widening at concave base to 3–4mm. Fruiting carpels unknown.

Bhutan: S – Sarbhang district (above Sarbhang and near Phipsoo); **Sikkim.** Subtropical forests, 350–1000m. May.

2. A. hexapetalus (L. f.) Bhandari; *A. odoratissimus* Ker, *A. uncinatus* (Lamarck) Baillon. Nep: *Katar Champa.*

Similar to *A. caudatus* but leaves up to 22 × 6cm; flowers greenish-yellow, strongly scented; petals spreading from clawed coherent base, ovate, outer ones c 2.3 × 1.3cm, inner c 1.5 × 0.7cm; fruiting carpels ovoid c 3 × 1.5cm, 1–2-seeded.

Bhutan: S – Cultivated, e.g. at Samchi. 500m. March.

Native of S India and Ceylon, cultivated for its sweet-scented flowers.

7. GONIOTHALAMUS Hook. f. & Thomson

Shrub or small tree. Leaves coriaceous, distinctly veined. Flowers solitary, usually axillary. Petals valvate, outer ± erect, inner clawed at base with upper part broadened and cohering in a narrow cone. Fruiting carpels ± sessile.

1. G. sesquipedalis (Wall.) Hook. f. & Thomson. Nep: *Sane* (34).

Leaves coriaceous, oblong, 20–40 × 4.5–10.5cm, acuminate, base cuneate, glabrous, lateral veins 16–20 pairs impressed above, interarching 5–8mm from recurved margin. Flowers greenish-yellow on pedicels up to 5mm, bracteate at base. Sepals ovate, c 6 × 5mm, acute. Outer petals ovate-lanceolate, 1.5–2 × 0.5cm; inner petals c 12 × 4mm, shortly clawed and free in lower part, cohering in ovate-acuminate upper part. Fruit ovoid, 2–2.5 × 1cm, acute, orange-red when mature.

Sikkim: Sivoke, Rungit etc. Subtropical forest, 300–1000m. May–September.

8. POLYALTHIA Blume

Trees. Flowers borne in axils of fallen leaves, few or numerous. Petals valvate, flat, inner slightly longer than outer. Fruiting carpels 1-seeded, stalked.

1. P. simiarum (Hook. f. & Thomson) Hook. f. & Thomson. Nep: *Khutti* (34), *Labshi* (34), *Lapchey, Lapche Kat*. Fig. 21f,g.

Tree 15–20m. Leaves ± coriaceous, elliptic-oblanceolate, 15–25 × 5–10cm, acuminate, base rounded or shallowly cordate, glabrous; petioles to 1cm. Flowers usually 2–3 on peduncles 3–6mm, pedicels 3–4cm with a rounded bract c 1mm near middle. Sepals broadly ovate, 3 × 3mm. Petals linear-lanceolate, outer 1.5–2 × 0.3–0.4cm, inner 2.5–3 × 0.3–0.4cm, yellowish. Fruit ellipsoid, c 3 × 1.75cm, orange at first, later black, borne on 3–4cm stalks.

Bhutan: S – Samchi, Sarbhang and Gaylegphug districts; **Sikkim**. Subtropical forests, 300–400m. May–June.

Wood tough and durable, used for axe and hoe handles (34).

2. P. longifolia (Sonnerat) Thwaites. Hindi: *Asok* (34).

Similar to *P. simiarum* but leaves pendulous, narrowly lanceolate, 9–19 × 2–3cm, gradually acuminate, base rounded, margin undulate; petioles 5–10mm; flowers 5–10 in subumbellate clusters on peduncles 3–10mm, pedicels 1.5–2.5cm; outer petals 6–7 × 2mm, inner 10–12 × 2.5mm; fruiting carpels ellipsoid, c 2.5 × 1.75cm, borne on stalks 1–1.5cm.

Bhutan: S – Cultivated, e.g. at Samdrup Jongkhar; **Sikkim**. 500m.

Native of S India and Ceylon, cultivated for its decorative habit and foliage.

9. ANNONA L.

Trees or shrubs. Flowers solitary or in few-flowered cymes, terminal, leaf-opposed or extra-axillary. Petals valvate, outer fleshy, ± erect, inner reduced or absent. Carpels becoming united into a fleshy fruit.

1. A. reticulata L. Eng: *Custard Apple, Bullock's Heart*. Fig. 21m.

Tree, 6–12m. Leaves oblong-lanceolate, 10–20 × 2.5–5cm, acuminate, base rounded, minutely pubescent at first, later glabrous; petioles 8–12mm. Cymes 2–9-flowered, extra-axillary or terminal. Sepals broadly ovate, 2–3 × 2–3mm. Outer petals oblong, 1.5–2 × 0.4–0.5cm, acute, pubescent, inner petals c 2mm. Fruit ovoid-globose, 5–10cm diameter, carpels completely fused, ± smooth but with slight lozenge-shaped projections, reddish green or brownish.

Bhutan: S – Phuntsholing district (Phuntsholing), **C** – Tashigang district (Dangme Chu). Cultivated on river banks and in gardens, 250–1050m. July.

Native of Central America, sometimes cultivated for its edible fruit.

2. A. squamosa L. Nep: *Shariphal* (34); Eng: *Sweetsop, Sugar Apple*.

Similar to *A. reticulata* but shrub or small tree 3–6m; leaves elliptic, 5–10 × 2–5cm, obtuse or acute, base cuneate, pubescent at first, soon glabrous; petioles 6–12mm; flowers solitary or few, leaf-opposed; outer petals yellow, oblong, 1.5–3cm, inner ones rudimentary or absent; fruit ovoid-globose, 8–9cm diameter, surface covered with the rounded tips of incompletely fused carpels, glaucous.

Sikkim: Cultivated in the Terai (34).

Native of Central America, cultivated for its edible fruit.

Family 36. MYRISTICACEAE

by A.J.C. Grierson & D.G. Long

Dioecious evergreen trees, often stellate-pubescent, inner bark exuding reddish juice. Leaves simple, alternate, pinnately veined, entire, exstipulate. Flowers in axillary clusters or panicles, unisexual, actinomorphic. Perianth shortly cup-shaped at base, 3-lobed; lobes thick, valvate. Male flowers with filaments connate into a column bearing anthers. Female flowers with superior 1-celled ovary; ovule solitary, basal; style short or absent, stigma 2-lobed. Fruit a dehiscent drupe, 2–4-valved. Seed partially or completely enveloped by a brightly-coloured, fleshy laciniate or subentire aril; endosperm ruminate.

1. Leaves not whitish beneath; flowers in panicles; staminal column globose bearing elongated anthers on surface ....**1. Horsfieldia**
\+ Leaves whitish beneath; flowers in short-stalked axillary clusters; staminal column a peltate disc bearing anthers around rim ....**2. Knema**

1. HORSFIELDIA Willdenow

Flowers numerous in axillary panicles, subglobose. Male flowers with globose staminal column bearing 18–20 elongated anthers on its surface. Female flowers with obovoid, tomentose ovary; stigma sessile, small. Fruit ovoid or ellipsoid, leathery; aril subentire.

1. H. kingii (Hook. f.) Warburg; *Myristica kingii* Hook. f. Nep: *Ramguwa, Runchepat* (34). Fig. 22 h–k.

Tree 10–12m, branches often crowded towards top of trunk. Leaves thinly coriaceous, oblong-obovate, 12–35 × 5–7cm, acute or shortly acuminate, base rounded or cuneate, glabrous; petioles 1.5–2cm. Male panicles lax, 9–15cm, flowers 3–4mm diameter; perianth segments triangular. Female panicles 3–7cm, flowers c 5mm diameter. Fruit 4–4.5 × 2.5cm.

Sikkim: Sivoke, Rungit valley, Tista valley, etc. Subtropical forests, 200–600m. June.

Fruit edible (34).

2. KNEMA Loureiro

Flowers in dense axillary clusters or on short thick branchlets. Male flowers obovoid, tomentose, with disc-like staminal column bearing 9–18 anthers around rim. Female flowers with tomentose ovary; stigma 2-lobed, on short thick style. Fruit ovoid or ellipsoid with fleshy pericarp; aril laciniate at apex.

1. K. tenuinervia de Wilde; *K. linifolia* auct. p.p. non (Roxb.) Warburg, *Myristica longifolia* sensu F.B.I. p.p. non Blume. Nep: *Ramguwa* (34), *Balaio, Rate Balayo*. Fig. 22a–g.

Tree 10–20m, twigs brownish stellate-tomentose when young. Leaves coriaceous, oblong, 20–65 × 8–16cm, acute, base rounded or shallowly cordate, glabrous above, whitish beneath, lateral veins 20–35 pairs, not prominent above; petioles 1.5–2cm, stout. Male flowers 7–10 × 5–6mm, perianth segments thick, red within. Female flowers 6mm; ovary tomentose bearing short 1.5mm style and bilobed stigma. Fruit ovoid or ellipsoid, c 3.5 × 2.2cm, tomentose.

Bhutan: S – Samchi, Sarbhang and Gaylegphug districts, widespread; **Sikkim**: Tista valley, etc. Subtropical forests, 300–800m. April–July.

Records of *K. linifolia* (Roxb.) Warburg from Sikkim probably all belong here; it differs in having the lateral veins of the leaf more prominent above.

2. K. erratica (Hook. f. & Thomson) Sinclair; *Myristica erratica* Hook. f. & Thomson, *M. longifolia* Blume var. *erratica* (Hook. f. & Thomson) Hook. f., *Knema cinerea* (Poiret) Warburg var. *andamanica* auct. p.p. non (Warburg) Sinclair, *K. angustifolia* auct. p.p. non (Roxb.) Warburg. Nep: *Ramguwa* (34).

Similar to *K. tenuinervia* but leaves smaller, 12–20(–25) × 2.5–5cm, base cuneate, lateral veins 18–24 pairs; fruit ellipsoid, 2–2.5 × 1.3–1.5cm.

Sikkim: Darjeeling duars, Jaldakha valley, Sivoke, Buxa, etc. Terai forests, 150–200m.

Family 37. SCHISANDRACEAE

by A.J.C. Grierson & D.G. Long

Monoecious or dioecious twining shrubs. Leaves alternate, exstipulate, simple, pinnately veined, glabrous. Flowers solitary, axillary; perianth parts all ± similar, several-seriate. Stamens numerous, variously aggregated, filaments entirely or

FIG. 22. **Myristicaceae, Schisandraceae** and **Illiciaceae. Myristicaceae.** a–g, *Knema tenuinervia*: a, portion of shoot with leaf and inflorescence; b, flower bud; c, female flower, two lobes of perianth removed to show ovary; d, male flower, half of perianth removed to show staminal column; e, staminal column; f, fruit; g, seed partially covered by aril. h–k, *Horsfieldia kingii*: h, portion of shoot with leaf and inflorescence; i, female flower; j, the same with perianth removed to show ovary; k, male flower seen from above. **Schisandraceae**. l–o, *Schisandra grandiflora*: l, portion of shoot with leaves and male flowers; m, male flower with perianth removed to show staminal column; n, female flower with perianth removed to show carpels; o, elongated fruiting column. p–s, *Kadsura heteroclita*: p, staminal column; q, dorsal view of carpel with style-like appendage; r, side view of carpel showing stigmatic crest along upper margin; s, developing fruit. **Illiciaceae**. t–v, *Illicium griffithii*: t, portion of shoot with leaves and flowers; u, flower; v, whorl of dehiscing follicles. Scale: h, o × ⅓; a × ½; f, g, l, s, t × ⅔; v × 1; b, u × 2½; m, n × 2⅔; i, j × 4; c, d, e, k × 4½; p × 5½; q, r × 8. Drawn by M. Bates.

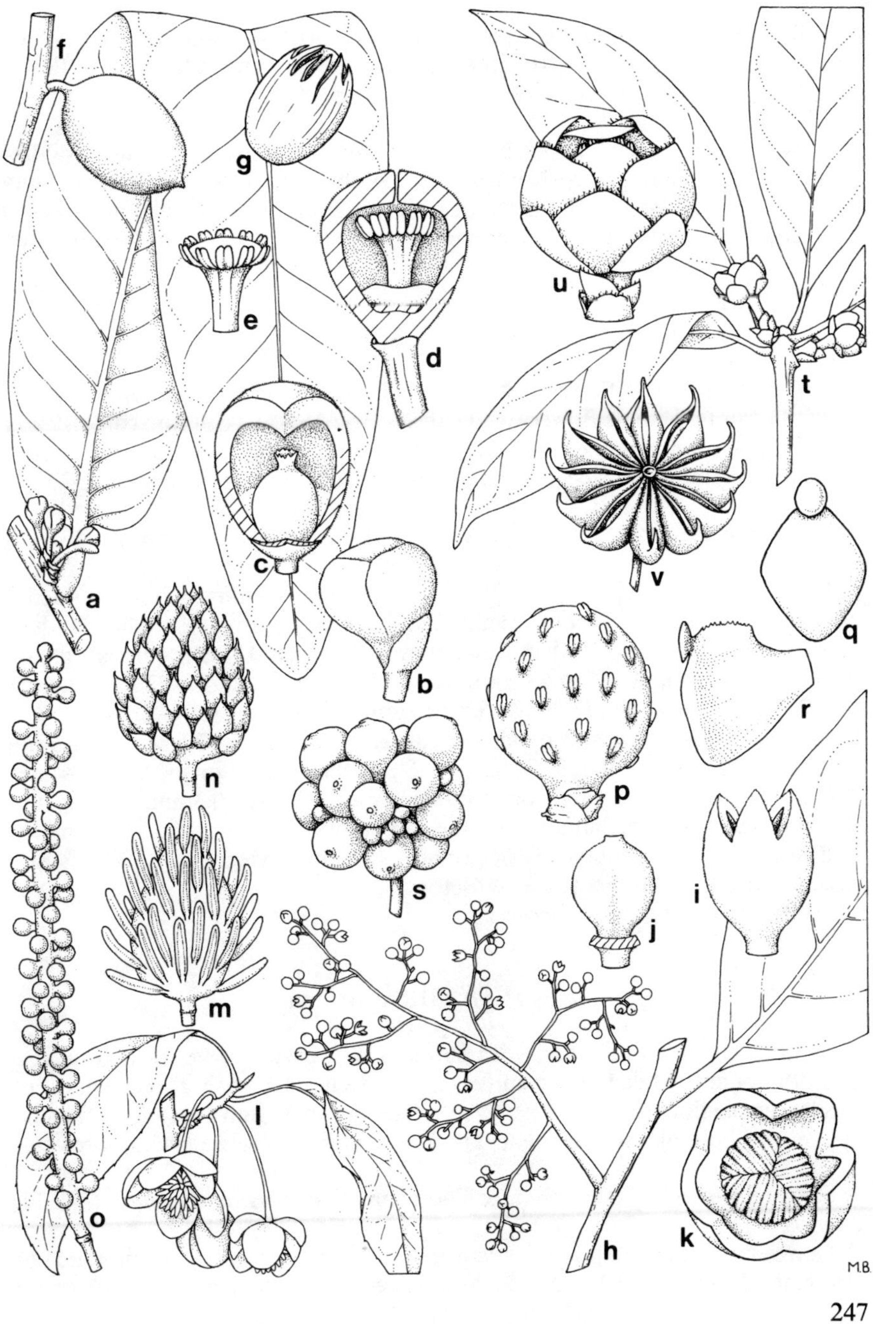
f
g
u
e
d
t
c
v
q
a
n
b
p
r
s
i
j
m
l
o
h
k
M.B.

basally fused into a column. Carpels numerous, borne on a column, ovaries each with two inner stigmatic crests and a distal style-like appendage. Fruit fleshy, subglobose, drupe-like, borne on an elongated or ellipsoid column, seeds 1–5.

1. Leaves entire; each carpel bearing an elliptic or suborbicular style-like appendage; fruiting carpels borne in subglobose clusters ....**1. Kadsura**
+ Leaves denticulate; each carpel bearing an elongated style-like appendage; fruiting carpels borne on an elongated column ....**2. Schisandra**

1. **KADSURA** Jussieu

Leaves entire. Male flowers with anthers in several series ± immersed in club-shaped column. Female flowers with carpels crowded on a column; carpels ovoid or ellipsoid, often angular, appendage elliptic or peltate. Column in fruit club-shaped, not elongating, bearing subglobose or obovoid carpels; seeds 2–5.

1. K. heteroclita (Roxb.) Craib; *K. roxburghiana* Arnott. Nep: *Pattiamlo* (34). Fig. 22p–s.

Leaves ovate-elliptic, 8–13(–17) × 3–6(–8.5)cm, acuminate, base rounded or cuneate, entire (rarely remotely denticulate near apex), minutely pale glandular beneath; petioles 0.5–2(–4)cm. Male flowers: pedicels 0.5–2cm; perianth parts white or yellowish, 11–15 in 4–5 whorls, outer rounded, 2–5 × 2–5mm, inner elliptic or obovate, 10–13 × 7–8mm; staminal column subglobose, 5–8 × 5–8mm. Female flowers as males but carpels angular, 1.5–2.5 × 1.5–2.5mm, densely packed into a subglobose head 6–8mm diameter; stigmatic crest obscurely ciliolate, terminated by elliptic or suborbicular appendage c 0.3mm; fruiting head subglobose, 2.5–5mm diameter, mature carpels obovoid, 1–2 × 0.5–1.5cm, red.

Bhutan: S–Gaylegphug district (Hatipoli near Gaylegphug); **Sikkim**. Subtropical forest, 270m. September–October.

Seeds edible (34).

2. **SCHISANDRA** Michaux

Leaves denticulate. Flowers with stamens or carpels borne on a conical or cylindric column. Male flowers: anthers free ± erect, outer on short filaments, inner sessile. Female flowers: carpels ovoid, each tapering above into an elongate flattened appendage; column much elongated in fruit, fruiting carpels obovoid; seeds 1–2.

1. S. grandiflora (Wall.) Hook. f. & Thomson. Nep: *Singhatta Lahara* (34). Fig. 22 l–o.

Leaves lanceolate, elliptic or oblanceolate, 7–15 × 2–6cm, acuminate, base attenuate; petioles 1–2.5cm. Male flowers: pedicels 1–4cm; perianth parts white or pale pink, 7–8 in 3 series, broadly elliptic or obovate, 1–1.8 × 0.5–1.5cm, staminal column ovoid, 7–15 × 7–15mm, anthers 2–3mm on filaments 1–2mm. Female flowers: pedicels 1.5–6cm, perianth as in males; carpels ovoid, c 1.5 × 0.75mm, appendage c 0.5mm; column in fruit 15–22cm long, 3–6mm thick, fleshy, usually bearing 30–80 obovoid carpels, 7–9 × 5–6mm, red.

Bhutan: **C**–Thimphu to Mongar districts, **N**–Upper Mo Chu district; **Sikkim**. Climbing on shrubs and trees in Cool broad-leaved and moist coniferous forests, 2440–4100m. May–June.

Fruit edible (34).

2. S. neglecta A. C. Smith; *S. elongata* Hook. f. & Thomson non Baillon. Nep: *Singhatta Lahara* (34).

Similar to *S. grandiflora* but smaller, leaves 5–15 × 2–7cm, acute or shortly acuminate, base rounded or cuneate, margins denticulate, sometimes sinuous; male flowers on pedicels 1–5cm, perianth parts yellowish, 5–10 × 5–10mm, anthers 1–2mm, outer ones on filaments c 0.5mm; female flowers as males but carpels ellipsoid, 1–2 × 1mm, on column 1.5–1.8mm; fruiting column 4–12cm, bearing 10–30 oblong-ellipsoid carpels 5–7 × 3–4mm, seeds 1–2.

Bhutan: **S**–Gaylegphug district (above Shamkhara), **C**–Tongsa district (Tsanka); **Sikkim**. Climber on shrubs in moist Cool broad-leaved forests, 2100m. June.

Records of *S. propinqua* (Wall.) Baillon from Sikkim (34) cannot be substantiated.

Family 38. ILLICIACEAE

by A.J.C. Grierson & D.G. Long

Glabrous shrubs or small trees. Leaves alternate, often clustered or whorled at branch ends, exstipulate, pinnately veined, entire. Flowers bisexual, solitary, axillary. Perianth parts usually numerous, several-seriate, hypogynous. Stamens numerous in several series. Carpels 12–14 in a single whorl, each laterally compressed, obliquely attached to torus, erect at first then spreading, unilocular, ovule solitary. Fruit a whorl of free spreading follicles.

1. ILLICIUM L.

Description as for Illiciaceae.

1. I. griffithii Hook. f. & Thomson. Fig. 22 t–v.

Leaves coriaceous, oblong-elliptic, 7–13 × 2–5cm, acuminate, base cuneate; petioles 0.5–2cm. Flowers on pedicels 1–2.5cm; perianth parts c 25, outer ones papery, broadly ovate to suborbicular, 3.5–10 × 3.5–10mm, ciliate, inner fleshy, oblong-obovate, 4–5 × 2.5–3mm. Stamens in 2 series, 2–2.5mm, filaments fleshy. Carpels in flower 4–4.5mm, style subulate. Follicles compressed ovoid, 1.5–2 × 0.8–1cm, 3–5mm thick, narrowed at apex to a stout point.

Bhutan: C–Punakha district (above Menchunang). Evergreen oak forest, 1800–2460m. April–May.

Family 39. LAURACEAE

by D.G. Long

Evergreen or deciduous trees or shrubs, often aromatic; twigs with or without conspicuous terminal vegetative buds and rings of bud scale scars. Leaves alternate, sometimes opposite or clustered at branch ends, simple, sometimes shallowly lobed near apex, pinnately veined or strongly 3-veined, exstipulate. Flowers in panicles, racemes or umbels, umbels often enclosed by subpersistent scales and arranged in fascicles or racemes; unisexual or bisexual, actinomorphic. Perianth segments usually 6, free, in two whorls of 3. Fertile stamens usually 9, rarely 6 or 12, in whorls of 3, innermost filaments usually bearing glands, often with an inner whorl of 3 staminodes; female flowers with 9 or 12 staminodes; anthers dehiscing by 2 or 4 valves. Ovary usually superior, ovule 1, apical, style short, rarely filiform. Fruit a drupe, usually borne on enlarged cup-shaped remains of perianth, rarely perianth completely absent or totally enclosing drupe.

1. Leaves 3-veined at or above base (sometimes pinnately veined above but basal lateral veins much stronger than others) 2
\+ Leaves pinnately veined throughout (lowermost lateral veins weaker than those above) 7
2. Flowers borne in panicles without conspicuous persistent bracts; fruit borne on enlarged perianth-cup 8–10mm across 3
\+ Flowers borne in ebracteate clusters or bracteate umbels (umbels sometimes arranged in panicles); fruit borne on little-enlarged perianth cup 2–4mm across 5
3. Leaves obovate, clustered at branch ends; panicles short and dense, bearing flowers to base ....**13. Actinodaphne** (*A. obovata*)
\+ Leaves elliptic, ovate or lanceolate, not clustered at branch ends; panicles elongate, slender, borne on naked peduncle 4
4. Leaves broadly ovate or broadly elliptic, coriaceous, opposite or subopposite but not distichous ....**3. Cinnamomum**
\+ Leaves ovate, elliptic or lanceolate but not broadly so, thinly coriaceous, alternate, distichous ....**4. Neocinnamomum**

5. Leaves suborbicular or broadly ovate; flowers in axillary ebracteate clusters containing a vegetative bud which develops into a leafy shoot **14. Parasassafras**
\+ Leaves ovate, elliptic or oblanceolate (broadly ovate or suborbicular only in *Lindera heterophylla*); flowers in bracteate umbels borne on old wood or on short leafless shoots, umbels not enclosing a vegetative bud 6

6. Leaves ovate, elliptic or suborbicular; basal lateral veins distinctly arching near base; umbels sessile or pedunculate but not densely clustered; anther cells 2 **9. Lindera**
\+ Leaves elliptic to oblanceolate; basal lateral veins almost straight in lower half of leaf; umbels sessile and densely clustered; anther cells 4 ....**11. Neolitsea**

7. Plants bearing flowers 8
\+ Plants bearing fruit 18

8. Flowers borne in panicles, not surrounded by conspicuous persistent scales (sometimes in racemes in *Cinnadenia*) 9
\+ Flowers borne in umbels, surrounded by conspicuous persistent scales, umbels sessile on leafy shoots or aggregated into short racemes or panicles (or flowers in ebracteate clusters in *Actinodaphne*) 15

9. Anthers 2-celled; flowers always bisexual 10
\+ Anthers 4-celled; flowers bisexual or unisexual 11

10. Terminal vegetative buds absent; leaves alternate, minor veins weakly reticulate only on lower surface; panicles elongate, 15–25cm ....**1. Cryptocarya**
\+ Terminal vegetative buds present; leaves alternate, opposite or subopposite, minor veins usually strongly reticulate on both surfaces; panicles short, 1–10cm **2. Beilschmiedia**

11. Panicles short, slender, 3–5cm, racemiform, borne in axils of lower leaves; leaves not clustered ....**6. Cinnadenia**
\+ Panicles short or long, often 10–20cm, diffuse, usually borne amongst upper leaves; leaves clustered at branch ends or not 12

12. Leaves ovate to elliptic-obovate, never oblanceolate, never clustered; panicles weak, naked peduncle usually shorter than panicle ....**3. Cinnamomum**
\+ Leaves often oblanceolate, obovate or elliptic, rarely ovate, often clustered at branch ends; panicles rigid, naked peduncle usually longer than panicle. .13

13. Perianth segments 1.5–2mm; leaves oblanceolate ....**5. Alseodaphne**
\+ Perianth segments mostly 3–6.5mm (2–2.5mm in *Persea minutiflora* but then leaves elliptic or ovate); leaves various 14

14. Perianth segments rigid, erect ....**7. Phoebe**
\+ Perianth segments soft, often spreading ....**8. Persea**

15. Umbels 1–2-flowered; flowers large, 14–18mm across, bisexual **12. Dodecadenia**
+ Umbels or clusters many-flowered; flowers small, usually 4–6mm across; unisexual 16

16. Leaves clustered at branch ends; flowers in clusters not subtended by persistent scales .... **13. Actinodaphne**
+ Leaves not clustered at branch ends; flowers in umbels subtended by persistent scales 17

17. Anthers 2-celled .... **9. Lindera**
+ Anthers 4-celled .... **10. Litsea**

18. Fruit completely enclosed by persistent perianth .... **1. Cryptocarya**
+ Perianth absent in fruit, or persisting completely or partly and enclosing base of fruit 19

19. Perianth remains completely absent in fruit 20
+ Perianth persisting as a minute rim or enlarged cup with or without persistent segments 21

20. Fruiting pedicel often stout but not swollen and fleshy ..... **2. Beilschmiedia**
+ Fruiting pedicel much enlarged and fleshy .... **5. Alseodaphne**

21. Perianth segments persisting completely in fruit, sometimes enlarged, sometimes on rim of enlarged fleshy cup 22
+ Perianth segments absent in fruit, sometimes persisting in part as shallow teeth on rim of enlarged fleshy cup 24

22. Fruit borne in umbels on old wood, perianth segments persisting on rim of enlarged fleshy cup .... **10. Litsea**
+ Fruit borne in panicles amongst leaves, perianth segments erect or reflexed, not borne on enlarged fleshy cup 23

23. Perianth segments thick, erect, appressed to fruit .... **7. Phoebe**
+ Perianth segments thin, spreading or reflexed in fruit .... **8. Persea**

24. Fruits few, on branches of racemes or panicles 25
+ Fruits solitary, on short pedicels on old wood, or several in umbels 26

25. Fruit globose, c 1cm, on elongate panicles 4–13cm .... **3. Cinnamomum**
+ Fruit obovoid, 2–2.5cm, on short racemes or panicles 3–4cm **6. Cinnadenia**

26. Leaves clustered at branch ends; fruit borne in clusters on old wood **13. Actinodaphne**
+ Leaves not clustered at branch ends; fruit solitary or several in umbels ... 27

27. Fruits 1–2 together on short pedicels on old wood .... **12. Dodecadenia**
+ Fruits several in sessile or peduncled umbels on old wood **9. Lindera & 10. Litsea**

1. CRYPTOCARYA Brown

Evergreen trees, shoots without winter buds. Leaves alternate, not clustered, pinnately veined. Flowers bisexual in axillary and terminal panicles. Perianth segments 6. Fertile stamens 9 with 2-celled anthers, outer 6 introrse, inner 3 extrorse. Fruit ellipsoid, completely enclosed by adnate, unlobed perianth with ring-like scar at apex.

1. Leaves oblong-elliptic, 5–9cm broad, green beneath; fruit 2–2.5cm **1. C. amygdalina**
+ Leaves ovate-elliptic, 7–11cm broad, glaucous beneath; fruit 3.5–4.5cm **2. C. bhutanica**

1. C. amygdalina Nees; *C. floribunda* Nees. Nep: *Kawla* (34), *Patmero* (34), *Gamda Tarsing*. Fig. 23 c–g.

Leaves coriaceous, oblong-elliptic, 10–22 × 5–9cm, bluntly apiculate or shortly acuminate, base broadly cuneate or rounded, lateral veins 4–7 pairs, glossy above, glabrous and green beneath; petioles 1–2cm. Panicles 15–25cm, puberulous. Perianth segments ovate, 1.5–2mm, puberulous. Fruit ovoid, 2–2.5 × 0.8–1cm, tapering to acute apex.

Bhutan: S–Samchi, Phuntsholing, Chukka, Sarbhang and Gaylegphug districts; **Sikkim**. Subtropical forests, 340–1050mm. March–April.

2. C. bhutanica Long

Similar to *C. amygdalina* but leaves ovate-elliptic, 14–23 × 7–11cm, acuminate, glaucous beneath, lateral veins 7–9 pairs; fruit ellipsoid, 3.5–4.5 × 1.8–2cm.

Bhutan: S–Sarbhang district (below Dara Chu) and Gaylegphug district (Surey and Chabley Khola). Warm broad-leaved forests, 1200–1800m.

Endemic to Bhutan; known only in fruit.

2. BEILSCHMIEDIA Nees

Nep: *Tarsing*

Evergreen trees, shoots with terminal vegetative buds. Leaves coriaceous, opposite, subopposite or alternate, not clustered, pinnately veined, lateral and minor veins prominent on both surfaces. Flowers bisexual in panicles which are solitary in leaf axils or clustered at base of new shoots. Perianth segments 5–6. Fertile stamens 9 with 2-celled anthers, outer 6 introrse, inner 3 extrorse. Fruit ovoid, oblong or globose, without perianth remains.

1. Terminal buds large, subglabrous, 6–12 × 2–8mm, leaving conspicuous scars at base of new shoots 2
+ Terminal buds small, tomentose, 2–5 × 1–2mm, leaving inconspicuous scars at base of new shoots 3
2. Terminal buds ovoid, 6–10 × 3–8mm; leaves ovate or ovate-lanceolate, lateral veins 10–15 pairs; fruit obovoid or subglobose, 2.5–3 × 2–2.5cm **1. B. gammieana**
+ Terminal buds lanceolate, 8–12 × 2–3.5mm; leaves elliptic, lateral veins 7–10 pairs; fruit ellipsoid, 3.5–4.5 × 1.6–1.9cm ....**2. B. assamica**
3. Leaves elliptic to lanceolate, 2.5–6cm broad; petioles 1–1.6cm 4
+ Leaves ovate, 5–8cm broad; petioles 1.5–3cm 6
4. Leaves alternate, elliptic-lanceolate; panicles 1–1.5cm; fruiting peduncles 1–1.5cm ....**3. B. dalzellii**
+ Leaves opposite or subopposite, elliptic; panicles 4–9cm; fruiting peduncles 4–6cm 5
5. Leaves coriaceous, base rounded ....**4. B. sikkimensis**
+ Leaves thinly coriaceous, base cuneate ....**5. B. clarkei**
6. Leaves opposite or subopposite, coriaceous; veins equally prominent on both surfaces; panicles 1–2cm; perianth segments oblong c 3mm **6. B. roxburghiana**
+ Leaves alternate, thinly coriaceous; veins very prominent beneath, weakly above; panicles 4–5cm; perianth segments ovate c 2mm ....**7. B. brandisii**

1. B. gammieana Hook. f. Sha: *Golo Shing*; Nep: *Koruke, Tarsing*. Fig. 23b.
Shoots bearing coriaceous, ovoid, acute, glabrous terminal buds, 6–10 × 3–8mm. Leaves opposite, coriaceous, ovate or ovate-lanceolate, 12–18 × 3.5–7cm, shortly acuminate, base cuneate, glabrous, lateral veins 10–15 pairs prominent on both surfaces, minor veins prominently reticulate on both surfaces;

FIG. 23. **Lauraceae**. a, *Beilschmiedia dalzellii*: upper portion of shoot with leaves, terminal bud and inflorescences. b, *Beilschmiedia gammieana*: upper portion of shoot with leaves, terminal bud and fruit. c–g, *Cryptocarya amygdalina*: c, upper portion of shoot with leaves and inflorescences; d, detail of inflorescence; e, dissected flower; f, fruit enclosed within persistent perianth tube; g, perianth partly removed to show fruit. h–k, *Lindera heterophylla*: h, portion of precociously flowering shoot with vegetative buds; i, outer stamen; j, inner stamen with basal glands; k, portion of shoot with leaf and developing fruit. l, *Dodecadenia grandiflora*: upper portion of shoot with leaves, terminal bud and umbels. m–q, *Neolitsea cuipala*: m, upper portion of shoot with leaves, terminal bud and umbels; n, female flower dissected; o, ovary; p, male flower dissected; q, portion of shoot with clusters of fruit. r–u, *Actinodaphne obovata*: r, upper portion of shoot bearing panicles on old wood and leaves on young; s, ovary; t, stamen with basal gland; u, fruit with persistent perianth cup. v, *Parasassafras confertiflora*: portion of stem bearing maturing fruit at base of young shoot. Scale: a × ¼; c, q, r × ⅓; h × 2/5; b, l, m × ½; f, g, k, u, v × ⅔; d × 1⅓; p × 2½; n, o × 3; e × 4; s × 4½; t × 5½; i, j × 8. Drawn by M. Bates.

a
b
c
d
e
f
g
h
i
j
k
l
m
n
o
p
q
r
s
t
u
v
M.B.

petioles 1–1.5cm. Panicles 2–3cm, clustered at base of new shoots. Perianth segments ovate, 2.5–3mm, glabrous. Fruit obovoid or subglobose, 2.5–3 × 2–2.5cm when ripe, apiculate, borne on a short stalk bearing abscission scar of caducous perianth.

Bhutan: S–Phuntsholing, Chukka, Sarbhang and Gaylegphug districts, **C**–Tongsa district; **Sikkim**. Cool broad-leaved and Evergreen oak forests, 1750–2050m. May.

2. B. assamica Meisner

Similar to *B. gammieana* but terminal buds lanceolate, 8–12 × 2–3.5mm; leaves elliptic, broader, 12–18 × 6–8cm; lateral veins 7–10 pairs; panicles longer 4–10cm; fruit ellipsoid 3.5–4.5 × 1.6–1.9cm.

Bhutan: S–Chukka district (Sinchu La), **C**–Tongsa district (near Pertimi, Mangde Chu); **Sikkim**. Subtropical and Warm broad-leaved forests, 300–1800m.

3. B. dalzellii (Meisner) Kostermans; *B. fagifolia* Nees var. *dalzellii* Meisner. Fig. 23a.

Terminal buds slender, 3–5 × 1–2mm, tomentose. Leaves alternate, membranous at first, becoming coriaceous, elliptic-lanceolate, 9–14 × 2.5–4cm, acuminate, base cuneate, glossy above, distinctly reticulate on both surfaces, lateral veins 10–12 pairs; petioles 1–1.6cm. Panicles very short, 1–1.5cm, greyish pubescent. Perianth segments oblong, 3–3.5mm. Fruit narrowly obovoid, c 2 × 1.3cm, borne on short stout peduncle 1–1.5cm.

Bhutan: S–Chukka, Sankosh, Sarbhang and Gaylegphug districts; **Sikkim**: Tista valley. Subtropical forest slopes, 200–1200m. March.

4. B. sikkimensis Hook. f. Nep: *Tarsing* (34).

Similar to *B. dalzellii* but leaves opposite or subopposite, elliptic, 6–9 × 3–4cm, acute or mucronate, base rounded; fruiting peduncles 4–6cm.

Sikkim: locality unknown.

5. B. clarkei Hook. f. Nep: *Sanu Tarsing* (34).

Similar to *B. dalzellii* and *B. sikkimensis* but leaves opposite, membranous when young becoming thinly coriaceous, elliptic, 11–16 × 4–6cm, acuminate, base cuneate, dull above; panicles 4–9cm; fruit c 3 × 1.5cm.

Sikkim: Lopchu and Reinak. Warm broad-leaved forests, 1400–1700m. May.

6. B. roxburghiana Nees. Nep: *Tarsing, Thulo Tarsing* (34).

Terminal buds 2–3mm, tomentose. Leaves opposite or subopposite, coriaceous, ovate, 12–19 × 5–8cm, shortly acuminate, base rounded or broadly cuneate, glossy above, prominently reticulate on both surfaces, lateral veins 7–10 pairs prominent on both surfaces; petioles 1.5–2.5cm. Panicles short, 1–2cm, pubescent. Perianth segments oblong, c 3mm. Fruit oblong, c 3 × 1cm, on peduncle 1.5–2cm.

Bhutan: S – Samchi district (Torsa Valley); **W. Bengal Duars**: Buxa; **Sikkim**: terai. Subtropical forests, 200–400m. March.

The record from Gaylegphug (117) refers to *Heliciopsis terminalis* (Kurz) Sleumer. Wood used for house building and tea boxes (48).

7. B. brandisii Hook. f.

Similar to *B. roxburghiana* but leaves alternate, more thinly coriaceous, lateral veins scarcely prominent above, strongly prominent beneath; petioles 2.5–3cm; panicles longer, 4–5cm; perianth segments broadly ovate, c 2mm; fruit oblong, 5–6cm.

Sikkim: Jalpaiguri terai. Subtropical terai forests. March.

3. CINNAMOMUM Schaeffer

Evergreen trees, shoots with terminal vegetative buds. Leaves alternate or opposite, not clustered, pinnately veined or strongly 3-veined. Flowers usually bisexual in axillary panicles. Perianth segments 6. Fertile stamens usually 9 with 4-celled anthers, outer 6 introrse, inner 3 extrorse. Fruit borne on an enlarged fleshy perianth cup, cup entire or with persistent perianth segments.

1. Leaves 3-veined, opposite or subopposite 2
\+ Leaves pinnately veined, alternate 5
2. Part or all of perianth segments persisting in fruit; leaves elliptic, lanceolate or ovate-lanceolate, 10–40cm 3
\+ Perianth segments completely deciduous in fruit; leaves ovate or ovate-elliptic, 6–12cm 4
3. Leaves elliptic, 15–40cm; panicles 10–20cm; fruit c 1.5...**1. C. bejolghota**
\+ Leaves lanceolate or ovate-lanceolate, 10–15cm; panicles 5–10cm; fruit 1.0–1.2cm ....**2. C. tamala**
4. Leaves ovate-elliptic or elliptic, 8–12cm, glossy above, veins strongly impressed above ....**3. C. impressinervium**
\+ Leaves ovate, 5–9cm, not glossy above, veins not impressed above **4. C. pauciflorum**
5. Lateral veins 5–7 pairs, without pit-like glands in axils; fruiting perianth cup 13–18mm long ....**7. C. tenuipilis**
\+ Lateral veins 3–5 pairs, with pit-like glands in axils; fruiting perianth cup 8–10mm long 6
6. Panicles glabrous, 3–5 per shoot, not clustered; fruiting perianth cup 6–7mm across ....**5. C. glanduliferum**
\+ Panicles brownish tomentose, 5–9 per shoot; densely clustered; fruiting perianth cup 10–12mm across ....**6. C. glaucescens**

1. C. bejolghota (Hamilton) Sweet; *C. obtusifolium* Nees. Dz: *Drongdo*; Sha: *Phramda, Tesparshing Chhilu*; Nep: *Bhale Sinkoli* (34), *Patrange, Sin Kaule*. Fig. 24h.

Tree to 20m. Leaves coriaceous, opposite or subopposite, elliptic, 15–40 × 6–12cm, obtuse or shortly and bluntly acuminate, base cuneate, glossy above, strongly 3-veined from above base, veins not impressed above; petioles 1.5–3cm. Panicles 12–20cm; perianth segments ovate, 2–3mm, pubescent. Fruit ellipsoid, c 1.5 × 1cm, borne on enlarged perianth cup 8–10mm diameter including completely persistent segments 3–4mm.

Bhutan: S – Phuntsholing to Deothang districts, common; **C** – Punakha and Tongsa districts; **Sikkim**. Subtropical, Warm and Cool broad-leaved forests, 400–2050m. March–May.

Two varieties occur in Bhutan: the widespread var. **bejolghota** with leaves, young shoots and main panicle branches quite glabrous but pedicels and flowers appressed-pubescent; and var. **pubescens** Grierson & Long with the young shoots, panicles and lower leaf surface densely pubescent. The latter variety is endemic to Bhutan, known from Chukka, Sarbhang, Gaylegphug and Tongsa districts.

2. C. tamala (Hamilton) Nees & Ebermaier. Dz: *Shingtsha*; Sha: *Tespar*; Nep: *Tespata, Tezpat* (34), *Sinkoli* (34); Hindi: *Tespata*.

Similar to *C. bejolghota* but leaves lanceolate or ovate-lanceolate, 10–15 × 3–6cm, shortly and bluntly acuminate; panicles shorter, 5–10cm; fruit 1–1.2cm, borne on enlarged perianth cup with lower part of segments persisting as short lobes 1–2mm.

Bhutan: S – Deothang district (Domphu); **C** – Punakha district (Mishichen, Changshingna and Tinlegang); **Sikkim**. Warm broad-leaved forests, 1400–1700m. April–May.

Similar to *C. impressinervium* but differs in its more bluntly pointed leaves with upper surface weakly glossy and veins not impressed. Leaves used for flavouring food; bark and leaves also used medicinally (13).

3. C. impressinervium Meisner. Nep: *Korsane, Sissi* (34).

Small tree to 15m. Leaves elliptic or ovate-elliptic, 7–12 × 3–5cm, finely acuminate, glossy above with strongly impressed veins; petioles 8–13mm. Panicles 6–10cm, appressed pubescent. Perianth segments c 2mm, deciduous. Fruit ellipsoid, c 12 × 7mm, borne on shallow entire perianth cup 6–7mm diameter.

Bhutan: S – Chukka district (Chimakothi and Jumudag), **C** – Punakha district (Ritang and below Norbding); **Sikkim**. Warm broad-leaved and Evergreen oak forests, 1800–2450m. May–August.

4. C. pauciflorum Nees

Similar to *C. tamala* and *C. impressinervium* but leaves ovate, smaller, 5–9 × 2.5–4cm, apex finely acuminate, base rounded or cuneate; panicles short, up to 4cm, 3–5-flowered; perianth segments completely deciduous.

Bhutan: C – Punakha district (above Tinlegang) (71). c 2000m.

5. C. glanduliferum (Wall.) Meisner. Dz: *Kipchu Shing*; Nep: *Kawla* (34), *Malagiri* (34). Fig. 24g.

Shrub or small tree to 15m, shoots glabrous. Leaves alternate, elliptic-obovate, 8–13 × 4–6cm, acuminate, base cuneate, pale beneath, lateral veins 3–4 pairs, with conspicuous pits in vein axils; petioles 1–2.5cm. Panicles 3–5 per shoot, distant in leaf axils, 6–12cm, glabrous. Perianth 2.5–3mm, segments ovate, sparsely pubescent, caducous. Fruit globose, c 1cm, borne on enlarged subentire obconical perianth cup 10–11 × 6–7mm.

Bhutan: S – Chukka district (Chukka), **C** – Thimphu district (Raidak Valley S of Confluence and Dobchu near Paro) and Tongsa district (Shamgong); **Sikkim**. Cool broad-leaved forest and dry scrub on river banks, 1900–2400m. May–June.

Records of *C. porrectum* (Roxb.) Kostermans (*C. parthenoxylon* (Jack) Meisner) from Bhutan and Sikkim (59a,135), a tropical Malaysian species without gland-pits in vein axils, refer to this species. Wood camphor-scented (126) and used for planking (48).

6. C. glaucescens (Nees) Drury; *C. cecicodaphne* Meisner. Nep: *Ghansaray, Gonserai, Malagiri* (34).

Similar to *C. glanduliferum* but leaves ovate-elliptic, base broadly cuneate or rounded; lateral veins 4–5 pairs; panicles brownish-tomentose, 5–9 densely clustered on young shoots; fruiting cup much broader, 8–10 × 10–12mm.

Bhutan: S – Sankosh district (Balu Khola), **C** – Punakha district (Punakha); **Sikkim**: Darjeeling district. Subtropical and Warm broad-leaved forests, 200–1200m. January–February.

Wood excellent for furniture and boat-building (16).

7. C. tenuipilis Kostermans

Similar to *C. glanduliferum* and *C. glaucescens* but leaf-bases always rounded; lateral veins 5–7 pairs, without pit-like glands in vein axils; panicles whitish-pubescent, 5–7 clustered on young shoots; fruiting cup elongate, 13–18 × 7–9mm.

Bhutan: S – Gaylegphug district (above Tama), **C** – Tongsa district (Mangde Chu valley); **Sikkim**: Kurseong and Lopchu. Warm broad-leaved and *Quercus griffithii/Q. glauca* forests, 1250–1550m. March.

4. NEOCINNAMOMUM Liou

Similar to *Cinnamomum* but shoots without conspicuous terminal buds; leaves distichous; fruit borne on much enlarged fleshy perianth cup with persistent enlarged perianth segments.

1. N. caudatum (Nees) Merrill; *Cinnamomum caudatum* Nees. Sha: *Bidaring Shing*; Nep: *Khorsane*. Fig. 24n.

Small tree to 8m. Leaves thinly coriaceous, broadly elliptic or ovate, 7–14 × 3.5–8cm, acuminate, base usually rounded, puberulous when young, strongly 3-veined from base to apex with numerous transverse reticulations beneath; petioles 7–15mm. Panicles slender, axillary, 2–3.5cm, terminal 3–9cm. Perianth segments ovate, appressed pubescent. Fruit c 2 × 1.5cm, red, pendulous, borne on fleshy obconical 6-lobed perianth cup c 1cm across.

Bhutan: S – Deothang district (Leborey), **C** – Punakha district (Rinchu) and Mongar district (Zimgaon, Shongar Chu); **Sikkim**: Gangtok, Rungit, Lopchu etc. Subtropical and Warm broad-leaved forests, 200–1500m. June–September.

5. ALSEODAPHNE Nees

Evergreen or(?) deciduous trees, terminal buds small. Leaves clustered at branch ends, pinnately veined. Flowers bisexual in axillary panicles. Perianth segments 6. Fertile stamens 9 with 4-celled anthers, outer 6 introrse, inner 3 extrorse. Fruit ellipsoid, borne on thick fleshy pedicel, perianth not persistent.

1. A. owdenii Parker. Fig. 24b–f.

Tree to 18m. Leaves membranous, oblanceolate, 7–12 × 2–3.5cm, acute or shortly acuminate, base cuneate, glabrous, lateral veins 5–9 pairs, pale beneath; petioles 8–12mm. Panicles 6–10cm, including peduncle 4–6cm, minutely puberulous. Perianth segments ovate, 1.5–2mm. Fruit solitary, ellipsoid, 3–3.3 × 1.5–1.7cm, on thickened obconical pedicel 6–7mm broad.

Bhutan: S – Sarbhang district (above Sarbhang); **Sikkim**: Kalijhora. Subtropical forests, 800m. May.

A valuable timber-tree (16).

6. CINNADENIA Kostermans

Evergreen trees, shoots without conspicuous terminal buds. Leaves alternate, not clustered, pinnately veined. Dioecious, flowers unisexual in short axillary panicles or subumbellate racemes. Perianth segments 6. Stamens 9–12 with 4-celled anthers, all introrse. Fruit obovoid, borne on enlarged entire fleshy perianth cup.

1. C. paniculata (Hook. f.) Kostermans; *Dodecadenia paniculata* Hook. f. Fig. 24a.

Tree to 30m. Leaves coriaceous, ovate-lanceolate, 11–22 × 4–7cm, acute, base broadly cuneate, glossy above, pale beneath, glabrous, lateral veins 9–11 pairs, impressed above; petioles 1–2cm. Panicles or racemes 3–5cm. Flowers c 8mm across, on pedicels 4–7mm. Perianth segments ovate, c 3mm. Fruit 1–2 per inflorescence, obovoid, 2–2.5 × 1–1.3cm, pink when cut, borne on fleshy entire cup 6–8mm diameter.

Bhutan: **S** – Chukka district (Chukka) and Deothang district (Sassi), **C** – Tongsa district (Shamgong and Pertimi); **Sikkim**: Mik. Warm broad-leaved forests, 730–2000m. October–January.

7. PHOEBE Nees

Evergreen trees, shoots with or without conspicuous terminal buds. Leaves alternate, clustered near branch ends or not, pinnately veined. Flowers bisexual or unisexual in axillary panicles often clustered near branch ends. Perianth segments 6, erect, rigid. Fertile stamens 9 with 4-celled anthers, outer 6 introrse, inner 3 extrorse. Fruit ellipsoid, enclosed at base by appressed hardened persistent perianth segments.

1. Leaves lanceolate or elliptic, acuminate; lateral veins 5–10 pairs; flowers glabrous or pubescent 2
\+ Leaves obovate or oblanceolate, apiculate; lateral veins 12–20 pairs; flowers densely pubescent or tomentose 3
2. Leaves lanceolate or narrowly elliptic; lateral veins 7–10 pairs; flowers glabrous **1. P. lanceolata**
\+ Leaves elliptic; lateral veins 5–7 pairs; flowers pubescent .... **2. P. cathia**
3. Leaves densely pubescent beneath; lateral veins 15–20 pairs; perianth segments 3–4mm .... **3. P. attenuata**
\+ Leaves sparsely pubescent or glabrous beneath; lateral veins 12–16 pairs; perianth segments c 5mm .... **4. P. hainesiana**

The record of *Phoebe pallida* (Nees) Nees from Sikkim (80) is based on a sterile specimen of *Persea glaucescens* (Nees) Long.

1. P. lanceolata (Nees) Nees. Nep: *Jakrikat, Jhankri Kath* (34).

Small tree to 16m; shoots with narrow rings of bud scale scars. Leaves usually clustered near branch ends, lanceolate or narrowly elliptic, 8–17 × 2–5.5cm, finely acuminate, base attenuate, lateral veins 7–10 pairs, not impressed above, glabrous or sparsely appressed pubescent beneath; petioles 6–15mm. Panicles 2–12 clustered at branch ends, 9–15cm. Perianth segments ovate, c 2.5mm, glabrous. Fruit ellipsoid, 10–11 × 5–7mm, enclosed at base by hardened perianth segments 3–4mm.

Bhutan: **S** – Phuntsholing, Chukka, Sarbhang, Gaylegphug and Deothang districts, **C** – Mongar district; **Sikkim**. Subtropical and Warm broad-leaved forests, 300–1500m. April–May.

Similar to *Persea minutiflora* when in flower; the latter differs in its acute leaves with lateral veins obscure beneath.

2. P. cathia (D. Don) Kostermans; *P. paniculata* (Nees) Nees

Similar to *P. lanceolata* but shoots without rings of bud scale scars; leaves not clustered, more broadly elliptic, 10–16 × 3–6cm, lateral veins 5–7 pairs, often impressed above, densely pubescent beneath; perianth pubescent.

Bhutan: C – Punakha district (between Bhotokha and Rinchu) and Mongar district (Zimgaon, Shongar Chu). Warm broad-leaved forests, 1250–1500m. June.

3. P. attenuata (Nees) Nees. Nep: *Angare* (34), *Aule Lapche Kawla* (34).

Tree to 20m; bud scale scars loosely aggregated into broad rings. Leaves clustered at branch ends, oblanceolate or obovate, 11–20 × 3.5–7cm, rounded and shortly apiculate at apex, base attenuate, lateral veins 15–20 pairs, not impressed above, pubescent and brown beneath when dry; petioles 1.5–2.5cm. Panicles stout, 4–8 clustered at branch ends, 12–22cm, peduncle pubescent. Perianth segments 3–4mm, densely pubescent. Fruit ellipsoid c 1.5 × 1cm, enclosed at base by hardened pubescent perianth segments 5–6mm.

Bhutan: S – Chukka district (Giengo), Sarbhang district (Lao Pani) and Gaylegphug district (Sher Camp); **W Bengal Duars**: Buxa; **Sikkim**: terai. Subtropical forests, 400–1300m. March–April.

Vegetatively similar to *Persea fructifera* which has only 10–15 pairs of lateral veins, the lowermost of which are strongly ascending, shorter panicles 6–10cm, and much larger globose fruit, c 4cm, with spreading or reflexed perianth segments.

Timber excellent for cabinet-making (16).

4. P. hainesiana Brandis; ? *P. goalparensis* Hutchinson. Nep: *Angare* (34). Fig. 24m.

Similar to *P. attenuata* but leaves 11–25 × 4.5–8cm, bluntly mucronate, lateral veins 12–16 pairs, pale beneath when dry, glabrous or sparsely pubescent beneath; peduncles subglabrous; perianth larger, segments c 5mm, densely greyish tomentose; fruit larger, broadly ellipsoid, c 3 × 1.5cm, surrounded at base by woody perianth segments c 7mm.

Sikkim: foothills and Gyabari. Terai forests, 200–500m. May.

A valuable timber tree, sometimes cultivated (34).

FIG. 24. **Lauraceae.** a, *Cinnadenia paniculata:* portion of shoot with leaves and infructescences. b–f, *Alseodaphne owdenii:* b, portion of shoot with leaves and inflorescence; c, outer stamen; d, inner stamen with basal glands; e, staminode; f, fruit on thickened pedicel. g, *Cinnamomum glanduliferum:* portion of shoot with leaf and inflorescence. h, *Cinnamomum bejolghota:* upper portion of shoot with leaves and inflorescences. i–k, *Persea duthiei:* i, upper portion of shoot with leaves and inflorescences; j, dissected flower; k, ovary. l, *Persea clarkeana:* upper portion of stem bearing maturing fruit at base of new new growth. m, *Phoebe hainesiana:* fruit. n, *Neocinnamomum caudatum:* portion of shoot bearing leaf and fruit borne on thickened peduncle. Scale: 1 × ⅓; a, b, f, h × ½; i × ⅔; g, m, n, × 4/5; j, k × 3; c, d, e × 8. Drawn by M. Bates.

a
b
c
d
e
f
g
h
i
j
k
l
m
n
M.B.

8. PERSEA Miller

Similar to *Phoebe* but branches usually with conspicuous terminal buds and rings of bud scale scars at base of each year's growth; leaves clustered at branch ends or not; perianth segments soft, often spreading; fruit usually globose, sometimes ellipsoid or obovoid; persistent perianth segments spreading or reflexed, not hardened.

1. Leaves broadly obovate, 18–25 × 8–10cm, with 15–18 pairs of lateral veins; petioles 3–4cm .... **1. P. bootanica**

\+ Leaves elliptic, ovate, oblanceolate or narrowly obovate, usually 7–20 × 2–8cm (up to 25 × 9cm in *P. americana*), with 6–15 pairs of lateral veins (up to 20 pairs in *P. clarkeana* and *P. duthiei*); petioles 0.7–2.5cm (up to 3 or 4cm in *P. americana, P. robusta* and *P. wallichii*) 2

2. Panicle branches and perianth segments glabrous or minutely puberulous; fruit ellipsoid, 1.2–1.7cm, or globose and large, 1.5–3cm 3

\+ Panicle branches and perianth segments silky-pubescent, villous or tomentose; fruit globose, small, 0.6–1.2cm, up to 4cm only in *P. fructifera*, or obovoid up to 15cm in *P. americana* 6

3. Leaves elliptic-obovate, 5–8cm broad; lateral veins strongly prominent beneath; petioles 1.5–4cm .... **2. P. robusta**

\+ Leaves elliptic, 2–5cm broad; lateral veins not or weakly prominent beneath; petioles 1–2.5cm 4

4. Leaves narrowly-elliptic, finely acuminate, with 14–20 pairs of lateral veins; fruit globose 1.5–3cm .... **3. P. clarkeana**

\+ Leaves elliptic, acute, subacute or shortly acuminate, with 7–11 pairs of lateral veins; fruit ellipsoid 12–17 × 7–10mm 5

5. Shoots with ring-like bud scale scars; leaves shortly acuminate; perianth segments oblong, 5–6mm, glabrous .... **4. P. odoratissima**

\+ Shoots without ring-like bud scale scars; leaves acute or subacute; perianth segments ovate, c 2.5mm, puberulous in bud .... **5. P. minutiflora**

6. Shoots with a distinct ring of closely aggregated bud scale scars at base of each year's new growth; leaves ± clustered at branch ends 7

\+ Shoots without a ring of bud scale scars at base of each year's new growth, budscale scars distant (sometimes loosely aggregated in *P. americana*); leaves ± distant on shoots 10

7. Leaves reticulate beneath; lateral veins strongly prominent beneath; leaves tomentose when young becoming villous beneath .... **6. P. fructifera**

\+ Leaves not reticulate beneath; lateral veins weakly prominent or obscure beneath; leaves appressed silky-pubescent when young, becoming glabrous or thinly appressed silky-pubescent beneath 8

8. Leaves elliptic, 12–20cm, with 12–18 pairs of lateral veins; perianth segments 4–5mm .... **7. P. duthiei**
+ Leaves usually obovate or oblanceolate, sometimes some elliptic, 7–13cm, lateral veins 8–12 pairs; perianth segments 3–4mm 9
9. Leaves pale brown beneath when dry; lateral veins 8–10 pairs; petioles 8–13mm **8. P. gamblei**
+ Leaves reddish-brown and glaucous beneath when dry; lateral veins 9–12 pairs; petioles 10–18mm .... **9. P. kurzii**
10. Leaves ovate or broadly-elliptic, 6–9cm broad; perianth segments 4.5–5mm; fruit obovoid, 5–15cm .... **12. P. americana**
+ Leaves oblong-elliptic, 3–5.5cm broad; perianth segments 3–3.5mm; fruit globose 8–12mm 11
11. Leaves shortly acuminate, red-brown beneath when dry; lateral veins very prominent beneath; panicles densely brownish tomentose; fruiting pedicels not thickened .... **10. P. glaucescens**
+ Leaves acute, green or brownish-green beneath when dry; lateral veins not very prominent beneath; panicles greyish pubescent; fruiting pedicels thick and fleshy **11. P. wallichii**

1. P. bootanica (Meisner) Kostermans; *Machilus bootanica* Meisner

Large tree; shoots with rings of bud scale scars. Leaves membranous, obovate, 18–25 × 8–10cm, apiculate, base attenuate, lateral veins 15–18 pairs, prominent beneath, brownish hirsute beneath especially on veins; petioles 3–4cm. Panicles 12–18cm, stout, densely pubescent. Perianth segments ovate-oblong, 5–6.5mm, tomentose. Fruit unknown.

Bhutan: S – Chukka district (near Chukka). Warm broad-leaved forest, 1200m. May.

Endemic to Bhutan; known only from Griffith's type collection.

2. P. robusta (W. W. Smith) Kostermans; *Machilus robusta* W. W. Smith

Tree to 15m; bud scale scars sometimes loosely aggregated but not forming distinct rings. Leaves coriaceous, elliptic-obovate, 13–24 × 5–8cm, apiculate or shortly acuminate, base cuneate, lateral veins 7–8 pairs, strongly prominent beneath, glabrous and often glaucous beneath; petioles 1.5–4cm. Panicles 8–13cm, stout, minutely puberulous or almost glabrous. Perianth segments oblong, 6mm, minutely puberulous. Fruit globose, 2.5–3cm.

Sikkim: Tong. Warm broad-leaved forest, 1200m. March.

3. P. clarkeana (Hook. f.) Kostermans; *P. gammieana* (Hook. f.) Kostermans, *Machilus clarkeana* Hook. f., *M. gammieana* Hook. f. Nep: *Chipli Kawla* (34), *Seto Kawla* (34). Fig. 24l.

Tree to 12m; shoots with rings of bud scale scars. Leaves often reddish-brown when dry, narrowly elliptic or oblanceolate, 10–22 × 2–4cm, finely acuminate, base attenuate, midrib impressed above, lateral veins 14–20 pairs, obscure beneath; petioles 1–2.5cm. Panicles short and stout, bright red, 4–7cm, glabrous. Perianth segments oblong, 5–6mm, glabrous or with a few hairs near apex. Fruit globose, 1.5–3cm.

Bhutan: S–Chukka district (near Chimakothi) and Deothang district (S of Riserboo); **C**–Mongar district (above Mongar); **Sikkim**. Cool broad-leaved and Evergreen oak forests, 2100–2400m. May–June.

4. P. odoratissima (Nees). Kostermans; *Machilus odoratissima* Nees. Dz: *Matchu Shing*; Nep: *Lali Kawla* (34).

Small tree to 10m; shoots with rings of bud scale scars. Leaves reddish-brown when dry, elliptic or broadly-elliptic, 8–13 × 3–5cm, shortly acuminate, base cuneate, glabrous, midrib slightly impressed above, lateral veins 7–11 pairs, not very prominent beneath; petioles 1–2cm. Panicles short, red, 5–8cm, glabrous. Perianth segments oblong, 5–6mm, glabrous. Fruit ellipsoid, 12–16 × 7–9mm.

Bhutan: S–Chukka district (Marichong), **C**–Punakha district (below Mendegang); **Sikkim**. Subtropical and Warm broad-leaved forests, 600–2000m. March–May.

5. P. minutiflora Kostermans; *Machilus parviflora* Meisner

Similar to *P. odoratissima* but bud scale scars not forming rings; leaves more narrowly elliptic, 9–16 × 2.5–3.5cm, acute or subacute, glaucous beneath; flowers much smaller with perianth segments c 2.5mm, sparsely silky pubescent becoming almost glabrous; fruit ellipsoid, c 17 × 10mm.

Sikkim: Latpanchor, Kalimpong and Darjeeling. Warm broad-leaved forest, 700–1200m. January.

Similar to *Phoebe lanceolata* which differs in its finely acuminate leaves with more prominent lateral veins.

6. P. fructifera Kostermans; *Machilus edulis* Hook. f. Nep: *Lapche Phal* (34), *Lapche Kawla* (34), *Lapchi*.

Tree to 12m; shoots with rings of bud scale scars. Leaves obovate or oblanceolate, 10–14 × 3–5cm, subacute, acute or shortly acuminate, base attenuate, reddish-brown beneath when dry, brownish villous beneath, lateral veins 10–15 pairs, strongly ascending and very prominent beneath, minor veins distinctly reticulate beneath; petioles 1–3cm. Panicles 6–10cm, densely brownish villous; pedicels stout, 2–5mm. Perianth segments 4–5mm, pubescent. Fruit globose, c 4cm.

Sikkim: Darjeeling district. Warm broad-leaved forests, 900–2000m. May.

Fruit edible (34). Similar in foliage to *Phoebe attenuata* which differs in having 15–20 pairs of lateral veins, the lowermost of which are spreading, longer panicles 12–22cm, and smaller ellipsoid fruit c 1.5cm, with appressed perianth segments.

7. P. duthiei (Hook. f.) Kostermans; *Machilus duthiei* Hook. f. Dz: *Mitsu Shing*. Fig. 24i–k.

Tree to 20m; shoots with rings of bud scale scars. Leaves coriaceous, elliptic, 12–20 × 2.5–4cm, acuminate, base cuneate or attenuate, minutely silky-pubescent beneath when young, pale brown and usually glaucous beneath when dry, lateral veins 12–18 pairs, not very prominent beneath; petioles 1–2cm. Panicles numerous, suberect, 5–15cm, minutely silky-pubescent. Perianth segments oblong, 4–5mm, minutely silky pubescent. Fruit globose 9–11mm.

Bhutan: S – Chukka district (Chukka and Chimakothi), **C** – Punakha district (Tinlegang and Tang Chu) and Tashigang district (Tobrang); **Sikkim**. Warm broad-leaved and Evergreen oak forests, 600–2100m. February–May.

8. P. gamblei (Hook. f.) Kostermans; *P. bombycina* (Hook. f.) Kostermans, *Machilus gamblei* Hook. f., *M. bombycina* Hook. f. Nep: *Kawla* (34).

Small tree; shoots with rings of bud scale scars. Leaves thinly coriaceous, obovate to oblanceolate, often with some leaves elliptic, 7–11 × 2.5–4.5cm, shortly acuminate, base attenuate, pale brown beneath when dry, minutely silky-pubescent beneath sometimes becoming glabrous; lateral veins 8–10 pairs, obscure beneath; petioles 8–13mm. Panicles 5–8cm, pubescent. Perianth segments 3.5–4mm, densely appressed pubescent. Fruit globose, c 1cm.

Sikkim: Darjeeling terai. Subtropical terai forests. March.

9. P. kurzii (Hook. f.) Kostermans; *Machilus kurzii* Hook. f. Nep: *Seto Kawla* (34).

Similar to *P. gamblei* but leaves usually oblanceolate, 7–13 × 2–4cm, dark reddish-brown when dry, strongly glaucous beneath, lateral veins 9–12 pairs, almost invisible beneath; panicles 4–7cm; perianth segments c 3.5mm, pubescent.

Sikkim: Darjeeling foothills. Warm broad-leaved forests, 1500–2100m. May.

10. P. glaucescens (Nees) Long; *P. villosa* (Roxb.) Kostermans, *Machilus glaucescens* (Nees) Wight, *M. villosa* (Roxb.) Hook. f. Nep: *Bhale Kaulo, Kawla* (34).

Tree to 20m; shoots without rings of bud scale scars. Leaves coriaceous, usually elliptic, 10–15(–20) × 3–6cm, acuminate, base cuneate, pale reddish-brown beneath, pubescent when young becoming almost glabrous; lateral veins 6–9 pairs, arching upwards, very prominent beneath, minor veins reticulate beneath; petioles 0.8–2cm. Panicles numerous, often spreading, 10–17cm, densely reddish-brown tomentose when dry. Perianth segments c 3.5mm, tomentose. Fruit globose, 10–12mm.

Bhutan: S – Sarbhang district (above Sarbhang); **Sikkim**: Sivoke, Jalpaiguri etc. Subtropical forests, 500–650m. March–April.

11. P. wallichii Long; *Machilus sericea* Blume

Similar to *P. glaucescens* but leaves elliptic-oblong, 10–18 × 3–5.5cm, acute, base broadly cuneate or rounded, green or brownish-green beneath when dry, lateral veins 7–9 pairs, not very prominent beneath; petioles 1.8–3cm; panicles greyish pubescent; fruit with much thickened fleshy pedicel.

Sikkim: Mal, Ryang and Mungpu (34). Warm broad-leaved forests. March–April.

Its presence in Sikkim is doubtful as no correctly determined specimens have been seen from the E Himalaya.

12. P. americana Miller; *P. gratissima* Gaertner. Eng: *Avocado Pear*.

Similar to *P. glaucescens* and *P. wallichii* but leaves ovate or broadly elliptic, 9–25 × 6–9cm, shortly acuminate, sparsely pubescent beneath, lateral veins 7–9 pairs, prominent beneath; petioles 1.5–3cm; panicles 6–11cm, pale villous; perianth segments 4.5–5mm, pale villous; fruit obovoid, 5–15 × 3–7cm.

Sikkim: Kalimpong, cultivated. 1500m. April.

Native of Central America, cultivated for its edible fruit.

9. LINDERA Thunberg

Deciduous or evergreen trees or shrubs; shoots with or without conspicuous terminal vegetative buds. Leaves alternate, not clustered, pinnately veined or 3-veined from base. Dioecious; flowers unisexual, in sessile or pedunculate, often clustered or racemose umbels, each umbel enclosed by large persistent decussate scales forming a subglobose bud when young. Perianth segments usually 6. Fertile stamens 9 with 2-celled anthers, all introrse. Fruit ovoid or globose, borne on a little-enlarged, entire or toothed perianth cup.

1. Leaves strongly 3-veined at base (basal lateral veins much stronger than others) ... 2
+ Leaves pinnately veined throughout (lowermost lateral veins weaker than those above) ... 5

2. Deciduous; leaves membranous, ovate, broadly elliptic or suborbicular, acute; lowermost lateral veins arising suprabasally ... 3
+ Evergreen; leaves coriaceous, narrowly elliptic or lanceolate, acuminate or caudate; lowermost lateral veins arising basally ... 4

3. Leaves broadly elliptic or suborbicular; umbels sessile on previous year's shoots; perianth segments deciduous in fruit ... **1. L. heterophylla**
+ Leaves ovate; umbels pedunculate on short leafless axillary shoots; perianth segments withering but persisting in fruit ... **2. L. neesiana**

4. Leaves long-acuminate or caudate, white beneath ... **3. L. pulcherrima**
+ Leaves shortly acuminate, green beneath ... **4. L. melastomacea**

5. Shoots glabrous, strongly striate when dry; umbels borne on short stout peduncles; fruit 15–18mm, borne on large perianth cup 10–12mm across ... **5. L. bootanica**
+ Shoots pubescent or tomentose when young, smooth; umbels sessile or on slender peduncles; fruit 6–12mm, borne on slightly enlarged perianth cup 3–4mm across ... 6

6. Umbels on slender peduncles 7
\+ Umbels sessile 8

7. Leaves elliptic, 3–4.5cm broad; umbel scales subglabrous; fruit broadly ellipsoid, c 10 × 8mm ....**6. L. assamica**
\+ Leaves obovate, 4–6cm broad; umbel scales appressed-pubescent; fruit globose, c 6mm ....**7. L. latifolia**

8. Leaves elliptic, 6–11 × 2–3cm, shortly acuminate; lateral veins 6–8 pairs; fruit globose, c 6mm ....**8. L. nacusua**
\+ Leaves obovate, 14–20 × 6–8cm, subacute or obtuse; lateral veins 14–16 pairs; fruit ellipsoid, c 12mm ....**9. L. hamiltonii**

1. L. heterophylla Meisner. Dz: *Chur Tego*; Nep: *Lekh Pipli* (34). Fig. 23h–k.

Deciduous shrub or small tree to 6m. Leaves membranous, appearing after flowering, broadly elliptic to suborbicular, 8–12 × 4–9cm, shortly acute or obtuse, base rounded or subcordate, margins entire or 3-lobed at apex, strongly 3-veined above base, silky hairy beneath when young becoming glabrous; petioles 1–2.5cm. Umbels silky pubescent, 1.5–2cm diameter, sessile on previous year's shoots; flowers on pedicels 8–12mm, perianth segments oblong, 3.5–4.5mm, deciduous. Fruit (immature) ellipsoid, c 8 × 4mm, perianth cup entire, scarcely enlarged, 2mm across, on pedicel 1.5–2cm which becomes thickened near apex.

Bhutan: C–Thimphu, Punakha, Tongsa, Bumthang and Mongar districts, **N**–Upper Mo Chu and Upper Kuru Chu districts; **Sikkim**. Amongst shrubs in Blue Pine and Spruce forests, 2500–3200m. April–May.

Plants with 3-lobed leaves are unknown in Bhutan but occur in Sikkim; in E Bhutan plants with broader suborbicular leaves are commoner, these have sometimes been wrongly reported as *L. cercidifolia* Hemsley (71) which is a synonym of *L. obtusiloba* Blume, a Chinese and Japanese species (58).

2. L. neesiana (Nees) Kurz. Sha: *Nenshing, Nengshing, Ru Neng*; Nep: *Siltimur* (34).

Deciduous shrub or small tree to 5m. Leaves membranous, strongly aromatic when crushed, ovate, 9–13(–18) × 3–8(–12)cm, acute, base rounded, 3-veined above base but not strongly so, pinnately veined and often reticulate in upper part, glabrous; petioles 1–2cm. Umbels 3–6, racemose on short leafless axillary shoots 1–1.5cm; peduncles slender 5–10mm, each umbel 4–8-flowered, pedicels 2–3mm. Perianth segments ovate, 1.5–2mm. Fruit broadly ellipsoid, c 5 × 4mm; perianth cup not enlarged, c 1mm across, bearing withered remains of segments.

Bhutan: S–Deothang district (S of Riserboo), **C**–Tongsa, Mongar and Tashigang districts; **Sikkim**. Cool broad-leaved and Evergreen oak forests, 1950–2450m. October–February.

Bark and fruits are used medicinally (13).

3. L. pulcherrima (Nees) Hook. f.; *L. thomsonii* Allen. Sha: *Sengkyermay Shing*; Nep: *Sissi* (34).

Evergreen shrub or small tree to 5m. Leaves elliptic to lanceolate, 6–12 × 2.5–6cm, finely acuminate to long-caudate, base rounded or cuneate, whitish sericeous beneath when young becoming subglabrous and glaucous beneath, strongly 3-veined from base to apex, minor veins transverse, not very prominent; petioles 8–12mm. Umbels solitary, subsessile, 3–8-flowered; pedicels 4–5mm. Perianth segments c 3mm, deciduous. Fruit ellipsoid, c 10 × 7mm, borne above entire perianth rim c 2mm across.

Bhutan: S – Deothang district, **C** –Thimphu, Punakha, Tongsa and Mongar districts, **N** – Upper Mo Chu district; **Sikkim**. Evergreen oak forests, 2100–2600m. April–June.

Two subspecies occur, subsp. **pulcherrima**, the common plant in Bhutan and Sikkim with leaves usually large, 8–12cm, narrowing gradually to caudate-acuminate apices with points 8–15mm, and subsp. **thomsonii** (Allen) Long (*Lindera thomsonii* Allen) known from Mongar district (Rudo La) which has shorter, rather abruptly caudate-acuminate leaves 6–8cm with longer points 15–25mm.

4. L. melastomacea (Nees) Villar

Similar to *L. pulcherrima* but leaves smaller, 4.5–7 × 2.5–3cm, shortly and bluntly acuminate, green beneath, minor transverse veins prominent.

Bhutan: C –Tongsa district (near Chendebi). Cool broad-leaved forest, 2300m. March–April.

5. L. bootanica Meisner; *L. venosa* (Meisner) Hook. f. Dz: *Ashi*.

Deciduous tree to 25m; terminal buds large, ovoid, 1–1.5cm; shoots glabrous, strongly striate when dry. Leaves membranous, elliptic, 15–26 × 4–9cm, acute, base cuneate, glossy above, densely greyish silky-pubescent beneath when young, becoming pubescent or glabrous, lateral veins 8–12 pairs, prominent beneath; petioles 2–3cm. Umbels 2–3 in axils of fallen leaves on stout pubescent peduncles 9–13mm, each umbel many-flowered, enclosed by large broad scales c 1cm; pedicels densely tomentose, in males 4–6mm, in females 6–9mm. Perianth segments 3–4mm. Fruit subglobose 1.6–1.8cm, borne on enlarged entire perianth cup 1–1.2cm across; pedicel stout, striate.

Bhutan: C – Punakha district (Tinlegang and Ngawang, Tang Chu) and Mongar district (Unjar), **N** – Upper Mo Chu district (Gasa). Warm broad-leaved forest, 1800–2150m. February-April.

A local species known only from E Nepal and Bhutan.

6. L. assamica (Meisner) Kurz. Nep: *Sanu Pahenle* (34).

Evergreen tree to 10m; young shoots smooth, tomentose. Leaves thinly coriaceous, elliptic, 9–16 × 3–4.5cm, acute or shortly acuminate, base cuneate, tomentose when young, becoming pubescent on veins beneath when mature, lateral veins 6–8 pairs, strongly prominent beneath, minor veins prominent beneath; petioles 6–10mm. Umbels 1–3 on very short axillary shoots, on slender peduncles 4–17mm, each umbel many-flowered with 4 broad subglabrous bracts 4–6mm; pedicels 2–4mm, silky pubescent; perianth segments oblong, c 2mm. Fruit broadly ellipsoid c 10 × 8mm, borne on slightly enlarged undulate perianth cup c 4mm across, on elongate pedicel 1.3–2cm.

Bhutan: S – Chukka district (near Takhti Chu); **Sikkim**: Darjeeling, Senchal, etc. Warm broad-leaved forest, 2100–2400m. October–December.

Records of *L. meissneri* Hook. f. from Sikkim refer to this species. Vegetatively very similar to *Litsea elongata* (Nees) Hook. f. which differs in its more prominent parallel leaf reticulations, silky pubescent umbel scales, 4-celled anthers and ellipsoid 10–14mm fruit borne on a larger perianth cup, 5–6mm across.

7. L. latifolia Hook. f.

Very closely related to *L. assamica* but more densely pubescent or tomentose throughout; leaves usually obovate, 9–15 × 4–6cm, acute or apiculate; umbel-scales appressed pubescent; fruit globose, c 6mm.

Sikkim: Gangtok (69). Warm broad-leaved forests, 1400–1600m. February–March.

The record of this species from Bhutan (71) refers to *Litsea elongata*; the Sikkim record requires confirmation.

8. L. nacusua (D. Don) Merrill; *L. bifaria* (Nees) Hook. f.

Similar to *L. assamica* but leaves smaller, more coriaceous, elliptic, 6–11 × 2–3cm, shortly acuminate; umbels sessile, densely clustered, scales silky pubescent, pedicels 2–3mm; fruit globose, 6mm, on persistent perianth cup 3–4mm across, on short pedicel c 4mm.

Bhutan: C – Tongsa district (between Rukubi and Neylong and between Tashiling and Tongsa). Cool broad-leaved forests, 2000–2200m. April.

These records (71) require confirmation, as in Nepal and NW Himalaya the plant occurs in warmer forests at lower altitudes.

9. L. hamiltonii Kostermans; *L. reticulata* (Nees) Hook. f. *nom. illeg.*

Similar to *L. assamica* and its allies, especially *L. latifolia*, but leaves obovate, 14–20 × 6–8cm, subacute or obtuse, base rounded, lateral veins 14–16 pairs; umbels sessile; fruits ellipsoid, c 12mm, on short thick pedicels 2–4mm.

Sikkim: Takdah, Darjeeling (69). Warm broad-leaved forests, 1600–1700m.

10. LITSEA Lamarck

Similar to *Lindera* but leaves always pinnately veined; anthers 4-celled; fruit borne on little- to much-enlarged, entire or rarely toothed perianth cup.

1. Deciduous; shoots with conspicuous winter buds; flowers precocious or appearing with young leaves 2
+ Evergreen; shoots without conspicuous winter buds (except in *L. elongata*); flowers not precocious 4
2. Young shoots red-villous; leaves obovate, obtuse or subacute **2. L. sikkimensis**
+ Young shoots glabrous or puberulous; leaves elliptic to lanceolate, acute or acuminate 3
3. Winter buds with membranous, obtuse outer scales; leaves silky-pubescent beneath, at least on veins; peduncles of umbel buds straight, 2–4mm **1. L. sericea**
+ Winter buds with thick, acuminate outer scales; leaves glabrous; peduncles of umbel buds deflexed, 6–10mm .... **3. L. kingii**
4. Umbels on peduncles arranged usually in axillary clusters, sometimes solitary, sometimes peduncles borne on a short stout axis (stalk) up to 4mm (rarely up to 8mm in *L. cubeba*) 5
+ Umbels on peduncles which are aggregated into racemes or corymbs with a conspicuous slender axis (stalk) 5–70mm 12
5. Shoots with scaly terminal vegetative buds and rings of bud scale scars; umbels usually solitary borne on slender peduncles (5–)10–15cm ... **8. L. elongata**
+ Shoots without scaly terminal vegetative buds and rings of bud scale scars; umbels borne on slender or stout peduncles 2–10mm, usually several in clusters 6
6. Leaves lanceolate or narrowly elliptic-oblong, 2–5cm broad (up to 9cm broad in *L. laeta* and *L. salicifolia*) 7
+ Leaves ovate, obovate or broadly elliptic, 5–11cm broad 10
7. Leaves rather membranous, dark green or brown above when dry; lateral veins 8–15 pairs; fruit 6–9mm long, on scarcely enlarged perianth cup 1.5–3mm across 8
+ Leaves coriaceous, pale or yellowish-green above when dry; lateral veins 5–9 pairs; fruit 15–17mm borne on much enlarged perianth cup 7–15mm across 9
8. Leaves lanceolate, 7–14 × 2–4cm, glabrous beneath except on veins; lateral veins 8–12 pairs, not prominent beneath; umbels in clusters of 2–4; fruit subglobose .... **4. L. cubeba**
+ Leaves elliptic-oblong, 12–30 × 2.5–8.5cm, minutely silky-pubescent beneath; lateral veins 10–15 pairs, prominent beneath; umbels in clusters of 6–18; fruit ellipsoid .... **5. L. salicifolia**
9. Leaves acute; lateral veins 5–7 pairs; umbel buds 6–8mm; fruit obovoid or subglobose, c 15 × 12mm, apex flattened with small mucro, on cup 11–15mm across with enlarged fleshy base 7–9mm long .... **6. L. laeta**

\+ Leaves shortly acuminate; lateral veins 7–9 pairs; umbel buds 4–6mm; fruit ellipsoid c 17 × 10mm, obtuse, on cup 7–9mm across borne on slender pedicel 10–15mm ....**7. L. chartacea**

10. Leaves long acuminate, glabrous; lateral veins 5–7 pairs .....**11. L. nervosa**

\+ Leaves obtuse, apiculate or shortly acuminate, tomentose or pubescent on veins beneath; lateral veins 7–15 pairs 11

11. Leaves elliptic, obtuse or apiculate, green and tomentose beneath; lateral veins 7–12 pairs; fruit subglobose c 7 × 6mm, on perianth cup c 5mm across ....**9. L. monopetala**

\+ Leaves elliptic-obovate, shortly acuminate, white and pubescent on veins beneath; lateral veins 10–15 pairs; fruit ellipsoid c 17 × 11mm, on perianth cup 15–17mm across ....**10. L. hookeri**

12. Leaves thickly coriaceous, obtuse, with 10–12 pairs of obscure lateral veins **15. L. nitida**

\+ Leaves thinly coriaceous, acuminate or acute, with 7–12 pairs of conspicuous lateral veins 13

13. Leaves acuminate, on petioles 4–6mm; umbel buds c 3mm; peduncles borne on short stalks 8–12mm; fruit oblong-ellipsoid, 17–23 × 9–10mm **14. L. albescens**

\+ Leaves acute, on petioles 15–35mm umbel buds 5–10mm; peduncles borne on stalks 15–70mm; fruit globose 8–9mm or depressed globose c 15 × 18mm 14

14. Leaves ovate-lanceolate, ovate, or broadly elliptic with 7–10 pairs of lateral veins; petioles 20–35mm; umbels on peduncles 7–10mm on stalks 15–30mm; fruit globose 8–9mm, on minute perianth cup 2–3mm across ..**12. L. glutinosa**

\+ Leaves elliptic-oblong with 10–12 pairs of lateral veins; petioles 15–20mm; umbels on peduncles 2–5mm on stalks 30–70mm; fruit depressed globose c 15 × 10mm, half surrounded by large perianth cup 20–25mm across **13. L. panamanja**

1. L. sericea (Nees) Hook. f.; *L. oreophila* Hook. f. *nom. illeg.*, *Lindera griffithii* Meisner, *Lindera ?hookeri* Meisner. Dz: *Geywe*; Nep: *Lekh Siltimur* (34).

Deciduous shrub or small tree to 15m, shoots black, smooth, leaf buds cylindric, 1–1.5cm, bluntly pointed, outer scales membranous, obtuse, pellucid-dotted. Leaves thinly coriaceous, elliptic, 5–10 × 2–3.5cm, shortly acuminate, base cuneate, sericeous when young, becoming silky pubescent on midrib when mature, prominently reticulate beneath, lateral veins 6–8 pairs, arching upwards, not prominent; petioles 7–10mm. Umbel buds ovoid, 3–4mm, on short straight peduncles 2–4mm. Umbels 1–2cm across, 5–10-flowered, pedicels 3–8mm, pubescent. Perianth segments c 3mm in male flowers, 2mm in females. Fruit ellipsoid, c 7 × 5mm, apiculate, above minute perianth rim 3mm across, sometimes bearing withered remains of segments, borne on elongated pedicels 7–12mm.

Bhutan: C – Ha to Tashigang districts, common, **N** – Upper Mo Chu and Upper Kuru Chu districts; **Sikkim**. Evergreen oak, Blue Pine, Hemlock and Spruce forests, 2400–3300m. April–May.

2. L. sikkimensis (Meisner) Long; *Lindera ?sikkimensis* Meisner

Similar to *L. sericea* but shoots striate; young parts and winter buds densely reddish-brown villous; leaf-buds without obtuse membranous scales; leaves (immature) obovate, 7 × 3.5cm or more, obtuse or subacute, appressed pubescent on midrib; umbels 3–6-flowered.

Sikkim: Lachen and Kanglassa. 3350–3650m. May–June.

A poorly-known plant which may prove to be only varietally distinct from *L. sericea*.

3. L. kingii Hook. f. Nep: *Siltimur* (34).

Similar to *L. sericea* but terminal buds sharply pointed, outer scales thicker, acuminate; leaves membranous, elliptic, 7–12 × 2–4cm, acute or shortly acuminate, glabrous except when very young, pale beneath, glossy above, lateral veins 11–16 pairs, spreading, prominent on both surfaces; petioles 9–12mm; umbels in bud borne on deflexed peduncles 6–10mm, flowers larger, perianth segments c 5mm in males, 3mm in females; fruit subglobose c 6mm, borne on pedicels 3–6mm on peduncles up to 1cm.

Bhutan: C – Tongsa district (Chendebi) and Sakden district (Taktoo and Choling La), **N** – Upper Mo Chu district (Tamji and Gasa); **Sikkim**: Senchal, Tonglo, Dumsong etc. Cool broad-leaved and Rhododendron forests, 2100–2900m. March–April.

Has been much confused with and sometimes (135) wrongly united with *L. cubeba*, which differs in its non-deciduous habit, absence of vegetative winter buds, lanceolate, acuminate leaves with longer petioles 1.5–2cm, and in its ecology, occurring in warmer forests at lower altitudes.

4. L. cubeba (Loureiro) Persoon; *L. citrata* Blume. Sha: *Nenshing*; Nep: *Timur* (117), *Siltimur* (34).

Evergreen shrub or small tree to 6m; winter buds absent. Leaves membranous, aromatic when crushed, lanceolate, 7–14 × 2–4cm, long acuminate, base cuneate, dark green above when dry, pale beneath, glabrous or sparsely pubescent on veins, lateral veins 8–12 pairs, not prominent, minor veins not reticulate; petioles slender, 1.2–2.4cm. Umbel buds 4–5mm, borne on slender peduncles 3–8mm, in sessile clusters of 2–4 or on a short slender stalk up to 4mm. Umbels c 5–10-flowered; pedicels 1–2mm, pubescent. Fruit subglobose 6–7mm, perianth rim c 1.5mm across, on thickened pedicel 3–5mm.

Bhutan: S – Samchi, Phuntsholing, Chukka, Gaylegphug and Deothang districts; **Sikkim**. Subtropical forest slopes, 300–1500m. December–February.

Fruit edible (34) and used medicinally (13). Silkworms are reared on the leaves (16).

5. L. salicifolia (Nees) Hook. f. Nep: *Sanu Pahenle* (34).

Similar to *L. cubeba* but leaves elliptic-oblong, variable in size, 12–30 × 2.5–8.5cm, dark brown above when dry, minutely pale silky pubescent beneath, lateral veins 10–15 pairs, prominent beneath sometimes impressed above; petioles stouter 8–12mm; umbels 6–18 in dense axillary clusters, on peduncles 2–6mm; fruit ellipsoid c 9 × 5mm, apiculate, on slightly enlarged entire perianth cup 2.5–3mm across.

Bhutan: S – Samchi, Chukka, Sankosh, Sarbhang, Gaylegphug and Deothang districts; **Sikkim**. Subtropical forest slopes and river banks, 300–1050m. February–April.

A very variable species especially in foliage; three varieties have been recorded: var. **salicifolia** (var. *ellipsoidea* (Meisner) Hook. f.) with pubescent shoots and leaves scarcely glossy above, the common plant in Bhutan and Sikkim; var. **attenuata** (Nees) Hook. f. with glabrous shoots and smaller, oblong leaves up to 15 × 3.5cm, recorded from Gaylegphug and Sikkim; and var. **polyneura** Hook. f. with pubescent shoots, large leaves glossy above, and lateral veins distinctly impressed above, from Sikkim.

6. L. laeta (Nees) Hook. f.

Shrub or small tree to 8m. Leaves coriaceous, oblong-elliptic, 10–20(–30) × 3–5(–9)cm, acute, base cuneate, glabrous, green above, white beneath when dry, lateral veins 5–7 pairs, strongly arching upwards; petioles 1–1.5cm, usually thickened. Umbels 6–8mm in bud, on peduncles 4–10mm, 3–6 in sessile clusters in leaf axils. Flowers subsessile. Fruit broadly obovoid or subglobose, c 15 × 12mm, apex flattened with small mucro, borne on enlarged fleshy cup-shaped perianth 11–15mm across, with thickened fleshy base 7–9mm long.

Bhutan: S – Samchi district (near Gokti), Sarbhang district (Phipsoo) and Deothang district (Satsalor); **Sikkim**. Subtropical forests, 290–670m. November-February.

7. L. chartacea (Nees) Hook. f.

Very closely allied to *L. laeta* but leaves shortly acuminate, lateral veins 7–9 pairs, more spreading; petioles slender; umbel buds 4–6mm; flowers on short pedicels 1–2mm; fruit ellipsoid, 17 × 10mm, obtuse but not mucronate, on shallow perianth cup 7–9mm across, borne on slender pedicel 10–15mm.

Sikkim: Lopchu, Mumsong and Rishap. Warm broad-leaved forests, 1500–1800m. May–June.

8. L. elongata (Nees) Hook. f. Nep: *Phamphal, Phusre, Thulo Pahenle* (34).

Tree to 20m, branchlets brownish tomentose, bearing conspicuous ovoid terminal vegetative buds and rings of bud scale scars. Leaves elliptic, oblanceolate or obovate, 8–18 × 2–6cm, acuminate, acute or obtuse, base cuneate, brownish tomentose and prominently reticulate beneath, lateral veins 6–13 pairs, very prominent beneath, ascending; petioles 6–16mm. Umbels usually solitary, rarely paired, on slender peduncles (5–)10–15mm (occasionally very short in male umbels), umbel scales silky pubescent, pedicels very short. Fruit ellipsoid, 10–14 × 6–7mm, with minute apical point, on persistent shallowly-lobed perianth cup 5–6mm across, on stout pedicels 3–6mm.

Bhutan: S – Chukka district (Bunakha, Gedu and Tala), **C** – Tongsa district (above Tongsa), **N** – Upper Mo Chu district (near Tamji); **Sikkim**. Evergreen oak forests, 1800–2600m. September–October.

A variable species, especially in foliage, in which two varieties are recognised: var. **elongata**, with leaves elliptic or oblanceolate, acuminate, peduncles short, up to 1cm, in Bhutan and Sikkim; and var. **latifolia** Cowan & Cowan, with leaves obovate, acute or obtuse and longer peduncles 1–2.5cm, from Darjeeling district, Vegetatively very similar to *Lindera assamica* and *L. latifolia* (see note under *L. assamica*) and also to *Phoebe attenuata* and *Persea fructifera* which differ in their paniculate inflorescences.

9. L. monopetala (Roxb.) Persoon; *L. polyantha* Jussieu. Sha: *Seychhanglu Shing*; Nep: *Bonsum, Kut Mero* (34), *Patmero* (34).

Tree to 12m, branchlets brownish tomentose. Leaves broadly elliptic, 9–24 × 5–11cm, obtuse or apiculate, base rounded, greenish beneath when dry, softly tomentose and prominently reticulate beneath, lateral veins 7–12 pairs; petioles 1–2.3cm. Umbels densely pubescent, in bud 5mm, on tomentose peduncles 3–10mm, forming dense clusters, peduncles sessile or borne on a short stout stalk 2–4mm. Fruit subglobose, c 7 × 6mm, apiculate, borne on perianth cup c 5mm across, on slender pedicels 9–12mm.

Bhutan: S – Chukka district (Marichong), Sarbhang district (near Phipsoo) and Gaylegphug district (Taklai Khola); **Sikkim**: terai. Subtropical and terai forests, 200–900m. March–April.

Silkworms are reared on the leaves of this tree (16).

10. L. hookeri (Meisner) Long; *L. khasyana* (Meisner) Hook. f. *nom. illeg.* Nep: *Dude Lampate* (34).

Similar to *L. monopetala* but leaves elliptic-obovate, 16–23 × 6–9cm, shortly acuminate, base cuneate, whitish beneath, pubescent on veins beneath; lateral veins 10–15 pairs; umbel buds larger, 6–7mm; fruit ellipsoid, c 17 × 11mm, borne on perianth cup 15–17mm across.

Sikkim: Rungbi, Darjeeling. Subtropical forests, 600–900m. May.

11. L. nervosa (Meisner) Grierson & Long; *L. wallichii* Hook. f. *nom. illeg.*

Tree, shoots glabrous. Leaves coriaceous, ovate, 12–20(–40) × 5–8.5cm, finely acuminate, base rounded, glossy above, pale beneath, glabrous, lateral veins 5–7 pairs, impressed above, very prominent beneath; petioles 1–1.4cm. Umbels 1–2 in leaf axils, c 4mm in bud, glabrous, on peduncle 4–5mm. Flowers subsessile. Fruit subglobose, c 8mm, borne on persistent 6-lobed perianth cup c 2mm across, on short thick pedicel 4mm.

Bhutan: locality unknown. Subtropical forests.

A poorly known species collected in Bhutan by Griffith in 1838 but not seen since that date.

12. L. glutinosa (Loureiro) Robinson; *L. sebifera* (Willdenow) Persoon. Nep: *Kawala* (34).

Tree to 15m, young shoots whitish pubescent. Leaves coriaceous, ovate-lanceolate, ovate or elliptic, 12–21 × 4.5–10cm, acute, base rounded or cuneate, glabrous or pale pubescent beneath, lateral veins 7–10 pairs, not very prominent beneath; petioles slender, 2–3.5cm. Umbels large, 7–10mm in bud, whitish pubescent, on slender peduncles 8–15mm, usually in a subumbellate cluster at apex of a slender stalk 1.5–3cm. Pedicels 3–5mm, white-pubescent. Fruit globose, 8–9mm, on scarcely enlarged perianth cup 2–3mm across, on short thick pedicel 4–5mm.

Bhutan: C–Tongsa district (Wangdegong near Shamgong), Mongar district (Mongar) and Tashigang district (Shali and Balfi); **Sikkim**: Darjeeling and Badamtam. Warm broad-leaved forests, 900–1700m. May–June.

Bark and fruits used medicinally (13).

13. L. panamanja (Nees) Hook. f. Nep: *Painle Champ, Dudhi Lampati*.

Tree to 25m, shoots glabrous or minutely puberulous. Leaves coriaceous, elliptic-oblong, 13–22 × 3–7cm, acute, base cuneate, pale beneath, lateral veins 10–12 pairs, slender, not impressed above, glabrous or pubescent on veins beneath; petioles 1.5–2cm. Umbel buds 5–6mm, puberulous, on short peduncle 2–5mm, peduncles racemosely arranged on puberulous axillary stalks 3–7cm in male inflorescences, shorter in females. Fruit depressed globose c 1.5 × 1.8cm, half enclosed by large fleshy perianth cup 2–2.5cm across.

Bhutan: S–Samchi district (Samchi) and Gaylegphug district (Lodrai Khola); **Sikkim**: Ryang and Mal Forest. Subtropical forests, 300–670m. March–May.

14. L. albescens (Hook. f.) Long; *L. oblonga* (Nees) Hook. f. var. *albescens* Hook. f. Nep: *Kalipahenle, Makai Kath* (34).

Similar to *L. panamanja* but young shoots appressed pale pubescent; leaves elliptic-lanceolate, 10–17 × 2.5–4.5cm, gradually acuminate, lateral veins 7–9 pairs, usually impressed above; petioles 4–6mm; umbel buds c 3mm, on short slender peduncles 2–8mm, arranged racemosely on axillary stalks 8–12mm (shorter in some axils); fruit oblong-ellipsoid, 17–23 × 9–10mm, on shallow perianth cup 6–7mm across, on short thick pedicel 4–7mm.

Bhutan: S–Chukka district (Sinchu La); **Sikkim**: Darjeeling, Dumsong, Kurseong, Lopchu etc. Warm broad-leaved forests, 1200–2100m. October.

Records of *L. doshia* (D. Don) Kostermans and *L. lancifolia* (Nees) Hook. f. from Bhutan and Sikkim probably all refer to this species. *L. doshia* is a Nepalese species with oblong-elliptic, acute leaves, brown beneath, lateral veins not impressed above, longer petioles 8–16mm, larger umbel buds 6–7mm diameter, umbels racemose on a stouter stalk 4–15mm. *L. lancifolia* from Sylhet is very different in its rusty brown tomentose shoots, opposite or subopposite leaves which are reddish-brown when dry, and brown-tomentose subsessile umbels not arranged in racemes. However *L. lancifolia* var. *alternifolia* Hook. f. from Sikkim may be a synonym of *L. albescens*.

L. albescens also shows some similarities to *L. chartacea* and *L. cubeba*. *L. chartacea* is vegetatively quite similar but has more coriaceous leaves, yellow-green when dry, only shortly acuminate, lateral veins less prominent beneath and obscure above, on longer petioles 10–15mm, larger umbel buds 4–6mm on peduncles which form sessile axillary clusters and fruit borne on a larger perianth cup 7–9mm across on long slender pedicel 10–15mm. *L. cubeba* has lanceolate leaves with 8–12 pairs of lateral veins, longer petioles 12–24mm and umbels clustered in leaf axils, smaller subglobose fruit 6–7mm with a minute perianth rim c 1.5mm across.

15. L. nitida (Nees) Hook. f.

Tree, branchlets strongly striate, glabrous. Leaves thickly coriaceous, oblanceolate or obovate, 11–25 × 5–10cm, obtuse, base attenuate, glabrous, lateral veins 10–12 pairs, obscure; petioles 1–1.5cm, stout. Umbel buds 4–5mm, on slender 5–16mm peduncles, arranged racemosely on stalks 1.5–3cm. Flowers subsessile. Fruit ellipsoid c 13 × 7mm, half-enclosed by much enlarged coarsely toothed perianth cup c 10 × 10mm.

Bhutan: locality unknown. Subtropical forests. September.

Known from Bhutan only from an unlocalised Griffith specimen.

11. NEOLITSEA (Bentham) Merrill

Similar to *Lindera* and *Litsea* but terminal buds always conspicuous; leaves sometimes clustered at branch ends, strongly 3-veined from above base; umbels sessile, crowded; anther cells 4; perianth cup not or slightly enlarged in fruit.

1. Leaves oblanceolate, 13–25cm, clustered at branch ends; flower clusters 1.5–2.5cm diameter; fruit ellipsoid 12–15mm .... **1. N. cuipala**
\+ Leaves ovate-elliptic, 7–14cm, not clustered at branch ends; flower clusters c 1cm diameter; fruit globose 8–9mm .... **2. N. foliosa**

1. N. cuipala (D. Don) Kostermans; *N. lanuginosa* (Nees) Gamble, *Litsea lanuginosa* Nees, *L. cuipala* (D. Don) Nees. Dz: *Akaru Sham.* Fig. 23m–q.

Evergreen or deciduous tree to 20m, terminal buds elongate, 2–5cm. Leaves clustered at branch ends, pendulous, coriaceous, oblanceolate, 16–24×3.5–6cm, gradually acuminate, base long attenuate, pale beneath, strongly 3-veined above base, basal lateral veins straight, extending to upper quarter of leaf; petioles 1.5–2.5cm. Umbel buds globose, in clusters on old shoots, flowering umbels 1.5–2.5cm diameter. Perianth segments 3mm. Fruit ellipsoid 12–15×7–8mm, borne on short pedicels 4–8mm, thickened throughout, with small perianth cup 3mm across.

Bhutan: S–Chukka district (Chukka Dzong), **C**–Punakha district (near Ngawang, Tang Chu); **Sikkim.** Warm broad-leaved forests, 1750–1850m. February–March.

2. N. foliosa (Nees) Gamble; *Litsea foliosa* Nees, *L. striolata* Blume. Dz: *Metsu Shing, Podzi Shing;* Nep: *Sisi Phul, Bhale Sissi* (34).

Similar to *N. cuipala* but terminal buds smaller, 8–12mm; leaves rather distant, not clustered, ovate-elliptic, 7–14×2–6cm, apex often abruptly acuminate, base cuneate; umbel clusters c 1cm diameter; fruit subglobose 8–9mm diameter, borne on slender pedicels 8–13mm becoming thickened only at apex, perianth cup 2.5–3mm across.

Bhutan: S–Chukka district (Gedu), **C**–Punakha district (Changshima and Tinlegang), **N**– Upper Kuru Chu district (Shawa); **Sikkim.** Cool broad-leaved forests, 1600–2080m. December–January.

N. foliosa belongs to a much-confused group of species, of which three others have been erroneously recorded from Bhutan and Sikkim (34, 80,135): *N. pallens* (D. Don) Momiyama & Hara (*Litsea consimilis* (Nees) Nees, *L. umbrosa* (Nees) Nees var. *consimilis* (Nees) Hook. f.), a W Himalayan speicies; *N. umbrosa* (Nees) Gamble (*Litsea umbrosa* (Nees) Nees), from Assam and Khasia; and *N. zeylanica* (Nees) Merrill (*Litsea zeylanica* Nees), from S India and Ceylon. These records probably all refer to *N. foliosa.*

12. DODECADENIA Nees

Evergreen trees; shoots with conspicuous terminal vegetative buds. Leaves alternate, not clustered, pinnately veined. Flowers bisexual, solitary or few in sessile umbels on previous year's wood, surrounded by persistent scales. Perianth segments 6. Fertile stamens 9–15, anthers 4-celled, all introrse. Fruit ellipsoid, seated on enlarged entire perianth cup.

1. D. grandiflora Nees; *D. griffithii* Hook. f. Fig. 23 l.

Tree to 15m, shoots with ellipsoid terminal vegetative buds. Leaves coriaceous, elliptic-oblanceolate, 8–11×2–3cm, shortly acuminate, base cuneate, almost glabrous, finely reticulate on both surfaces, lateral veins 7–10 pairs, slender, scarcely prominent; petioles 5–12mm. Flower buds subsessile, globose, scales appressed silky, containing 1–2 large flowers 1.4–1.6cm across. Perianth segments oblong 7–9mm. Fruit c 1.5×1 cm, reddish, borne on thick woody perianth cup 8–9mm across, on stout pedicel 3–5mm.

Bhutan: locality unknown; **Sikkim:** Tonglo; **Arunachal Pradesh:** Nyam Jang Chu. Evergreen oak forests, 2400-2560m. March–April.

The Nyam Jang Chu plant belongs to the typical var. **grandiflora** with pubescent young shoots and ovary, whilst the Sikkim and Bhutan plants belong to var. **griffithii** (Hook. f.) Long (*Dodecadenia griffithii* Hook. f.) with glabrous young shoots and ovary.

13. ACTINODAPHNE Nees

Evergreen trees, shoots with conspicuous terminal vegetative buds. Leaves clustered at branch ends, pinnately veined or strongly 3-veined at base. Dioecious; flowers unisexual in sessile or shortly pedunculate umbels or clusters, without persistent scales, or in short panicles, usually in axils of fallen leaves on previous year's wood. Perianth segments 6. Fertile stamens 6 or 9, anthers 4-celled, all introrse. Fruit ellipsoid or subglobose, or borne on usually enlarged fleshy entire perianth cup.

1. Leaves obovate, 9–17cm broad, strongly 3-veined near base; flowers in panicles **1. A. obovata**
\+ Leaves elliptic, lanceolate or oblanceolate, 2–7cm broad, pinnately veined; flowers in umbels or clusters 2
2. Leaves 3–7cm broad, with 5–7 pairs of lateral veins ......**2. A. angustifolia**
\+ Leaves 1.5–4cm broad, with 8–16 pairs of lateral veins 3
3. Leaves membranous; lateral veins 8–11 pairs; fruiting perianth cup scarcely enlarged, c 2mm across ....**3. A. sikkimensis**
\+ Leaves coriaceous; lateral veins 12–16 pairs; fruiting perianth cup much enlarged, 9–10mm across ....**4. A. longipes**

1. A. obovata (Nees) Blume. Nep: *Runchey, Runchey Kath, Runche* (34), *Runchepat* (34). Fig. 23 r–u.

Tree to 20m. Young growth red-brown tomentose. Leaves drooping, obovate, 25–40×9–17cm, acute or apiculate, base cuneate, glaucous and thinly pubescent beneath, lateral veins 5–7 pairs, strongly prominent beneath, basal pair strongest, minor veins prominent beneath; petioles 2.5–5.5cm. Male panicles 2.5–4.5cm, perianth segments c 5mm. Female panicles shorter, 2–3cm, perianth segments c 2.5mm. Fruit ellipsoid, 1.5–2cm, borne on enlarged perianth cup c 1cm across.

Bhutan: S–Samchi, Sarbhang and Gaylegphug districts, **C**–Tongsa district (lower Mangde Chu valley); **Sikkim.** Subtropical and Warm broad-leaved forests, 320–1400m. March–April.

2. A. angustifolia (Blume) Nees

Similar to *A. obovata* but leaves elliptic or oblanceolate, 15–25×3–7cm, acuminate, lateral veins 5–7 pairs, lowermost weaker than upper, minor veins almost invisible; flowers in shortly pedunculate clusters; perianth segments 2–3mm; fruit subglobose, 8mm diameter, on perianth cup 5–6mm across.

Sikkim: Darjeeling. Subtropical forests. October.

3. A. sikkimensis Meisner. Nep: *Rudilo* (117), *Sik Siki* (34), *Phurke Sissi* (34).

Small tree to 6m. Leaves membranous, lanceolate, 10–14×2–4cm, finely acuminate, base cuneate, glabrous, pale and not reticulate beneath; lateral veins slender, 8–11 pairs, not very prominent beneath; petioles 8–10mm. Flowers in small subsessile clusters c 1cm diameter, perianth segments c 3mm. Fruit ellipsoid, c 13×7mm, borne on small entire perianth cup c 2mm across, fruiting pedicels 6–8mm.

Bhutan: S – Phuntsholing district (above Kamji) and Gaylegphug district (Sham Khara); **Sikkim.** Warm broad-leaved forests, 1500–2000m. November–February.

4. A. longipes Kostermans; *A. reticulata* Meisner var. *glabra* Meisner

Similar to *A. sikkimensis* but leaves coriaceous, narrowly elliptic-lanceolate, 11–15×1.5–3cm, somewhat reticulate on both surfaces, lateral veins 12–16 pairs; fruit broadly ellipsoid, c 15×10mm, borne on enlarged perianth cup 9–10mm across, fruiting pedicels 2–3cm.

Sikkim: Tonglu and Manibanjan. Cool broad-leaved forests, 1700–2000m. October.

14. PARASASSAFRAS Long

Small evergreen tree, shoots with conspicuous terminal and axillary vegetative buds. Leaves alternate, not clustered, strongly 3-veined from above base. Dioecious; flowers unisexual in ebracteate shortly pedunculate umbels borne around base of young leafy shoots. Perianth segments 6. Fertile stamens 9, anthers 4-celled, all introrse. Fruit subglobose, perianth cup entire, slightly enlarged.

1. P. confertiflora (Meisner) Long; *Actinodaphne confertiflora* Meisner, *Neocinnamomum confertiflorum* (Meisner) Kostermans, *Litsea confertiflora* (Meisner) Kostermans. Sha: *Selung Shing, Sey Shing;* Nep: *Kalo Bori* (117). Fig. 23v.

Small tree 4–6m, branches smooth, glabrous, with ovoid terminal and axillary buds and broad rings of bud scale scars. Leaves coriaceous, broadly ovate or suborbicular, 9–17×6–10cm, apiculate or shortly acuminate, base rounded, juvenile leaves sometimes acutely 2–3-lobed near apex, glabrous, pale beneath, strongly 3-veined 6–11mm above base; petioles 1.5–2.5cm. Flowers 3–6 in ebracteate umbels on short peduncles 1–3mm, several umbels clustered around a terminal bud which develops into a new leafy shoot, pedicels slender, 6–9mm. Perianth segments ovate, 3–4mm. Fruit subglobose, 7–8mm borne on flat entire perianth cup c 4mm across, on slightly thickened pedicels 6–7mm.

Bhutan: S – Gaylegphug district (Sham Khara) and Deothang district (Deothang, Wamrung and Riserboo), **C** – Mongar district (near Mongar) and Tashigang district (Shali). Warm broad-leaved forests, 900–2150m. November–January.

A monotypic genus known only from Bhutan, Upper Burma and W China.

Family 40. TETRACENTRACEAE

by A.J.C. Grierson & D.G. Long

Deciduous trees with branches bearing alternately arranged short shoots covered with crowded scars of fallen leaves and bud scales, terminated in late season by a bud at first enclosed by sheathing base of petiole but later free. Leaves borne singly at apex of short shoots, simple, serrate, palmately veined. Inflorescence a slender pendulous catkinate spike borne close to petiole near apex of short shoot. Flowers bisexual, numerous, borne in clusters of 4; sepals 4; petals 0; stamens 4, borne opposite the sepals; carpels 4, united at base; styles subulate, erect at first then recurved. Fruit a cluster of 4 laterally united follicles, styles ± basal due to unequal growth of carpels; seeds c 5 per follicle, compressed ellipsoid, shortly winged at each end.

1. TETRACENTRON Oliver

Description as for Tetracentraceae.

1. T. sinense Oliver. Fig. 20h–k.

Tree up to 30m, glabrous, short shoots up to 4cm long. Leaves ovate, 7–15×5–12cm, acuminate, base cordate, rarely rounded, palmately 5–7-veined at base; petioles 2–4cm. Spikes 7–20cm on peduncles 0.5–2cm. Sepals broadly ovate, 1–1.5mm long and broad, obtuse. Stamens 2–3.5mm. Carpels c 1.5mm at flowering time. Follicles 2.5–4.5mm; seeds 2–3.5mm including wings.

Bhutan: C – Punakha district (near Lometsawa and Norbding) and Tongsa district (between Neylong and Tashiling). Cool broad-leaved forests, 2200–2900m. April–May.

The Himalayan plants are sometimes treated as a distinct variety, var. **himalense** Hara & Kanai.

Family 41. EUPTELEACEAE

by A.J.C. Grierson & D.G. Long

Deciduous trees or shrubs. Leaves alternate, simple, pinnately veined, finely serrate. Inflorescence of 6–12 flowers in axils of deciduous bracts at base of lateral branches, opening before development of leaves, bisexual; pedicels slender, broadened distally into a flattened disc. Perianth none. Stamens usually 10–15, borne in a single whorl near margin of disc, anthers bearing a slender appendage at apex. Carpels 8–18, free, borne on slender stalks in a single whorl within the stamens, compressed, obliquely oblong or elliptic bearing a stigmatic crest along upper margin, ovules 1–3; carpels rapidly changing shape after pollination due to unequal growth. Fruit a cluster of oblong or obovate samaras, seeds 1–3 surrounded by a papery wing, rounded at apex, tapering at base into stalk, concave and with stigmatic crest still visible on inner side, outer side ± straight.

1. EUPTELEA Siebold & Zuccarini

Description as for Eupteleaceae.

1. E. pleiosperma Hook. f. & Thomson. Fig. 20 l–o.

Tree or shrub 2–15m. Leaves ovate or elliptic, 7–16(–19) × 4–12(–15)cm, acuminate, base cuneate, sparsely pubescent beneath; petioles 1–4.5cm. Flowers borne on minute pedicels, soon elongating to 0.5–2cm in fruit; disc 0.5–1.5mm diameter, pubescent at margin. Stamens with filaments 2–8mm, anthers 5–10mm, crimson or brownish, apical appendage 0.5–2mm. Carpels at flowering 0.5–1.5 × 0.3–0.6mm. Samaras 5–11 × 3.5–6mm, reddish or purplish, on stalks 5–15mm; persistent stigmatic crest 1–4mm.

Bhutan: C – Mongar district (Dengchung N of Lhuntse). Margins of Evergreen oak forests, 2100m. March.

Family 42. RANUNCULACEAE

by A.J.C. Grierson

Annual or perennial herbs, erect or stoloniferous, rarely shrubs or woody climbers. Leaves basal and on stems, alternate, rarely opposite, entire or palmately, ternately, or pinnately dissected, exstipulate or petioles sometimes broadened into stipule-like auricles at base. Hairs when present simple. Inflorescence 1-flowered, cymose, racemose or paniculate. Flowers actinomorphic or zygomorphic, bisexual or rarely unisexual (plants dioecious). Sepals 3–8, mostly 5, free, often showy and petaloid. Petals sometimes absent, 1–many, free, often nectariferous, sometimes inconspicuous. Stamens many, rarely 8–20. Carpels 1–many, superior, free or shortly connate at base (in *Dichocarpum*), ovules 1, basal or few–many, marginal. Fruit a cluster of 1-seeded indehiscent achenes or few- to many-seeded follicles dehiscing along dorsal suture, rarely a berry.

1. Plants woody, either erect shrubs or climbers, sometimes slender and almost herbaceous but then climbing by means of twining petioles from opposite leaves 2
\+ Plants herbaceous, erect, decumbent or prostrate, rarely scrambling but never climbing by means of twining petioles 4

2. Erect shrubs; leaves alternate; fruits follicular, styles short, not hairy **22. Paeonia**
+ Climbing shrubs or subshrubs; leaves opposite; fruits achenial, achenes with elongated hairy styles 3

3. Plants climbing by means of twining petioles; sepals showy; petals absent ....**1. Clematis**
+ Plants climbing by means of tendrils (from modified terminal leaflets); sepals not showy, caducous; petals small, spathulate ....**2. Naravelia**

4. Flowers zygomorphic; fruit always follicular 5
+ Flowers actinomorphic; fruit achenial or follicular, rarely a berry 7

5. Petal solitary (± concealed in showy, hooded upper sepal); follicle solitary ....**20. Consolida**
+ Petals 2 or 4 (± concealed within showy, hooded upper sepal); follicles 3–10 6

6. Petals 4, the upper pair elongated, attached near middle, the upper section extending into the spur of upper sepal, the lower pair with basal claw and oblique lamina ....**19. Delphinium**
+ Petals 2, with basal claw and S- or T-shaped head divided into a ± bulbous or coiled spur and flattened lamina ....**21. Aconitum**

7. Flowers usually numerous in compact racemes or panicles; fruit a cluster of follicles or a solitary berry 8
+ Flowers solitary or few; fruit achenial or follicular, sometimes (in *Thalictrum*) flowers in diffuse racemes or panicles but then fruit achenial 10

8. Flowers in slender racemes 7–15cm, usually forming terminal panicles; petals bearing 2 rounded glandular discs at apex ....**18. Cimicifuga**
+ Flowers in short terminal racemes 2–5(–10)cm, never paniculate; petals without glandular discs at apex 9

9. Leaflets serrate; stigma rounded, disc-like, sessile; fruit a solitary berry ....**17. Actaea**
+ Leaflets pinnatisect; stigma elongated, borne on short thickened style; fruit of 1–3 follicles ....**16. Souliea**

10. Leaves all basal; scapes leafless or with a whorl of leafy bracts or reduced leaves below the flowers 11
+ Leaves basal and with several on stems 18

11. Leaves compound 12
+ Leaves simple 14

12. Rootstock surrounded at apex by stiff bristly remains of petioles; scapes bearing a pair of linear-lanceolate or ternate bracts below flowers; fruit of 5–7 follicles **13. Paraquilegia**
+ Roostock fibrous with leaf remains but not bristly; scapes leafless or with a whorl of leafy bracts; fruit achenial 13

13. Scapes with a whorl of leafy bracts; petals absent .... **3. Anemone**
+ Scapes leafless; petals present .... **5. Callianthemum**

14. Scapes bearing a whorl of leafy bracts or minute leaves 15
+ Scapes naked 16

15. Leaves suborbicular, peltate; scapes bearing 3 minute 2–3mm bracts **15. Asteropyrum**
+ Leaves ovate or reniform, attached to petiole at base; scapes bearing a whorl of leafy bracts .... **3. Anemone**

16. Sepals 12–15 × 3–10mm; petals absent; fruit a cluster of follicles **10. Caltha** (*C. scaposa*)
+ Sepals less than 7 × 4mm; petals present; fruit a cluster of achenes 17

17. Flowers bisexual; sepals 3–7 × 2–4mm; petals 10–15 .... **8. Oxygraphis**
+ Flowers unisexual; sepals c 2 × 1.5mm; petals 5 .... **9. Paroxygraphis**

18. Petals absent; sepals showy 19
+ Petals present, sometimes smaller than showy sepals 21

19. Leaves simple, reniform; flowers several, corymbose; fruit a cluster of follicles .... **10. Caltha** (*C. palustris*)
+ Leaves compound; flowers solitary or numerous in lax racemes or panicles; fruit a cluster of achenes or follicles 20

20. Leaves palmatisect; flowers solitary; fruit follicular .... **11. Calathodes**
+ Leaves repeatedly ternately divided; flowers in lax racemes or panicles; fruit achenial .... **4. Thalictrum**

21. Leaves repeatedly ternately divided into rounded leaflets; follicles 2, divergent **14. Dichocarpum**
+ Leaves ternately or bipinnately divided, leaflets or segments not rounded; fruit achenial or with 5–20 follicles 22

22. Leaves bipinnatisect; flowers white; fruit achenial .... **6. Adonis**
+ Leaves ternatisect; flowers yellow (or if white then plants aquatic with filamentous leaves); fruit achenial or follicular 23

23. Sepals shorter than petals, sometimes deciduous; fruit achenial **7. Ranunculus**
+ Sepals showy, longer than very small petals; fruit follicular ...... **12. Trollius**

1. CLEMATIS L.

The Nepali name *Kanari Lahara* has been applied to several species (34).

Climbing or rarely erect shrubs or subshrubs. Leaves opposite, simple or more often ternate, biternate or pinnate; petioles and petiolules often twining. Flowers actinomorphic, solitary or in panicles, axillary or terminal. Sepals 4(–8), petaloid, valvate. Petals absent. Stamens numerous. Carpels many, each with one ovule. Fruit a head of achenes each with an elongated slender persistent style; style plumose with long straight hairs, or rarely short, stout and without plumose hairs.

1. Flowers solitary or in axillary fascicles 2
\+ Flowers in axillary or terminal racemes or panicles 3
2. Pedicels naked ....**Species 1–3**
\+ Pedicels bearing a pair of leafy bracts near or above the middle **Species 4 & 5**
3. Leaves simple or ternate (rarely pinnate in *C. buchananina)*.....**Species 6–9**
\+ Leaves pinnate or biternate 4
4. Petiole bases connate, becoming broadly winged ....**10. C. connata**
\+ Petiole bases not connate or winged ....**Species 11–13**

1. C. montana DC. Fig. 25g.

Woody climber. Leaves ternate, leaflets ovate, 1.5–7 × 1–3.5cm, acute, base rounded, margins coarsely toothed or lobed, sparsely appressed-pubescent especially beneath; petioles 4–7cm. Flowers 2–3 in axillary fascicles; pedicels 3–5(–8)cm. Sepals 4, elliptic, 1.5–2 × 0.7–1.25cm, white or pinkish, obtuse or acute, pubescent outside, glabrous within. Filaments glabrous, anthers yellow. Achenes ovate, c 5 × 4mm, compressed, glabrous; styles up to 4.5cm at maturity.

Bhutan: C – Ha to Bumthang districts, common, **N** – Upper Mo Chu to Upper Kulong Chu districts; **Sikkim**. On shrubs in dry valleys, in Blue Pine and Spruce forests and Juniper/Rhododendron scrub, 2100–4100m. April–July.

A variable species; some specimens approach *C. tongluensis* in having larger sepals and longer peduncles.

2. C. tongluensis (Bruehl) Tamura; *C. montana* DC. var. *tongluensis* Bruehl. Nep: *Indrakamal*.

Similar to *C. montana* but flowers usually 1 per axil; pedicels longer, 12–30cm; sepals 4–6 × 1–2cm, acuminate; anthers purple.

FIG. 25. **Ranunculaceae**. a & b, *Aconitum novoluridum*: a, flower in side view; b, petal. c, *Aconitum spicatum*: vertical section of flower to show arrangement of petals within uppermost sepal, also stamens and carpels. d, *Delphinium ludlowii*: vertical section of flower to show arrangement of two upper petals which extend into spur of uppermost sepal, and two lower petals with slender basal claw and oblong lamina, also arrangement of stamens and carpels. e, *Caltha palustris*: dehisced follicles. f, *Cimicifuga foetida*: petal. g, *Clematis montana*: achene with persistent elongated plumose style. h, *Dichocarpum adiantifolium*: petal. i, *Ranunculus trichophyllus*: petal with basal nectar-pit. j, *Paraquilegia anemonoides*: petal. k, *Thalictrum elegans*: winged achene with appressed style. l, *Thalictrum javanicum*: ribbed achene with persistent hooked style. m & n, *Trollius pumilus*: m, flower; n, petal with nectar-pit above basal claw. Scale: c × 1; m × 1½; e, g × 2; a, d × 2½; b × 4½; n × 5½; f × 6½; k × 6⅔; h × 7; i, j × 7½; l × 10. Drawn by J. Chisholm.

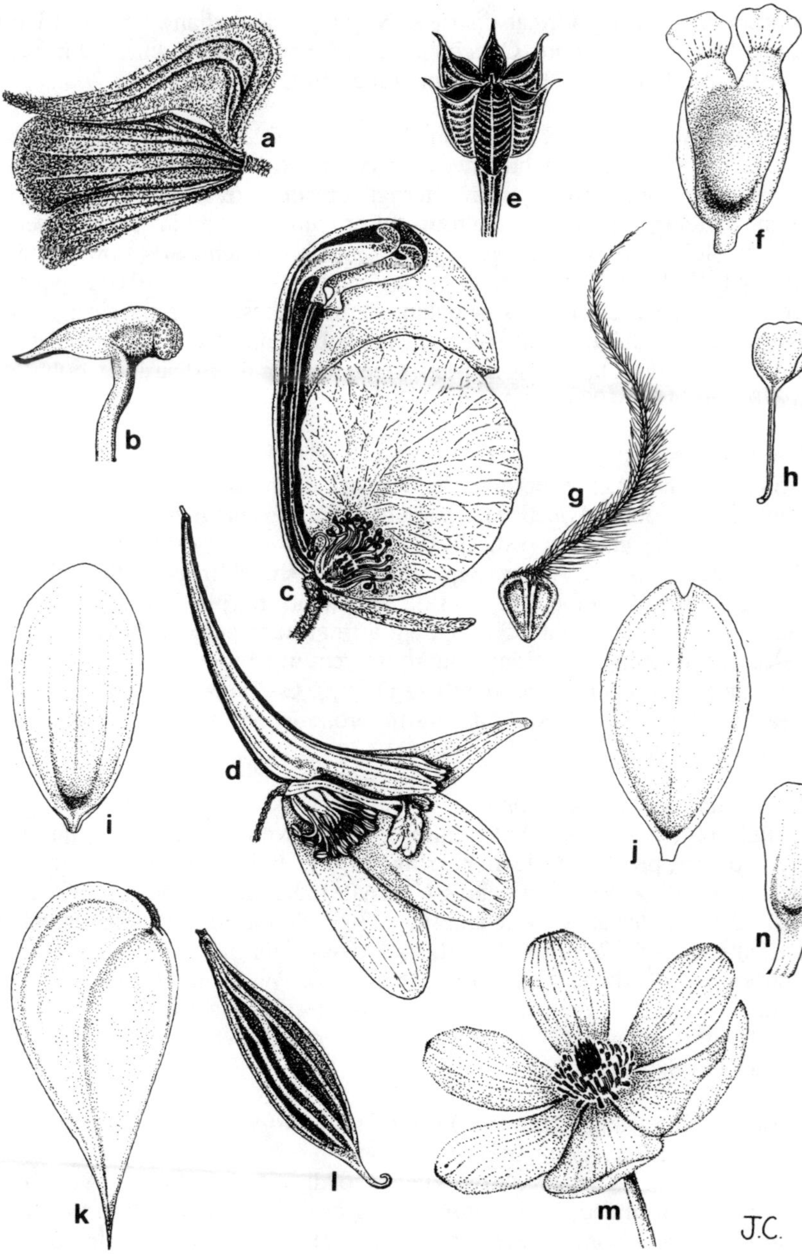
a
b
c
d
e
f
g
h
i
j
k
l
m
n
J.C.

Bhutan: **C** – Ha to Mongar districts, **N** – Upper Bumthang Chu and Upper Kulong Chu districts; **Sikkim**. Cool broad-leaved, Spruce, Hemlock and Fir forests, on shrubs and bamboo, 2300–3500m. June–July.

3. C. acutangula Hook. f. & Thomson; *C. ranunculoides* Franchet. Med: *Emong*.

Similar to *C. montana* but stems erect or weakly scandent, 0.3–3m, ± herbaceous or slightly woody; leaves ternate or biternate, leaflets ovate, 2–5 × 1–3cm, acute or acuminate, base rounded or truncate, margins coarsely serrate and often 3-lobed, sparsely or densely pubescent with straight hairs; flowers solitary or few, pedicels 2–5(–20)cm; sepals oblong-elliptic, 1–1.5 × 0.4–0.6cm, purplish, narrowly 3-winged on the back, subglabrous but pubescent at margins; stamens densely covered with long white hairs; achenes pubescent, styles c 2.5cm at maturity.

Bhutan: **C** – Ha to Tashigang districts. Open scrub in Blue Pine and Cool broad-leaved forests, 1500–2740m. August–October.

4. C. napaulensis DC.

Woody climber. Leaves ternate, leaflets ovate-lanceolate, 2–5 × 0.5–2cm acuminate, base rounded or cuneate, margins entire or sometimes 3-lobed, glabrous. Flowers in axillary clusters; pedicels 3–4cm with a pair of ovate connate bracts 5–10mm near base of flower. Sepals 4, yellowish, oblong-elliptic, 1.5–2 × 0.5–0.7cm, densely pubescent outside. Filaments purple, glabrous. Achenes obovate, densely pubescent; styles c 5cm at maturity; pedicel above the bracts pubescent, thickening and elongating up to 2cm in fruit.

Bhutan: **S** – Chukka district (Chukka Dzong), **C** – Punakha district (between Tsarza La and Ratsoo); **Sikkim**. Warm broad-leaved forests, 1760–2000m. December–January.

5. C. cadmia Hook. f. & Thomson

Similar to *C. napaulensis* but stems slender, scarcely woody; leaves biternate; flowers solitary, pedicels 7–15cm bearing near the middle a pair of sessile, ovate, acute bracts 1–3 × 0.5–3cm; sepals 4–6, pale blue, elliptic-oblanceolate, 2–4.5 × 0.6–2.25cm; achenes ovate-elliptic, c 7 × 5mm, compressed, appressed-pubescent; styles 5–7cm, stout, without long straight hairs.

Bhutan: **S** – Deothang district (N of Narfong). Warm broad-leaved forest, ?c 1500m.

The Bhutan record (117) requires confirmation, elsewhere known only from Manipur and Sylhet.

6. C. smilacifolia Wall.; *C. munroiana* Wight, *C. loureiriana* auct. non DC. Nep. *Pinasi Lahara* (117), *Halure Lahara*.

Woody climber. Leaves simple or ternate, broadly ovate, 6–14 × 3–9cm, acute, base cordate or rounded, 5- or 7-veined, margins entire or remotely and shallowly serrate, glabrous, or sometimes ternate with more narrowly-ovate leaflets 6–15 × 3–8cm, acuminate, base rounded or slightly cordate. Panicle branches often whorled and bearing a number of linear-spathulate bracts c 2cm. Sepals 4–8, ovate 2–2.5 × 0.5–0.75cm, densely and finely brownish pubescent outside, purple within, becoming reflexed. Filaments white, glabrous, connective elongated into an acuminate point above anthers. Achenes pubescent; styles up to 5.5cm at maturity.

Bhutan: S – Chukka district (Marichong and Jumudag) and Gaylegphug district (Sureylakha), **C** – Punakha district (Kabjisa), Mongar district (Lhuntse) and Tashigang district (Dangme Chu and Ghunkarah); **Sikkim**. On shrubs in Subtropical, Warm broad-leaved and Evergreen oak forests, 1200–2060m. October–November.

The Bhutan plants represent typical *C. smilacifolia* with simple, entire leaves. The Sikkim plants often have ternate leaves with serrate leaflets and are sometimes separated as *C. munroiana* Wight.

7. C. acuminata DC.

Similar to *C. smilacifolia* but more slender; leaves ternate, leaflets ovate or lanceolate, 2.5–10 × 1.5–4.5cm, acuminate, rounded and 5-veined at base, margins serrate, glabrous or sparsely pubescent; panicles sometimes few-flowered, branches slender, glabrous or sparsely pubescent; sepals ovate, yellowish, 1–1.5 × 0.4–0.6cm, erect or recurved at tip, subglabrous or densely and shortly pubescent; filaments densely hairy; achenes with styles up to 2cm at maturity.

Bhutan: S – Samchi, Chukka and Sarbhang districts, **C** – Tongsa and Tashigang districts; **Sikkim**. On shrubs in Subtropical and Warm broad-leaved forests, 200–2400m. November–December.

Some plants from the Darjeeling area have been distinguished as var. **sikkimensis** Hook. f. & Thomson (*C. sikkimensis* (Hook. f. & Thomson) Burkill), on account of their glabrescent sepals which are pubescent at the margins, but the suggested correlation (69) between larger leaves and smaller flowers seems not to be well-founded.

8. C. buchananiana DC.; *C. tortuosa* (Hook. f. & Thomson) Fischer. Nep: *Pinasi Lahara* (117).

Similar to *C. smilacifolia* but usually brownish pubescent throughout; leaves ternate or rarely pinnate, leaflets 3(–5), broadly ovate, 6.5–12 × 4–10cm, acute or shortly acuminate, base rounded or cordate, margins coarsely serrate or 3-lobed, sparsely pubescent above, denser beneath; petioles sometimes broadened and thickened at base and narrowly connate; panicles bearing a pair of coarsely toothed leafy bracts at each node; sepals greenish-yellow, 4–6, ovate, 2–3 × 0.5–0.7cm, brownish pubescent externally; filaments hairy.

Bhutan: S – Phuntsholing, Chukka and Gaylegphug districts, **C** – Ha to Tashigang districts, common, **N** – Upper Mo Chu and Upper Kuru Chu districts; **Sikkim**. Warm and Cool broad-leaved, Hemlock and Fir forests, on shrubs, 1000–3650m. July–October.

A variable species; two varieties have been reported from Bhutan and Sikkim, but are scarcely distinct: var. *rugosa* Hook. f. & Thomson with leaves prominently veined and var. *tortuosa* Hook. f. & Thomson (*C. tortuosa* (Hook. f. & Thomson) Fischer) with more glabrous leaves. Forms with pinnate leaves resemble *C. connata* but lack the broadly-winged petiole-bases of the latter.

9. C. grewiiflora DC.

Similar to *C. buchananiana* but stems, leaves and sepals very densely brown tomentose; sepals broader, 8–11mm broad.

Bhutan: S – Samchi and Chukka districts, **C** –Thimphu, Punakha, Tongsa and Tashigang districts; **Sikkim**. Dry scrub in Subtropical and Warm broad-leaved forests, 1000–1525m. November–February.

10. C. connata DC.; *C. trullifera* (Franchet) Finet & Gagnepain

Slender woody climber. Leaves pinnate; leaflets 5–7, distant, ovate, 4.5–13 × 3–10cm, acuminate, base rounded or cordate, margins serrately toothed or sometimes coarsely lobed, sparsely pubescent; petioles connate at base and broadened into a collar 5–12mm broad around stem. Panicles cymosely branched. Sepals ovate, c 2 × 0.5–1cm, acute, yellowish pubescent on both surfaces. Filaments hairy. Achenes pubescent, styles up to 5cm at maturity.

Bhutan: S – Chukka district (Chukka), **C** – Punakha district (Lometsawa, Wangdu Phodrang and Ritang), Bumthang district (Shabjethang and Ura) and Tashigang district (Bamri Chu and Pintsogong); **Sikkim**. On shrubs in dry valleys and in Blue Pine and Spruce forests, 1370–3050m. September–November.

11. C. gouriana DC.

Similar to *C. connata* but leaves pinnate with 7 leaflets or sometimes biternate, rarely ternate in the inflorescence; leaflets ovate, 2.5–8 × 1.5–3.5cm, acuminate, base rounded, margins entire or with 1 or few teeth near apex on either side, glabrous or sparsely pubescent along veins; petioles not connate and broadened at base, petiolules slender up to 2.5cm; panicles cymose, branches rather stout, pubescent, flowers numerous; sepals elliptic, pubescent; styles up to 4cm at maturity.

Bhutan: S – Samchi district (Dorokha and Torsa valley) and Phuntsholing district (Phuntsholing), **C** – Paro district (117) and Mongar district (Unjar) (117); **Sikkim**. On shrubs at margins of Subtropical and Warm broad-leaved forests, 150–2000m. October–November.

12. C. puberula Hook. f. & Thomson

Similar to *C. gouriana* but leaves often biternate, leaflets usually toothed or 3-lobed, pubescent on both surfaces; flowers few in leafy panicles; sepals elliptic-spathulate, 5–15 × 3–5mm, whitish pubescent with soft hairs outside.

Bhutan: C – Punakha district (Kencho). Climbing on shrubs in open dry valley, 1830m. August.

13. C. zemuensis W.W. Smith

Similar to *C. gouriana* but leaves bipinnate or biternate, leaflets ovate, smaller, 1.5–3 × 1–2.5cm, acute or acuminate, base rounded, margins coarsely toothed or lobed, sparsely and finely pubescent on both surfaces; panicles branching into 5–7 straight, erect pedicels 5–6cm; sepals 4–6, elliptic, 1.5 × 0.3–0.5cm, acute, yellowish-white, sparsely pubescent; filaments glabrous; achenes pubescent.

Bhutan: C–Ha district (Tegong La); **Sikkim**: Zemu valley. Juniper/Rhododendron scrub, 2745–3800m. July–August.

2. NARAVELIA DC.

Climbing shrubs similar to *Clematis* but leaves ternate with terminal leaflet modified into a tendril; sepals 4–5, caducous; petals 6–12; achenes linear, thickly stalked.

1. N. zeylanica (L.) DC. Nep: *Rashgagri* (34).

Leaflets ovate, 7–12 × 4–8cm, acuminate, base rounded or cordate, margins entire, glabrous above, densely pubescent beneath. Flowers numerous in divaricately branched panicles. Sepals elliptic, c 6 × 3mm, densely appressed pubescent, falling away as petals expand. Petals spathulate, slender, c 7 × 1mm, greenish-yellow, glabrous. Connective prolonged into a short point above anthers. Achenes c 7 × 1mm, sparsely hairy, borne on stalks c 1mm, styles 3.5–4cm at maturity.

Bhutan: S–Gaylegphug district (Gaylegphug) and Deothang district (Deothang); **Sikkim**. Margins of Subtropical forests, 270m. November–January.

3. ANEMONE L.

Rhizomatous perennial herbs. Leaves all basal, petiolate, lobed or divided, often ternately. Flowers actinomorphic, solitary, or several in umbellate cymes on scapes, with an involucre of 2–3 leafy bracts below flowers. Sepals 5–6(–8), petaloid. Petals absent. Stamens many. Carpels numerous, ovule one. Fruit a head of glabrous, pubescent or woolly achenes; styles persistent, minute or short.

1. Scapes bearing an umbellate cyme of flowers (rarely flowers solitary) 2
\+ Flowers usually solitary, rarely 2–3 3
2. Branches of inflorescence (4–)10–15(–20)cm, some at least usually again branched or bearing one or more bracteoles near middle ......**Species 1 & 2**
\+ Branches of inflorescence 2.5–5(–13)cm, simple, rarely again branched or bracteate ....**Species 3–5**
3. Bracts of involucre divided like leaves ....**Species 6–8**
\+ Bracts of involucre not divided like leaves, entire or shallowly toothed (rarely deeply lobed in *A. rupestris*) ....**Species 9 & 10**

1. A. vitifolia DC.

Leaves simple, broadly ovate, 7–20cm long and broad, 3–7-lobed, acute, base cordate, margins serrate, sparsely pubescent above, whitish tomentose beneath; petioles 15–50cm. Scapes 20–50cm, in upper part bearing an involucre of bracts smaller than but similar to leaves. Umbellate cyme usually with 3 branches, one bearing a single flower, the others again ternately divided and involucrate. Sepals obovate-elliptic, 1.5–2 × 0.7–1.2cm, obtuse or acute, white within, purplish and pubescent outside. Achenes ellipsoid, c 1mm, densely surrounded by white woolly hair.

Bhutan: S – Chukka and Deothang districts, **C** –Thimphu to Tashigang districts, **N** – Upper Mo Chu district; **Sikkim**. Clearings in Cool broad-leaved, Evergreen oak and Fir forests, 1525–2745m. July–September.

2. A. rivularis DC. Med: *Supka*.

Similar to *A. vitifolia* but leaves ternate; leaflets broadly elliptic or rhombic, 4–7 × 3–5cm, acute, base cuneate often deeply trilobed, margins serrate, stiffly pubescent; petioles 10–30cm; scapes 7–35cm, involucres similar to leaves but segments narrowly linear sometimes only 2mm broad; sepals elliptic, 10–15 × 4–10mm; achenes broadly elliptic, 7 × 5mm, compressed, glabrous, style persistent, c 1mm, hooked.

Bhutan: S – Chukka district (Chima Kothi), **C** – Ha to Tashigang districts, common, **N** – Upper Mo Chu district (Laya) and Upper Pho Chu district (Chojo Dzong); **Sikkim**. Streamsides and clearings in Cool broad-leaved, Evergreen oak, Blue Pine and Hemlock forests, more rarely in alpine meadows, 1980–3960m. April–August.

3. A. demissa Hook. f. & Thomson

Rootstock surrounded by fibrous leaf remains. Leaves ternate, leaflets broadly ovate or rhombic in outline, 1.5–2.5cm long and broad, coarsely lobed and toothed, ± densely villous with long white hairs; petioles 3–10cm, usually with dense white spreading hairs. Scapes 10–30cm, densely white-hairy, 3–6-flowered (rarely 1-flowered), pedicels 2–4cm; involucral bracts 1–2cm, similar to leaves. Sepals elliptic, 7–15 × 4–10mm, whitish, yellow or purple. Stamens blackish. Achenes elliptic, 5–7 × 3.5–5mm, glabrous, compressed and surrounded by a wing up to 2mm broad; style minute, hooked.

Bhutan: C – Ha, Thimphu, Punakha and Sakden districts, **N** – Upper Mo Chu to Upper Kulong Chu districts, widespread; **Sikkim**. Alpine meadows and clearings in Juniper/Rhododendron scrub, also on screes and cliff ledges, 3350–4570m. May–July.

A common species, variable in dissection of leaf, indumentum, number and colour of flowers; the following two species are uncertainly distinguished from it.

4. A. polyanthes D. Don

Very similar to *A. demissa* but rootstock enlarged, woody, not usually covered with leaf remains; leaflets broader and less deeply lobed; sepals white, or bluish outside, 2–3cm; style straight.

Bhutan: C – Ha, Tongsa, Bumthang and Sakden districts, **N** – Upper Mo Chu district (Chomo Lhari); **Sikkim**. Clearings in Fir forests and on rocky alpine slopes, 3300–4110m. May–July.

5. A. smithiana Lauener & Panigrahi

Similar to *A. demissa* but rootstock densely surrounded by blackish fibrous leaf remains; sepals whitish, pink or purple, c 2.5 × 1cm; style hooked.

Bhutan: C –Thimphu district (Pajoding); **Sikkim**. Alpine rock-ledges and turf and in Juniper/Rhododendron scrub, 3960m. August.

6. A. obtusiloba D. Don; *A. geum* Léveillé. Tongsa: *Ugala*.

Leaves simple, broadly ovate in outline, 1.5–5cm long and broad, cordate, ternately divided almost to base, lobes ± obtusely toothed and divided, usually densely villous with long pale hairs; petioles 3–15cm. Scapes 5–20cm, involucral bracts similar to leaves but smaller, pedicels 1–12cm. Sepals elliptic-obovate, 1.5–2 × 0.7–1.2cm, white on both surfaces or purplish outside. Achenes elliptic, ± compressed, 2.5 × 1.5mm, densely hairy, style hooked.

Bhutan: C – Ha district (Ha Dzong) and Tongsa district (Pele La and Black Mountain), **N** – Upper Mo Chu district (between Nala and Tzatogang); **Sikkim**. Open glades in Fir forests and alpine slopes, 3200–3960m. April–May.

7. A. rupicola Cambessedes

Similar to *A. obtusiloba* but rootstocks slender bearing few leaf remains; leaves ternate, leaflets broadly ovate or rhombic in outline, 1.5–5 × 1–4cm, each divided into 3–5 sharply toothed lobes, sparsely long-hairy especially beneath; petioles 4–20cm; scapes 3–20cm, involucral bracts very similar to leaves; pedicels 4–15cm; sepals broadly-elliptic, 1.5–3 × 1–2cm, white within, purplish outside; achenes linear, c 2mm, surrounded by a wing c 1mm on either side, densely woolly at base, style minute.

Bhutan: C –Thimphu and Tongsa districts, **N** – Upper Mo Chu to Upper Kulong Chu districts, widespread; **Sikkim**. Alpine meadows amongst Juniper/ Rhododendron scrub, and on cliff edges, screes and moraines, 2830–4800m. May–July.

8. A. griffithii Hook. f. & Thomson

Similar to *A. obtusiloba* and more especially to *A. rupicola*, but leaflets ovate, 1.5–3 × 1–3cm, acute, margins subacutely toothed and lobed, sparsely pubescent on both surfaces; petioles 5–15cm; scapes 5–40cm, involucral bracts similar to leaves but smaller, pedicels 2–7cm; sepals obovate, 0.7–1.25 × 0.5–0.7cm, white, sometimes purple on margins or at base, subglabrous; achenes ovoid, c 4 × 2mm, glabrous, styles minute.

Bhutan: S – Chukka district (near Chukka), **C** – Ha, Thimphu, Tongsa, Bumthang and Mongar districts, **N** – Upper Mo Chu and Upper Kuru Chu districts. Streamsides and damp clearings in Hemlock and Fir forests, 1980 – 2800m. April – June.

9. A. rupestris Hook. f. & Thomson; *A. polycarpa* Evans, *A. bhutanica* Tamura

Rootstocks slender with few fibrous leaf remains. Leaves divided into 3 or 5 broadly ovate leaflets, 5 – 15mm long and broad, each 3-lobed or divided into toothed segments, sparsely pubescent with long fine hairs; petioles 3 – 7cm. Scapes 1.5 – 15cm, involucral bracts 5 – 12mm, pedicels 1.5 – 7cm. Sepals elliptic, 6 – 10 × 2 – 6mm, white or purplish. Achenes elliptic, c 2mm, compressed, glabrous; styles c 1mm, straight.

Bhutan: C – Ha district (Kyu La and Ha) and Tongsa district (Padima Tso and Black Mountain), **N** – Upper Mo Chu, Upper Bumthang Chu, Upper Kuru Chu and Upper Kulong Chu districts; **Sikkim**. Alpine cliff ledges, screes and open grassy slopes, 3500 – 4880m. April – August.

10. A. trullifolia Hook. f. & Thomson

Similar to *A. rupestris* but rootstocks stout, bearing fibrous leaf remains; leaves simple, broadly obovate, orbicular or oblong-elliptic, up to 6cm long and broad, shallowly divided (not more than half-way to midrib) into 3 coarsely toothed lobes, or oblong-elliptic, 1.5 – 5 × 0.7 – 1.5cm, entire or with 3 coarse subacute teeth at apex; base cuneate into petiole up to 15cm, pubescent with silky hairs on both surfaces; scapes 3 – 20(– 50)cm, involucral bracts 10 – 30 × 3 – 10mm, pedicels 1 – 6(– 12)cm; sepals obovate, 7 – 10(– 15) × 3 – 8(– 10)mm, yellow or often purplish outside; achenes narrowly ovate, c 3 × 1mm, compressed, densely brownish hairy, styles c 2mm, hooked.

Bhutan: C – Ha and Thimphu districts, **N** – Upper Mo Chu, Upper Pho Chu and Upper Bumthang Chu districts; **Sikkim**. Alpine pastures, also in marshes and open glades in Fir forests, 3200 – 4730m. May – July.

Two varieties occur in both Bhutan and Sikkim, the commoner, var. **trullifolia**, with leaves broadly ovate or orbicular, distinctly divided into 3 toothed lobes, from Ha, Thimphu and Upper Mo Chu districts, and var. **linearis** (Bruehl) Handel-Mazzetti, with oblong-elliptic leaves, entire or 3-toothed at apex, a more local plant from Upper Pho Chu and Upper Bumthang Chu districts.

4. THALICTRUM L.

Erect perennial herbs. Leaves pinnate or ternate with 3 – 7 leaflets, or more usually repeatedly ternatisect with numerous leaflets; petioles sheathing at base, sometimes with stipule-like auricles. Flowers actinomorphic, erect or pendent in diffuse racemes or panicles, rarely few-flowered, axillary. Sepals 4–5, petaloid. Petals 0. Stamens 5–many. Carpels (1–)5–10 or more, ovule 1. Fruit a head of stalked or sessile achenes, styles persistent or deciduous.

1. Stemless herbs; leaves all or mostly basal ....**Species 1 & 2**
+ Herbs with well-developed leafy stems (rarely only 2–5cm in *T. saniculiforme)* 2

2. Leaves simply ternate or biternate with up to 9 leaflets ....**Species 3 & 4**
+ Leaves repeatedly ternately or pinnately dissected, leaflets 15–many 3

3. Leaflets small, 2–3(–5)mm 4
+ Leaflets longer, 0.5–5cm 5

4. Achenes 2–4mm broad, strongly compressed ....**Species 5 & 6**
+ Achenes 1–1.5mm broad, not or slightly compressed ....**Species 7 & 8**

5. Stems 2–20cm ....**9. T. saniculiforme**
+ Stems 25–400cm 6

6. Flowers large and showy; sepals 8–15mm ....**10. T. chelidonii**
+ Flowers insignificant; sepals 2–6mm 7

7. Leaves glandular-pubescent beneath ....**Species 11 & 12**
+ Leaves glabrous or bearing minute sessile glands 8

8. Slender erect herbs; lower leaves 6–13cm, leaflets suborbicular mostly 5–10mm **Species 13 & 14**
+ Robust diffuse herbs; lower leaves 20–50cm, leaflets ovate or suborbicular, mostly 10–20mm ....**Species 15 & 16**

1. T. alpinum L.

Stemless herb. Leaves 4–10cm, pinnate or bipinnate, leaflets broadly obovate, often 3-lobed or divided, lobes subacute or rounded, margins and veins prominent beneath, glabrous. Flowers in racemes on peduncles 8–20cm. Sepals elliptic, 2–3 × 1mm. Stamens 5–10, anthers apiculate. Achenes 5–7, elliptic, 2–2.5mm, ± compressed, obscurely ribbed, borne on stalks ± as long, styles triangular at first.

Bhutan: N – Upper Mo Chu district (Lingshi and Laya), Upper Mangde Chu, Upper Bumthang Chu (Lubsing La) and Upper Kulong Chu (Me La) districts; **Sikkim**. Grassy alpine slopes and in Juniper/Rhododendron scrub, 3650–4570m. May–July.

2. T. setulosinerve Hara; *T. alpinum* L. var. *acutilobum* Hara

Similar to *T. alpinum* but leaves larger, up to 20cm, leaflets 3.5–10mm deeply and acutely cut into narrow lobes, coarsely pubescent on the veins beneath.

Bhutan: N – Upper Kulong Chu district (Me La). Alpine grassy hillsides, 4570m. June.

3. T. virgatum Hook. f. & Thomson

Stems 30–45cm. Leaves ternate, subsessile, leaflets broadly obovate or suborbicular in outline, 1–3cm long and broad, deeply and obtusely toothed or lobed above, base rounded or cordate, glabrous, finely reticulate; petioles 0.5–1cm. Flowers erect in terminal panicles. Sepals narrowly elliptic, 5–15 × 3–6mm. Anthers rounded at apex. Achenes narrowly elliptic, c 2 × 0.5mm, glabrous, ± compressed, shortly (c 1mm) stalked, tapering into style c 0.5mm.

Bhutan: C–Thimphu district (Barshong) and Tashigang district (Tobrang); **Sikkim**: Changu etc. On wet cliffs and in gorges, 2280–3650m. July.

4. T. punduanum Wall.

Similar to *T. virgatum* but leaves ternate or biternate with up to 9 broadly ovate or suborbicular leaflets, 1.5–4cm long and broad, obtuse, base cordate, shallowly 3-lobed, margins coarsely toothed, veins prominent and glandular-pubescent beneath; petioles up to 10cm; flowers in corymbose panicles; sepals elliptic, c 5 × 3mm, white or purplish; anthers acute at apex; achenes ellipsoid, c 2.5mm, strongly ribbed and glandular pubescent, sessile, style 0.5–1mm.

Bhutan: S–Chukka district (Chukka). Warm broad-leaved forest, 1400–1800m. July–August.

5. T. elegans Royle. Fig. 25k.

Stems slender, 15–50cm. Leaflets elliptic-obovate, 2–4 × 1–2mm, bluntly 2–3-lobed, glabrous or minutely glandular on veins beneath. Flowers in racemes or panicles. Sepals elliptic, c 3 × 1.75mm, greenish red. Anthers reddish, mucronate. Achenes obliquely obovate, 4–7 × 2.5–4mm, each with a strong median and two finer marginal ribs, narrowly winged on back, purplish, glabrous, borne on slender stalk half as long.

Bhutan: C–Tongsa district, **N**–Upper Pho Chu, Upper Mangde Chu, Upper Bumthang Chu and Upper Kulong Chu districts, local; **Sikkim**. On alpine cliff ledges, rocky slopes and moraines, and in Juniper/Rhododendron scrub, 3960–4265m. June–July.

6. T. platycarpum Hook. f. & Thomson

Similar to *T. elegans* but leaflets usually more acutely lobed, glandular-pubescent especially beneath; achenes 2–2.5mm broad, subsessile, strongly 3–4-ribbed on each side, minutely glandular-pubescent, not winged.

Sikkim: locality unknown. Mountain slopes.

7. T. rutifolium Hook. f. & Thomson

Stems 10–60cm. Leaflets 3–5mm, deeply and bluntly lobed, glaucous. Flowers in racemes or narrow panicles. Sepals greenish, c 2 × 1mm, ovate. Stamens usually 5, anthers apiculate. Achenes usually 5, narrowly ellipsoid, 5–6mm, curved, strongly deflexed against pedicel, prominently 7–8-ribbed.

Sikkim: Temu La. Alpine slopes, 4880m.

8. T. squamiferum Lecoyer; *T. glareosum* Handel-Mazzetti

Similar to *T. rutifolium* but smaller, stems 10–15cm surrounded by scales at base; leaflets 2–3mm, glaucous; flowers 2–3, axillary; sepals purplish, 2–3mm; stamens 5–10, connective elongated above anther-cells; style triangular, almost winged at base; achenes spreading, ellipsoid, 3-nerved on each side.

Bhutan: N – Upper Mo Chu district (Lingshi) and Upper Pho Chu district (Chojo Dzong); **Sikkim**: Chulong. Alpine slopes, 4570m. June–July.

9. T. saniculiforme DC.

Leaves 1–3, 15–25cm, upper ones smaller; leaflets broadly ovate or obovate, rounded or subcordate at base, margins roundly toothed or lobed, glabrous. Peduncle ± as long as leaf, divaricately branched, few-flowered. Sepals elliptic, c 5 × 2mm, white. Achenes ellipsoid, c 2mm, sessile, style ± as long, persistent, hooked at apex.

Sikkim: Lachen and Cheuntong. On mossy rocks and tree trunks, 2440m. August.

10. T. chelidonii DC. agg.; *T. reniforme* Wall.

Stems 40–200cm. Leaves 8–30cm, leaflets broadly ovate or suborbicular, 1–4cm long and broad, margin coarsely and bluntly toothed or lobed, minutely pubescent. Flowers in racemes or panicles. Sepals ovate, 8–15 × 5–10mm, mauve or purplish. Achenes obliquely obovate, 5–6 × 3–4mm, compressed, usually with 3 branching veins on each side, glabrous or minutely puberulous; stalks 1–4mm.

Bhutan: C – Ha to Mongar districts, common, **N** – Upper Mo Chu, Upper Pho Chu, Upper Bumthang Chu and Upper Kuru Chu districts; **Sikkim**. Fir forests and Juniper/Rhododendron scrub, 2130–4570m. June–September.

The distinction between *T. chelidonii* and *T. reniforme* is unclear; Hooker (80) distinguished *T. reniforme* by its smaller greenish-white flowers. In Bhutan and Sikkim flowers are apparently always mauve or purplish, but are very variable in size.

11. T. foetidum L.; *T. minus* L. var. *foetidum* (L.) Hook. f. & Thomson

Stems up to 75cm. Leaflets obovate, 5–10 × 3–8mm, acutely toothed or lobed, rounded or cuneate at base, veins very prominent beneath and glandular-pubescent. Sepals elliptic, c 3.5 × 2mm. Anthers apiculate. Achenes elliptic, 2–2.5mm, slightly compressed, finely glandular-pubescent.

Bhutan: C – Ha district (Ha) and Thimphu district (Kyapcha and Paga), **N** – Upper Mo Chu district (Lingshi). Evergreen oak and moist coniferous forests, often on streamsides or edges of cultivation, 2130–3960m. June–July.

12. T. cultratum Wall.

Similar to *T. foetidum* but leaflets bluntly-toothed, veins not very prominent beneath, glandular hairs minute, scattered over undersides; sepals elliptic, 4–6 × 1.5–2mm, greenish, tinged purple; achenes obliquely obovate, broadest near apex,

4–6 × 2–5mm, compressed with 3 ribs on each side, glabrous or glandular-pubescent, shortly stalked.

Bhutan: C – Ha, Thimphu and Tongsa districts, **N** – Upper Mo Chu, Upper Bumthang Chu and Upper Kulong Chu districts; **Sikkim**. Margins of moist conifer forests, also alpine slopes, marshes and cliff-ledges, 2300–4200m. May–August.

13. T. rostellatum Hook. f. & Thomson

Stems slender, erect, 40–75cm. Leaflets mostly suborbicular, 7–10(–15)mm, each with 3–5 blunt teeth at apex, pale and glabrous or minutely glandular beneath; petiole base broad and sheathing; leaf divisions with minute stipels c 1mm. Flowers few, terminal or on short lateral shoots. Stamens c 15, anthers acute, filaments filiform. Achenes 5–6, narrowly ellipsoid, c 3mm, sessile or shortly stalked, ± compressed, distinctly 6–7-ribbed, tapering into the persistent hook-tipped style ± as long.

Bhutan: C – Ha, Thimphu, Tongsa and Bumthang districts, **N** – Upper Mo Chu, Upper Pho Chu and Upper Bumthang Chu districts; **Sikkim**. Streamsides and margins of cultivation in Blue Pine, Spruce and Fir forests, 2600–3650m. May–August.

14. T. leuconotum Franchet

Similar to *T. rostellatum* but leaflets more deeply dissected into 3–5 obtuse lobes; petiole-base auriculate; leaf divisions with conspicuous whitish stipels 2–4mm; flowers in slender racemes with short branches; sepals greenish-purple, c 4 × 1.5mm; anthers mucronate; achenes 1–3, elliptic, compressed, c 5mm, 3–4-veined; style c 1.5mm, deciduous.

Bhutan: N – Upper Kulong Chu district (Me La). Amongst shrubs on alpine slopes, 3800m. June.

15. T. foliolosum DC.

Robust bushy herb, 1–2(–4)m. Leaflets broadly ovate, 1.5–4 × 1–3cm, acute or obtuse, bluntly toothed or lobed. Sepals obovate, c 4 × 2mm, white outside, mauve within, caducous. Anthers 2–3mm, acute or mucronate, ± as long as filaments which are not broadened at apex. Achenes ellipsoid, c 3mm, strongly ribbed, sessile, glabrous, style deciduous.

Bhutan: S – Deothang district (N of Narfong), **C** – Punakha, Tongsa, Mongar and Tashigang districts; **Sikkim**. Clearings in Cool broad-leaved and Evergreen oak forests, 1370–2130m. June–August.

An extract of the root is used medicinally (Caius in *J. Bomb. Nat. Hist. Soc.* 39: 729, 1937).

16. T. javanicum Blume. Fig. 25 l.

Similar to *T. foliolosum* but anthers c 1mm, obtuse at apex, much shorter than filaments which are broadened at apex; achenes ellipsoid, 2.5–3mm, distinctly ribbed, glabrous, style c 0.75mm, persistent and hooked.

Bhutan: C –Thimphu, Tongsa and Bumthang districts, **N** – Upper Mangde Chu and Upper Kulong Chu districts. On banks in Hemlock and Fir forests, 2440 – 3800m. May – August.

5. CALLIANTHEMUM Meyer

Stemless perennial herbs. Leaves pinnate or bipinnate. Scapes leafless, 1-flowered. Flowers actinomorphic. Sepals 5, deciduous. Petals 5 – 10, each with a nectar pit in lower half. Stamens numerous. Carpels many, styles minute, ovule 1. Fruit a cluster of subglobose achenes.

1. C. pimpinelloides (D. Don) Hook. f. & Thomson; *C. cachemirianum* Cambessedes

Leaves 2 – 12cm at flowering time (later up to 30cm), leaflets 5 – 7, broadly ovate, 0.5 – 3 × 0.5 – 2cm, deeply and subacutely toothed or lobed or the lower pair pinnately divided, glabrous. Scapes 3 – 7cm. Sepals elliptic, 4 – 6 × 2 – 3mm, obtuse, often reddish. Petals obovate, 0.7 – 1.5 × 0.2 – 0.5cm, white tinged purple outside. Stamens greenish. Achenes ovoid, c 3.5 × 2mm, glabrous.

Bhutan: C – Ha, Thimphu and Tongsa districts, **N** – Upper Mo Chu and Upper Kulong Chu districts; **Sikkim**. Alpine cliff-ledges, damp turf and sandy ground, 2750 – 4000m. April – June.

6. ADONIS L.

Perennial herbs with erect stems. Leaves basal and alternate on stems, much divided into linear segments. Flowers actinomorphic, solitary, terminal. Sepals 5 – 6. Petals 8 – 16, nectar-pits absent. Stamens numerous. Carpels many, style short. Fruit a cluster of achenes.

1. A. brevistyla Franchet

Stems erect, 20 – 40cm. Leaves ovate in outline, 5 – 10 × 3 – 8cm, tripinnatisect into narrow acuminate lobes, glabrous; petioles up to 10cm. Flowers solitary at branch or stem ends, pedicel (i.e. portion of stem above uppermost leaf) 1.5 – 6cm. Sepals elliptic, 10 – 20 × 3 – 4mm, greenish. Petals obovate, 1.2 – 2.5 × 0.5 – 1cm, white or yellow, sometimes tinged with blue outside. Stamens orange-yellow. Achenes ovoid, c 4 × 3mm, sparsely pubescent, tapering above into style c 1mm.

Bhutan: C – Ha district (Ha), Punakha district (Lao La and Pele La), Tongsa district (Black Mountain and Chendebi) and Bumthang district (Byakar). Wet ground in Hemlock forests, open mountainsides and in wet ravines, 2500 – 4110m. April – June.

2. A. nepalensis Simonovicz

Similar to *A. brevistyla*, but smaller, stems 6–8(–15)cm in flower (up to 20cm in fruit); leaves oblong-ovate, 2–10 × 1.5–3cm, apex of segments acute or subacute; pedicels up to 1cm; sepals c 7 × 4mm, sometimes purplish; petals yellow or rarely dark blue, 8–13 × 4–5mm; achenes glabrous.

Sikkim: Eumtso La. Alpine grassy slopes and in ravines, 4570m. April–June.

7. RANUNCULUS L.

Annual or perennial, terrestrial or aquatic herbs. Leaves basal or alternate on stems, often ternately lobed or divided, sometimes dissected into filamentous segments, rarely entire, auricles membranous or absent. Flowers actinomorphic, solitary or few in a corymb. Sepals 3–5, deciduous. Petals usually 5, each with a nectar-pit near base. Stamens many. Carpels many, style short, ovule 1. Fruit a head or spike of achenes.

1\. Water plants with floating leaves and stems; leaves divided into filamentous segments .... **Species 1 & 2**
\+ Terrestrial herbs, sometimes growing in water but leaves never divided into filamentous segments 2

2\. Plants with slender stolons regularly producing rosettes of leaves **Species 3–5**
\+ Plants without slender stolons, usually erect (but *R. diffusus* and *R. ficariifolius* prostrate though not producing rosettes of leaves) 3

3\. Basal leaves entire or margins broadly and shallowly toothed (sometimes more deeply divided into 3 lobes in *R. pulchellus* var. *stracheyanus*) **Species 6 & 7**
\+ Basal leaves deeply divided into 3 segments or leaflets which are sometimes further divided 4

4\. Small herbs up to 15(–25)cm tall; leaves 1–2(–3)cm broad ..**Species 8–10**
\+ Large herbs, usually 20–30cm or more tall, rarely only 5cm but then leaves 3–5cm broad 5

5\. Flowers small, 5–7(–10)mm across .... **Species 11 & 12**
\+ Flowers medium-sized, (1–)1.5–2.5cm across 6

6\. Sepals spreading .... **Species 13 & 14**
\+ Sepals reflexed .... **Species 15 & 16**

1. R. trichophyllus Chaix; R. *aquatilis L.* var. *trichophyllus* (Chaix) Hook. f. & Thomson. Fig. 25i.

Annual or perennial. Leaves 2–3cm long and broad, divided into filamentous segments; petioles c 5mm, each sheathed by an auricle for most of its length.

Flowers solitary in leaf axils, pedicels 2.5 – 3(– 4.5)cm. Sepals elliptic, c 3.5mm, reflexed. Petals obovate, 5 – 6 × 2 – 2.5mm, limb white, yellow near base, nectar-pit lunate. Achenes obovoid, c 1.5mm, glabrescent, receptacle hairy.

Bhutan: C – Tongsa district (Maruthang), **N** – Upper Bumthang Chu district (Pame Tso). In ponds and slow-flowing streams, 3800 – 4500m. August.

Records of *R. aquatilis* L. from E Himalaya (117) probably refer to this species.

2. R. flavidus (Handel-Mazzetti) Cook

Similar to *R. trichophyllus* but leaves 1.5 – 3cm; sepals 4 – 5mm; petals 7 – 8 × 4 – 5mm, yellow, nectar-pit lunate; achenes obovoid, c 2.5mm, glabrous, borne on pedicels up to 8cm.

Chumbi: locality unknown. In standing and running water.

3. R. tricuspis Maximowicz; *R. cymbalariae* Pursh, *R. hyperboreus* sensu F.B.I. non Rottboell, *R. salsuginosus* auct. non Georgi, *Halerpestes tricuspis* (Maximowicz) Handel-Mazzetti

Leaves all basal, broadly ovate or orbicular in outline, 0.5 – 1.5cm long and broad, shallowly or deeply tripartite, segments oblong or oblanceolate, obtuse, sometimes lobed, glabrous; petioles 1 – 2cm. Scapes 2 – 2.5cm bearing solitary flowers. Sepals narrowly ovate, c 4mm, spreading, greenish. Petals obovate, 4 – 5 × 1.75mm, yellow. Achenes 40 – 50, obovoid, c 1mm, weakly striate, style persistent, straight or slightly hooked, receptacle subglobose.

Bhutan: C – Bumthang district (Byakar), **N** – Upper Pho Chu district (Leji); **Sikkim**: Nathu La. Margins of shallow pools, 3650m. June.

4. R. pegaeus Handel-Mazzetti

Similar to *R. tricuspis* but smaller, leaves 0.5 – 0.7cm on petioles up to 2cm; sepals 1.5 – 2mm; petals c 2.5 × 1.5mm; achenes 10 – 15.

Sikkim: locality unknown (73). Mossy alpine streamsides.

5. R. sarmentosus Adams; *Halerpestes sarmentosa* (Adams) Komarov

Similar to *R. tricuspis* but leaves ± orbicular, broadly 3 – 5-toothed, achenes 80 – 130 borne on an oblong receptacle.

Sikkim: Laghep and Naku Chu. 4000 – 5000m.

6. R. ficariifolius Léveillé & Vaniot; *R. flaccidus* Hook. f. & Thomson non Persoon, *R. microphyllus* Handel-Mazzetti

Prostrate or decumbent herb often rooting at nodes. Leaves broadly ovate or suborbicular, 3 – 12(– 20)mm long and broad, acute or subacute, base truncate or cordate, margins with 5 – 7 broad rounded teeth, glabrous; petioles 0.3 – 5cm, sheathing at base. Flowers solitary, leaf-opposed. Sepals elliptic, 1.5 – 3mm, reflexed. Petals obovate, 2.5 – 3.5 × 1.5 – 3mm, yellow. Achenes ellipsoid, c 1.75 × 1mm, compressed, surrounded by a narrow rim, glabrous; style weakly hooked, deciduous.

Bhutan: **S** – Gaylegphug district, **C** –Thimphu, Bumthang and Tashigang districts, **N** – Upper Mo Chu district; **Sikkim**. In marshes and by streams in Warm broad-leaved, Blue Pine, Spruce and Hemlock forests, 1200 – 3200m. March – June.

7. R. pulchellus Meyer; *R. lobatus* Jacquemont

Erect perennial herb, stems 5 – 30cm. Basal leaves variously elliptic or ovate, 20 – 35 × 3 – 10mm, entire, or broadly ovate in outline, 1.5cm long and broad, 3-lobed, glabrous or densely silky; petioles 1.5 – 9cm; stem leaves narrowly elliptic or deeply 3-lobed, sheathing at base; sometimes all leaves linear-elliptic up to 9 × 0.4cm. Flowers solitary or 2 – 3 per leaf axil. Sepals elliptic, 4 – 6 × 3 – 5mm, blackish at tip. Petals broadly obovate, 6 – 10mm long and broad, yellow. Achenes obovoid, c 1.5mm, style straight or curved.

Bhutan: **C** –Thimphu district, **N** – Upper Mo Chu, Upper Bumthang Chu and Upper Kulong Chu districts; **Sikkim**. Damp sandy soil and cliff-ledges, 3650 – 4420m. May – July.

Three varieties occur in Bhutan and Sikkim: the widespread var. **pulchellus** with leaves elliptic or ovate, entire or slightly toothed, completely glabrous or pedicels pubescent; var. **sericeus** Hook. f. & Thomson, with leaves as in var. *pulchellus* but plants white silky, from Upper Mo Chu district (Phile La) and Sikkim; and var. **stracheyanus** (Maximowicz) Handel-Mazzetti (*R. lobatus* Jacquemont), with leaves broadly ovate in outline, deeply 3-lobed, truncate or cordate at base and sparsely pubescent, from Thimphu, Upper Mo Chu, Upper Bumthang Chu and Upper Kulong Chu districts and Sikkim.

8. R. adoxifolius Handel-Mazzetti; *R. affinis* sensu F.B.I. p.p. non Brown

Suberect herb, stems 2 – 12cm. Basal leaves reniform in outline, 5 – 10mm; leaflets broadly ovate, dentate, sparsely silky pubescent, petioles slender at base; stem leaves often with less-divided segments. Flowers usually solitary, pedicels sparsely pubescent. Sepals broadly ovate, 2.5 – 3.5mm, sparsely pubescent, spreading. Petals obovate, 3.5 – 4.5 × 2.5 – 3.5mm, yellow. Achenes subglobose, c 1mm, forming oblong heads c 5mm, glabrous.

Bhutan: **C** – Bumthang and Tongsa districts, **N** – Upper Mo Chu and Upper Bumthang Chu districts; **Sikkim**. In damp alpine grassland, boggy ground and moist sand, 3650 – 4110m. May – August.

Records of *R. hirtellus* D. Don from E Himalaya probably refer to this species.

9. R. sikkimensis Handel-Mazzetti; *R. nivalis* sensu F.B.I. non L.

Similar to *R. adoxifolius* but leaves glabrous; petioles broadly membranous at base; pedicels and sepals densely yellowish-pubescent.

Sikkim: Tunkra La. 4880m.

10. R. brotherusii Freyn; *R. affinis* sensu F.B.I. p.p. non Brown. Med: *Chetsha, Cheytsha*.

Similar to *R. adoxifolius* but stems sometimes up to 25cm; basal leaves 1.5 – 3cm long and broad, leaflets deeply cut into linear-lanceolate segments 1–1.5(–2)mm broad, sparsely hirsute beneath, glabrous above; petioles 2–6cm; petals 4–6mm long and broad; achenes glabrous or pubescent.

Bhutan: C – Ha, Thimphu and Mongar districts, **N** – Upper Mo Chu, Upper Bumthang Chu and Upper Kulong Chu districts; **Sikkim**. Damp clearings in Spruce, Hemlock and Fir forests, in alpine meadows and scrub, 2560–4570m. April–August.

11. R. diffusus DC.

Prostrate or decumbent perennial, stems usually densely hirsute with spreading hairs. Leaves deeply divided into 3 obovate lobes, margins shallowly and sharply toothed, pubescent especially beneath; petioles up to 10cm, densely hirsute, auricles brown scarious, 8–10mm. Flowers solitary, axillary. Sepals elliptic, c 3.5 × 1.5mm, spreading. Petals obovate, 3.5–6 × 2–3mm, yellow. Achenes suborbicular, compressed, 1.5–2mm diameter, pitted at middle, margin smooth, acute, intramarginal rib present; style hooked; receptacle globose.

Bhutan: S – Phuntsholing, Chukka and Sarbhang districts, **C** –Thimphu, Punakha and Tongsa districts; **Sikkim**. On damp ground, in ditches and by waterfalls in Warm and Cool broad-leaved and Evergreen oak forests, 1300–2700m. April–August.

12. R. sceleratus L. Dz: *Damgi Metog*.

Similar to *R. diffusus* but an erect annual, glabrous or sparsely pubescent above; basal and lower stem leaves reniform or suborbicular in outline, 2.5–5cm long and broad, divided almost to base into 3 obovate segments, bluntly 3–5-toothed or-lobed; petioles 2–7cm, auricles scarious, 5–7mm; upper stem leaves obovate, unlobed, bluntly 3–5-toothed at apex; flowers several or many, diffusely racemose; sepals elliptic, c 3 × 1.5mm, pubescent, somewhat reflexed; petals ± as long as sepals, yellow; achenes obovoid, c 0.75mm, ± compressed, pale at margin, style minute, receptacle oblong-ovoid, 7–8mm.

Bhutan: C – Punakha district (Wangdu Phodrang and Lobesa). Weed of irrigation ditches and river banks, 1400–1510m. April.

13. R. silerifolius Léveillé

Erect perennial, stems up to 60cm, spreading-pubescent. Basal and lower stem-leaves 3–6cm long and broad, ternately divided into 3 deeply trilobed leaflets, lobes broadly elliptic, appressed-pubescent. Sepals elliptic, 5–6 × 2–3mm, hirsute. Petals obovate, c 7 × 5mm, yellow. Achenes ovate, compressed, c 3 × 2mm, tapering into a persistent hooked style at apex; receptacle subglobose, hirsute.

Bhutan: S – Chukka district, **C** – Ha, Tongsa, Mongar and Tashigang districts. Fields and Cool broad-leaved forest clearings, 1060–2770m. May–October.

14. R. laetus Royle

Similar to *R. silerifolius* but leaves 7–10cm long and broad, divided into narrow oblong sharply-toothed lobes, softly appressed-pubescent; sepals c 6 × 4mm; petals 8–12 × 7–10mm; achenes compressed, dark in centre, margin thin and paler, style persistent, ± straight, receptacle glabrous.

Bhutan: C –Thimphu district (Dochu La); **Sikkim**. Damp fields and ditches, 3000m. May.

15. R. cantoniensis DC.; *R. pensylvanicus* sensu F.B.I. p.p. non L.f., *R. napaulensis* DC.

Erect annual or biennial, stems 20–40cm. Leaves ternately divided into 3 leaflets, each leaflet deeply cut into oblanceolate lobes, appressed-pubescent. Sepals elliptic, c 5 × 3mm, margins membranous. Petals obovate, 7–8 × 3–5mm, yellow. Achenes broadly elliptic, c 2mm long and broad, compressed, surrounded by a narrow marginal rim; style persistent, ± straight; receptacle subglobose.

Bhutan: C –Thimphu district (Tsalimarphe and Paro) and Tashigang district (Tashigang). Weed of irrigated fields, 2200–2500m. April–May.

16. R. chinensis Bunge

Similar to *R. cantoniensis* but achenes broadly elliptic, c 2.5 × 2mm, pale and thickened at margins, depressed and darker at middle, style minute, ± straight, receptacle elongated up to 1.5cm, hairy.

Bhutan: C –Thimphu district (common in Thimphu and Paro valleys) and Punakha district (Lobesa). Paddy fields and cultivated ground, 1400–2500m. April–May.

8. OXYGRAPHIS Bunge

Dwarf perennial stemless herbs. Leaves all basal, entire. Scapes 1-flowered, leafless; flowers actinomorphic, bisexual. Sepals 5, persistent. Petals 10–15, each with a nectar-pit at base. Stamens many. Achenes many, laterally compressed.

1. O. endlicheri (Walpers) Bennet & Chandra; *O. polypetala* (Royle) Hook. f. & Thomson *nom. illeg., O. glacialis* sensu F.B.I. non (Fischer) Bunge

Rootstock surrounded by fibrous leaf remains. Leaves broadly ovate or suborbicular, 5–15(–25)mm long and broad, obtuse or subacute, base truncate or subcordate, margins crenate with a few rounded teeth, glabrous; petioles 1–3(–7)cm. Scapes 1.5–4(–10)cm. Sepals ovate, 3–7 × 2–4mm, rounded, base subcordate. Petals 4–10 × 1.5–2mm, yellow. Achenes obovate, 1.5–2mm, strongly ribbed along outer and inner angles and with a distinct rib on each side.

Bhutan: C –Thimphu, Tongsa, Bumthang and Tashigang districts, **N** –Upper Mo Chu district; **Sikkim**. Damp clearings in Fir forests, and in alpine meadows, flushes, mossy rocks and grassy slopes, 3350–4900m. May.

9. **PAROXYGRAPHIS** W. W. Smith

Similar to *Oxygraphis* but smaller, dioecious; sepals and petals each 5; stamens 10; achenes 10–15.

1. P. sikkimensis W. W. Smith

Flowering stems 5–20mm. Leaves ovate, 1.5–3 × 1–2.5mm, obtuse, base weakly cordate, margins entire, glabrous; petioles 5–20mm. Sepals elliptic, c 2 × 1.5mm, obtuse. Petals elliptic, c 2 × 0.75mm, clawed at base, yellow. Achenes obovate, compressed, c 2.5mm.

Bhutan: C–Thimphu district (Pajoding and Nala to Tzatogang); **Sikkim**: Changu. Alpine marshes and damp rocks, 3400–3950m. May.

10. **CALTHA** L.

Glabrous perennial herbs with creeping rhizomes. Leaves basal or alternate on stems, simple. Flowers solitary or few in corymbs, actinomorphic. Sepals usually 5, petaloid. Petals absent. Stamens numerous. Carpels 5–10(–25). Fruit a cluster of subcompressed follicles.

1. C. palustris L.; *C. govaniana* Royle. Eng: *Marsh Marigold*. Fig. 25e.

Stems 10–60(–80)cm. Leaves reniform or deltoid, 3–12(–20)cm long and broad, obtuse, base cordate, margins sharply or crenately dentate; petioles of basal leaves 8–30(–60)cm, those of stem leaves shorter. Flowers 1 or few(–12). Sepals elliptic, 12–15 × 3–10mm, yellow. Filaments filiform. Follicles oblong, 6–8 × 3–4mm, sessile, bearing a straight persistent style 1.5–3mm at apex.

Bhutan: C–Ha, Thimphu, Tongsa, Bumthang, Mongar and Sakden districts, **N**–Upper Bumthang Chu district; **Sikkim**. Damp ground in Fir forests, and in alpine meadows and marshes, 3050–4570m. April–July.

2. C. scaposa Hook. f. & Thomson

Similar to *C. palustris* but smaller, stems 7–15(–20)cm; leaves broadly ovate, 1.5–2.5(–4.5) × 1–2(–4)cm; scapes leafless or with one small leaf, 1–2-flowered; filaments flattened; follicles shortly (1.5–3mm) stalked.

Bhutan: C–Thimphu, Tongsa and Bumthang districts, **N**–Upper Mo Chu, Upper Bumthang Chu and Upper Kulong Chu districts; **Sikkim**. Damp alpine meadows, marshes and streamsides, 3800–4570m. May–October.

11. **CALATHODES** Hook. f. & Thomson

Erect perennial herbs. Leaves basal and on stems, palmatisect. Flowers solitary, terminal, actinomorphic. Sepals 5, petaloid. Petals absent. Stamens numerous. Carpels many. Fruit follicular.

1. C. palmata Hook. f. & Thomson

Stems 20–40cm. Leaves reniform or suborbicular in outline, 3–6 × 4–10cm, ternately divided into rhombic, sharply serrate segments, glabrous; petioles up to 20cm, auricles membranous, 1–3cm, sheathing at base. Sepals elliptic, 10–15 × 4–7mm.

Bhutan: C –Tongsa district (Sephu) and Mongar district (Lhuntse); **Sikkim**. 2300–3050m.

12. TROLLIUS L.

Erect glabrous perennial herbs. Leaves palmatisect, mostly basal. Flowers solitary, terminal, actinomorphic. Sepals 5–6, petaloid. Petals (honey leaves) 10–12 (rarely 3–4), very small, obovate, with nectar pit above basal claw. Stamens numerous. Carpels 5–20, becoming follicles.

1. T. pumilus D. Don; *T. gammieanus* Stapf. Fig. 25m,n.

Stems 10–20cm in flower, up to 30cm in fruit, bearing up to 2 leaves in lower half. Basal leaves broadly ovate or reniform in outline, 3–5cm long and broad, divided to base into 3–5 rhombic primary segments, these deeply pinnatisect into linear-lanceolate, acuminate lobes c 4–7 × 2mm; petioles 5–15cm, auricles narrow at base; stem leaves similar, petioles shorter, sometimes with well-developed auricles. Sepals obovate, 15–20 × 6–10mm, yellow. Petals narrowly obovate, 3.5–5mm. Follicles oblong, 7–8 × 2mm, slightly compressed; styles persistent, c 2mm.

Bhutan: C –Thimphu, Bumthang, Tashigang and Sakden districts; **N** –Upper Mo Chu and Upper Mangde Chu districts; **Sikkim**. Damp alpine meadows and peaty ground, 3000–4300m. June–July.

2. T. sikkimensis (Bruehl) Doroszewska

Similar to *T. pumilus* but usually smaller, stems 5–10cm, rarely up to 25cm; primary leaf-segments obovate, less deeply divided, ultimate divisions 3–5 × 2–4mm, teeth acute, suddenly narrowed to finely acuminate points; petals oblong-spathulate.

Bhutan: C –Thimphu, Tongsa and Sakden districts, **N** –Upper Mo Chu, Upper Bumthang Chu and Upper Kulong Chu districts; **Sikkim**. Alpine meadows and marshes, 3000–4570m. May–August.

3. T. vaginatus Handel-Mazzetti

Similar to *T. pumilus* but always small, stems 5–12cm; leaves more similar to those of *T. sikkimensis* but primary segments more deeply divided, ultimate divisions 4–5 × 2–3mm, sharply toothed with acuminate points; petioles 4–6cm with large sheathing basal auricles about half as long; petals narrowly obovate.

Bhutan: C –Tongsa district (Black Mountain). Open grassy alpine meadows, 3960–4265m. June.

4. T. bhotanicus (Bruehl) Mukerjee

Similar to *T. pumilus* but taller, stems 30–40cm; leaves similar to those of *T. sikkimensis*, lobes with acuminate teeth; petals 3–4; staminodes present, resembling stamens but with shorter empty pollen sacs.

Chumbi : Dotha.

13. PARAQUILEGIA Drummond & Hutchinson

Densely tufted perennials. Rootstocks densely surrounded by persistent petioles of previous years. Leaves ternatisect. Flowers solitary, actinomorphic. Sepals 5, petaloid. Petals 5, small, obovate, emarginate at apex, base concave. Stamens numerous. Carpels 5–7, becoming ± erect follicles.

1. P. anemonoides (Willdenow) Ulbrich; *P. microphylla* (Royle) Drummond & Hutchinson, *Isopyrum grandiflorum* Fischer. Fig. 25j.

Leaves glabrous, glaucous, leaflets obovate in outline, 3–8mm long and broad, ± ternately lobed, ultimate divisions 1–2mm broad, obtuse; petioles 3–10cm. Scapes 6–10cm, bearing a pair of oblanceolate or ternate bracts 1–1.5cm above middle. Sepals obovate, purplish, 1.5–3 × 1–2cm, obtuse or acute. Petals obovate, c 6 × 3mm. Stamens yellow. Follicles oblong, c 7 × 3mm, ± compressed, shortly (c 2mm) stalked.

Bhutan: C–Thimphu, Punakha, Tongsa and Tashigang districts; **N**–Upper Kulong Chu district; **Sikkim**. Mountain cliff ledges and in ravines, 3500–4880m. May–August.

The Himalayan plants are sometimes segregated as *P. microphylla* (Royle) Drummond & Hutchinson.

14. DICHOCARPUM Wang & Hsiao

Perennial herbs with leafy branched stems surrounded at base by fleshy scales. Leaves ternately 5–15-foliate. Flowers in terminal cymes, actinomorphic. Sepals 5, petaloid. Petals (nectariferous staminodes?) 2–5 with blade shorter than stalk. Stamens 8–20. Carpels 2, connate at base. Follicles erect at first, later divaricate, spreading, borne on a slender pedicel thickened at apex.

1. D. adiantifolium (Hook. f. & Thomson) Wang & Hsiao; *Isopyrum adiantifolium* Hook. f. & Thomson. Fig. 25h.

Stems 10–35cm. Leaflets broadly obovate, 5–10mm long and broad, rounded and crenate at apex, cuneate at base, glabrous; petiolules up to 2cm; petioles up to 6cm. Sepals elliptic or obovate, 7–9 × 2–4mm, white. Blade of petals broadly obovate, c 0.7mm, on stalk 2–3mm. Follicles cylindrical, 7–9 × 1.5mm.

Bhutan: S–Sarbhang district (Loring Falls), **C**–Tongsa district (Tashiling and

Chendebi), Mongar district (Lhuntse and Saleng) and Tashigang district (Tobrang and Donga La); **Sikkim**. On mossy rocks by streams and waterfalls in Warm and Cool broad-leaved and Hemlock forests, 1310–2900m. April–July.

15. ASTEROPYRUM Drummond & Hutchinson

Rhizomatous perennials. Leaves all basal, peltate, petiolate. Flowers solitary or rarely 2 on erect scapes, actinomorphic. Sepals 5, petaloid. Petals 6–8, small, suborbicular, long-stalked. Stamens many. Carpels usually 10, erect in flower, later spreading as follicles.

1. A. peltatum Drummond & Hutchinson

Leaves suborbicular or weakly lobed, 2–4.5cm diameter, sparsely pubescent on upper surface near margins; petioles 5–9cm. Scapes 5–10cm, bearing an involucre of 3 minute bracts c 3mm. Sepals white, obovate, 5–7 × 3.5–5mm. Petals c 1mm diameter, stalks 2–3mm. Follicles ovoid, ± compressed, c 8 × 2.5mm, with persistent style c 2mm at apex.

Bhutan: C – Mongar district (Pangkar near Lhuntse). In dense rain forest, 2745m. April.

16. SOULIEA Franchet

Rhizomatous, erect, perennial herbs. Leaves usually only 2, on stems, bipinnate. Flowers in terminal racemes. Sepals 4–5, petaloid. Petals (staminodes?) 4–5, smaller than sepals. Stamens many. Carpels 1–3, styles elongate, thickened. Fruit of 1–3 stalked follicles.

1. S. vaginata Franchet; *Coptis ospriocarpa* Bruehl

Stems at first 10–15cm, reddish, surrounded at base by 2 large (2–4cm) sheathing membranous scales, in fruit up to 55cm. Leaves 15–20cm, ternately divided into 3 primary divisions, these then pinnate into ovate, acutely-toothed leaflets, glabrous. Sepals obovate, 5–6 × 3–4mm, white, tinged with purple. Petals suborbicular, 3mm. Stamens dark purple. Follicles oblong, 3.5–4 × 0.4cm, prominently veined.

Bhutan: C – Ha, Thimphu, Tongsa and Mongar districts, **N** – Upper Mo Chu and Upper Kulong Chu districts; **Sikkim**. Fir and Hemlock forests, in clearings and on damp banks, 2895–4000m. May–July.

17. ACTAEA L.

Erect perennial herbs. Leaves alternate, 3–4 times ternately compound. Flowers actinomorphic in short terminal racemes. Sepals 3–5, petaloid, caducous. Petals 3–5(–10), spathulate. Stamens many. Carpels solitary, stigma rounded, sessile. Fruit a berry.

1. A. acuminata Royle; *A. spicata* sensu F.B.I. non L., *A. spicata* L. var. *acuminata* (Royle) Hara

Rhizomatous. Stems 45–100cm, sheathed by ovate scales at base. Leaves 18–35cm, leaflets ovate, 4–8 × 1.5–3cm, acuminate, base rounded, margins irregularly serrate, pubescent at first; petioles 2–10cm. Racemes 2–6cm, pedicels up to 1.5cm. Sepals elliptic, c 5 × 2mm, white. Petals c 3 × 1mm, whitish, weakly 3-nerved. Berry ellipsoid, 7–8 × 5mm, blackish-purple.

Bhutan: C –Thimphu district (Drugye Dzong), **N** –Upper Mo Chu district (Tzatogang). Ravines and streamsides in Hemlock and Fir forests, 2900–3100m. May.

18. CIMICIFUGA L.

Erect perennial herbs. Leaves alternate, 2–5 times ternately compound. Flowers in slender axillary racemes or terminal panicles, actinomorphic. Sepals 4–5, petaloid, caducous. Petals (or ?staminodes) 4–5, fleshy, bifid with 2 rounded lobes at apex. Stamens many. .Carpels 3–5(–8), styles elongate, thickened. Fruit a cluster of follicles.

1. C. foetida L. Fig. 25f.

Stem 1.5–4m. Lower leaves up to 30cm, leaflets broadly ovate or somewhat trilobed, 4–8 × 2.5–6cm, acute, base rounded, margins coarsely and irregularly serrate, sparsely pubescent; petioles 2–20cm. Sepals obovate, 3–4 × 2–3mm, greenish. Petals elliptic or ovate, c 4 × 2mm, terminal lobes pale, c 1mm. Stamens white. Follicles oblong, c 10 × 5mm, papery.

Bhutan: C – Ha, Thimphu, Punakha and Bumthang districts, **N** – Upper Mo Chu district; **Sikkim**. In Spruce forests and on alpine hillsides, by streams and in scrub, 2400–3600m. July–September.

19. DELPHINIUM L.

Erect perennial herbs. Leaves basal and on stems, palmately lobed or dissected. Flowers few or many in terminal racemes, zygomorphic. Sepals 5, petaloid, the uppermost (posterior) elongated into a spur, the laterals in two similar pairs. Petals 4, the upper pair with linear, sometimes oblique, blades and produced at the base into nectariferous spurs which project into the sepal spur (and subequal to it), the lower pair with slender basal claw and oblong lamina usually at right angles to it. Stamens numerous, filaments broadened at base. Carpels 3–5 becoming follicles. Seeds winged or wingless, bearing transverse rows of scales.

1. Sepal spur 10–15mm, broad (7–10mm) at point of insertion (sometimes only 5mm broad in *D. nepalense* and *D. vestitum*) .... **Species 1–3**
\+ Sepal spur 15–25mm, slender (3–5mm broad) at point of insertion (sometimes up to 7mm broad in *D. candelabrum*) 2
2. Plants small, flowering stems 10–25(–40)cm bearing 1–3 flowers 3
\+ Plants tall (20–)40–150cm; flowers several or numerous in racemes 4
3. Leaves divided to middle into broad (7–12mm) segments ..... **Species 4 & 5**
\+ Leaves divided to base into narrow (1–3mm broad) segments .. **Species 6–8**
4. Bracteoles inserted at base of receptacle, closely appressed to sepals **Species 9–11**
\+ Bracteoles inserted above middle of pedicel but not closely appressed to sepals **Species 12–14**

1. D. nepalense Kitamura & Tamura

Stems 12–30cm, deflexed pubescent. Leaves suborbicular in outline, 3–6cm diameter, palmatisect; segments cuneate, divided into subacute teeth or lobes 2–4mm broad; petioles up to 10cm. Flowers few (up to 10) in short corymbose racemes, pedicels up to 7cm, bearing a pair of lanceolate bracteoles 8–10 × 2–3mm close below flower. Sepals blue or purplish, papery, covered with yellowish hairs, veins black; blade of upper sepal broadly ovate, 1.5–2 × 1.2–1.8cm, spur downwardly curved, 8–14 × 7–8mm; lateral sepals broadly ovate, 15–20 × 12–15mm. Petals darker than sepals, blade of upper pair straight, 12–15 × 2–3mm, bilobed at apex; blades of lower pair oblong, c 7 × 4mm, bifid to middle, densely pubescent near base.

Bhutan: C–Tongsa district (Thampe La), **N**–Upper Bumthang Chu district (Tolegang and Larko La). Alpine screes and gravel and on open grassy slopes, 4575–4730m. August–October.

2. D. glaciale Hook. f. & Thomson. Med: *Jagoid Poy, Jagoe Poi.*

Similar to *D. nepalense* but smaller, stems not more than 15cm, pubescent with spreading hairs some of which are glandular with swollen bases; leaves more deeply divided into lanceolate, acute lobes 2–3mm broad, sparsely pubescent; flowers few (up to 5) in a corymb; blade of upper sepal broadly ovate, 25–30 × 20mm, acute, spur 12–15 × 7–8mm; lateral sepals broadly ovate, 18–23 × 15–17mm; lamina of upper petals upwardly curved, 12–13 × 2–3mm, bilobed at apex; lower petals oblong c 6 × 3mm, bifid to middle into acute lobes, pubescent near base.

Bhutan: **C** –Thimphu district (near Tremo La), **N** – Upper Mo Chu district (Chhew La) and Upper Pho Chu district (Kangla Karchu La and Gafoo La); **Sikkim**. Alpine screes and gravel slopes, 4730 – 5100m. July – October.

3. D. viscosum Hook. f. & Thomson; *D. conocentrum* Chatterjee

Similar to *D. nepalense* but taller; stems (15 –)20 – 60(– 80)cm, glandular-pubescent above; leaves suborbicular, 3 – 10cm diameter, divided to middle into 5 – 7 broad (7 – 15mm) lobes, toothed at apex; petioles up to 20cm; flowers few, greenish to blue or purple, in lax racemes; bracts oblong, 3 – 5cm, entire or lobed; bracteoles similar, 1 – 2cm, borne near flowers; lamina of upper sepal broadly ovate, 18 – 23mm long and broad, spur c 15 × 5 – 8mm; lateral sepals 22 – 25mm long and broad; lamina of upper petals 12 – 15 × 3mm, divided into 2 narrow lobes c 3mm at apex; lamina of lower petals oblong, 10 × 3mm, bifid to below middle into linear lobes, stiffly pubescent.

Bhutan: **C** –Thimphu and Tongsa districts, **N** – Upper Mo Chu, Upper Bumthang Chu, Upper Kuru Chu and Upper Kulong Chu districts; **Sikkim**. Alpine cliff ledges, screes and gravel, 3660 – 4730m. July – October.

D. vestitum Royle, a species ranging from W Himalaya to Nepal has been doubtfully recorded (117) from Thimphu district (Chelai La). It differs from *D. viscosum* as follows: stems taller, 40 – 100cm, with long, spreading or deflexed hairs; leaves suborbicular up to 12cm diameter, shallowly lobed and toothed; flowers in well-developed racemes; spur 10 – 15 × 5 – 7mm.

4. D. cooperi Munz. Med: *Jakang*.

Stems slender, 10 – 45cm. Leaves palmately reniform, 3 – 5cm long and broad, divided into 3 – 5 broadly obovate segments each with a few ovate 4 – 5mm broad teeth; petioles 2 – 6(– 10)cm, spreading-villous. Pedicels 3 – 8cm, villous with some hairs glandular; bracteoles elliptic or oblanceolate, 4 – 6 × 1 – 2mm, borne above middle. Sepals purplish violet with yellowish hairs; lamina of upper sepal obovate, 12 – 13 × 7 – 8mm, spur spreading or decurved, 15 – 17 × 3mm; lateral sepals elliptic-ovate, 12 – 13 × 8 – 10mm. Petals dark blue-black, much darker than sepals; lamina of upper pair oblong, curved at apex, entire; lamina of lower pair ovate, obtuse, 4 × 3mm, notched at apex, ciliate.

Bhutan: **C** –Thimphu district (Paro and Dotanang), Punakha district (Ritang) and Tongsa district (Chendebi). Dry banks and in Evergreen oak forest, 2440 – 2600m. August – October.

5. D. wardii Marquand & Shaw

Similar to *D. cooperi* but stems not above 20cm, leaf-lobes more shallowly toothed at apex; bracts spathulate or 3-lobed, 10 – 30 × 3 – 7mm, bracteoles similar, up to 20mm, borne close below flower, pedicels villous, eglandular; lamina of upper sepal ovate, 17 – 20 × 14 – 16mm, spur 22 – 25mm, spreading or slightly decurved; lateral sepals ovate-elliptic, 18 – 20 × 12 – 15mm; lamina of upper petals ± straight, c 13mm, shortly bilobed at apex, sparsely pubescent; lamina of lower petals oblong, c 8 × 2–3mm, bifid to middle, pubescent.

Chumbi: near Bhutan border E of Phari. Alpine slopes, 4600m. September.

6. D. muscosum Exell & Hillcoat. Med: *Jakang*.

Stems branched from base, 7–10cm, deflexed-pubescent, each bearing one flower. Leaves suborbicular in outline, 2–3cm diameter, finely dissected; ultimate segments linear, 3–6 × 1–2mm, pubescent. Pedicels usually bearing 2 leaf-like bracteoles above middle or close to flower. Sepals violet-purple, pale hirsute, lamina of upper sepal broadly ovate, 18–20mm long and broad, spur 14–15 × 3–4mm slightly decurved; lateral sepals ovate-elliptic, 20–22 × 16–18mm. Lamina of upper petals oblong, c 10 × 2–3mm, subacute, minutely bilobed; lamina of lower petals obovate, c 9 × 7–8mm, margin ± crenate, bearing a tuft of yellow hair near base.

Bhutan: C –Tongsa district (Rinchen Chu and Omta Tso), **N** –Upper Mo Chu district (Chhew La), Upper Mangde Chu (Saga La) and Upper Bumthang Chu (Marlung, Pangothang and Tolegang) districts. Alpine screes and sandy slopes, 4570–4875m. July–September.

7. D. caeruleum Cambessedes

Similar to *D. muscosum* but stems 5–15(–40)cm, leaves with fewer linear-lanceolate segments 5–20 × 2–3mm; flowers solitary or 2–3 in racemes, pedicels 2–8cm, bearing linear bracteoles 5–8mm near or above middle; sepals often densely pubescent, lamina of uppermost ovate, 14–16 × 8–9mm, spur 18–22 × 2–3mm, spreading; lateral sepals oblong-ovate, 15–17 × 5–6mm; lamina of upper petals c 7 × 3mm, rounded at apex; lamina of lower pair ovate, c 7 × 5mm, entire or notched at apex, pubescent at base.

Bhutan: C –Thimphu district (Kumathang, Paro Chu), **N** –Upper Mo Chu district (Soe and Yale La); **Sikkim**. Alpine cliff-ledges, and on grassy and rocky slopes, 3960–4570m. August–October.

8. D. candelabrum Ostenfeld. Med: *Jakang*.

Similar to *D. muscosum* but ultimate segments of leaves elliptic-lanceolate, 3–5 × 2–3mm; pedicels 4–6cm, with 3-partite bracteoles 5–7mm near middle; lamina of upper sepal broadly ovate, 20–25 × 18–20mm, spur straight or upcurved, 15–22 × 5mm; lateral sepals ovate-elliptic, 18–20 × 10–12mm; lamina of upper petals ± straight, 10–15 × 3mm, curved and emarginate at apex; lamina of lower petals broadly elliptic, 7–8 × 5–6mm, shallowly bifid, sparsely pubescent.

Chumbi: E of Phari; **Sikkim**. Alpine slopes, 4570m. August.

9. D. ludlowii Munz. Fig. 25d.

Stems 75–100cm, pubescent with deflexed hairs especially above. Leaves suborbicular in outline, 10–15cm diameter, ternately palmatisect; segments divided to about 1cm from base, laterals deeply bifid, coarsely toothed, teeth 2–3mm broad; petioles up to 30cm. Inflorescence lax, up to 20-flowered, pedicels up to 3cm, bracteoles linear-spathulate, 5–7 × 1mm. Sepals blue or purple, minutely pubescent; lamina of uppermost broadly ovate, 12–15 × 10mm, spur spreading, 12–17 × 4–5mm; lateral sepals obovate, 11–13 × 5–7mm. Lamina of upper petals oblong, upwardly curved, 10 × 3mm, narrowed and notched at apex; lamina of lower petals oblong, 3–4 × 2–3mm, shallowly bilobed, sparsely pubescent.

Bhutan: C – Punakha, Tongsa, Bumthang and Tashigang districts, **N** – Upper Kuru Chu and Upper Kulong Chu districts; **Sikkim**. Streamsides and shady ravines in Spruce, Hemlock and Fir forests, 2740–3650m. June–October.

10. D. scabriflorum D. Don

Similar to *D. ludlowii* but pubescent with spreading hairs, some glandular; lamina of lower petals 5–6 × 3–4mm, deeply bifid into lanceolate lobes, strongly ciliate and hirsute.

Bhutan: C – Punakha district (Ritang) and Tongsa district (Yuto La and Chendebi), **N** – Upper Kuru Chu district (Khoma Chu valley). Mossy rocks by streams in moist conifer forests, 2130–3200m. June–July.

11. D. drepanocentrum (Bruehl) Munz

Similar to *D. ludlowii*, but pubescent with long, deflexed, yellowish hairs; lamina of upper sepal ovate, 12–15 × 10mm, spur downwardly curved almost into a half-circle, 23–25mm, nearly twice as long as lamina; lamina of lower petals ovate, 6–7 × 4mm, deeply bifid into two lanceolate lobes, ciliate and sparsely pubescent at base.

Sikkim: near Wallanchun. Alpine slopes.

12. D. bhutanicum Munz

Stems up to 1m, pubescent with deflexed hairs. Leaves suborbicular in outline, 6–17cm diameter, palmatisect, acutely 5-lobed, ultimate segments lanceolate, 5–20 × 2–5mm. Racemes lax, 2–4-flowered; pedicels 3–5cm, with linear 4–6mm bracteoles borne above middle. Lamina of upper sepal obovate, 12–15 × 10–12mm, spur 15–16 × 3mm, spreading; lateral sepals obovate, 13–16 × 7–9mm. Petals not darker than sepals, lamina of upper pair oblong, upwardly curving, c 9 × 3mm, entire; lamina of lower petals obovate, 8–9 × 5–7mm, shallowly bifid into 2 broad lobes, white-hairy especially near base.

Bhutan: N – Upper Bumthang Chu district (Kurmathang and Pangothang) and Upper Kulong Chu district (Me La). Alpine grassy slopes and in Juniper/Rhododendron scrub, 3650–3960m. August–September.

13. D. stapeliosmum Huth

Similar to *D. bhutanicum* but sometimes taller, up to 1.5m; leaves with shorter, broader ultimate segments 5–7 × 3–5mm; bracteoles oblanceolate, 6–10 × 2–3mm; sepal-spur 15–20mm, spreading; petals blue-black, much darker than sepals; lamina of upper pair c 10 × 3mm, bidentate at apex; lamina of lower petals ovate, 6–7 × 3mm, deeply bifid into lanceolate lobes, pubescent.

Bhutan: C–Thimphu district (E side of Thimphu valley); **Chumbi**: Lingmuthang. Moist conifer forests and alpine grassy slopes, 2440–3050m. September.

14. D. altissimum Wall. Med: *Jakang*.

Similar to *D. bhutanicum* but stems sometimes up to 1.5m; bracteoles linear-lanceolate, 4–5 × 0.5mm; lamina of upper sepal obovate, 13–15 × 7–8mm, spur 18–20 × 3–4mm; petals rich purple, somewhat darker than sepals; lamina of upper pair oblong, c 12 × 3mm, lobed in upper half, bidentate at apex; lamina of lower petals oblong-ovate, 5–6 × 4–5mm, deeply bifid, ciliate.

Bhutan: C–Punakha district (Punakha, Wangdu Phodrang, Hinglai La and Lometsawa), Tongsa district (Tongsa) and Mongar district (Lhuntse, Unjar and Kuru Chu valley). Oak forest, open slopes amongst cultivation and on streamsides, 1400–2100m. July–September.

The records of the W Himalayan *D. denudatum* Hook. f. & Thomson from Bhutan (117) may refer to this species.

20. CONSOLIDA Gray

Similar to *Delphinium* but annual; petal solitary, lobed, produced at base into a nectariferous spur which extends into the sepal spur; carpel 1, becoming a follicle.

1. C. ambigua (L.) Ball & Heywood; *Delphinium ambiguum* L. Eng: *Larkspur*.

Stems erect, 20–30(–60)cm, pubescent, glandular in part, some hairs with swollen bases. Leaves bipinnatisect, ovate in outline, 2–4 × 1.5–4cm; segments linear-lanceolate, c 2mm broad; lower leaves with petioles up to 2.5cm. Flowers blue, pink or white, in terminal racemes; pedicels 0.5–4cm with two linear-elliptic 2–5mm bracteoles above the middle. Lamina of upper sepal ovate, c 10 × 5–6mm, spur 10–15mm, spreading but upwardly curved, lateral sepals broadly elliptic, 12–15 × 5–10mm. Lamina of petal c 12 × 14mm, 3-lobed; median lobe longest, itself shortly bilobed.

Bhutan: S–Gaylegphug district (Gaylegphug), **C**–Mongar district (Mongar). Cultivated as an ornamental in gardens, 300–1800m. March–June.

21. ACONITUM L.

The medicinal name *Bonga Nakpo* has been applied to the tubers of several species.

Erect or scandent herbs, perennial or rarely annual. Leaves basal and on stems, suborbicular or broadly ovate in outline, ternately palmatisect with the lateral segments again deeply divided, or rarely pinnatisect. Flowers zygomorphic, solitary or few to numerous in racemes, usually blue or purplish, sometimes white or yellowish. Sepals 5, petaloid, the uppermost larger and broader, concave, boat-shaped or helmet-shaped, rarely spurred; upper lateral sepals obovate or suborbicular; lower lateral sepals oblong or elliptic. Petals 2, enclosed within and ± as long as uppermost sepal, having a slender grooved basal claw broadening into a sac-like head, often S- or T-shaped in outline, with terminal nectariferous spur and at lower end a flattened, often bilobed lamina; anterior petals much reduced or absent. Carpels usually 3–5, rarely 9–12, becoming ± erect follicles in fruit.

1. Annual with slender simple tap root; leaves pinnatisect; carpels 9–12 **1. A. gymnandrum**
+ Perennials; rootstock rhizomatous or more often with paired tubers, one of which is replaced each year; leaves palmatisect, ± suborbicular in outline; carpels usually 3–5 (6–8 in *A. fletcherianum*) 2
2. Scandent herbs ....**Species 2 & 3**
+ Erect or decumbent herbs, never scandent 3
3. Plants small, 5–35cm tall; leaves 1.5–5cm across; flowers solitary or up to 4 in a raceme 4
+ Plants tall, commonly 0.5–1.5m; leaves 5–15cm across; flowers few or numerous 6
4. Petal head funnel-like with a finger-shaped spur and a rounded, entire lamina **4. A. fletcherianum**
+ Petal head S- or T-shaped with a ± bulbous spur 5
5. Petal head 3–6mm, glabrous ....**Species 5 & 6**
+ Petal head c 13mm, hairy..**7. A. sherriffii**
6. Uppermost sepal with a dorsal spur (short and bulbous in *A. novoluridum*); rootstock rhizomatous ....**Species 8 & 9**
+ Uppermost sepal not spurred; roots tuberous 7
7. Flowers whitish, tinged yellow or blue ....**Species 10 & 11**
+ Flowers blue or purple (sometimes whitish in *A. spicatum* but then stems robust, c 1.5m, and leaves larger, c 15cm across) 8
8. Leaves very deeply divided, base of sinus between principal segments to top of petiole 2–3mm 9
+ Leaves less deeply divided, base of sinus between principal segments to top of petiole 5–10mm 11
9. Flowers numerous in dense racemes; uppermost sepal boat-shaped, c 0.5cm broad ....**12. A. parapolyanthum**
+ Flowers few at branch ends or in diffuse racemes; uppermost sepal helmet-shaped, 0.8–1.5cm broad 10

10. Petals hairy .... **Species 13 & 14**
\+ Petals glabrous .... **Species 15 & 16**

11. Petals hairy (sometimes glabrous in *A. spicatum*) 12
\+ Petals glabrous 13

12. Racemes laxly flowered; pedicels 6–10cm .... **Species 17 & 18**
\+ Racemes densely flowered; pedicels up to 3cm .... **Species 19 & 20**

13. Racemes laxly flowered .... **21. A. bisma**
\+ Racemes densely flowered .... **22. A. hicksii**

1. A. gymnandrum Maximowicz

Stems 15–50cm, villous with long white downwardly spreading hairs. Leaves broadly ovate in outline, 3–6cm long and broad, dissected into linear oblong lobes 2–3mm broad, pubescent above, white villous beneath. Racemes densely flowered. Sepals blue; the uppermost narrowly boat-shaped, 15–20 × 4–5mm; lateral sepals c 15 × 5–7mm. Petals glabrous, head c 10mm, the upper half an oblong spur, the lower half expanded into a reniform lamina 6–7mm broad. Follicles cylindrical, c 10 × 3mm.

Chumbi : Phari. Alpine grassy slopes, and open plains, 4265–4880m. September.

2. A. bulbilliferum Handel-Mazzetti

Stems scrambling, 1.5–2.5m. Leaves 6–12cm diameter, segments ± rhombic, deeply-cut or lobed, glabrous; petioles bearing 1–4 bulbils in the axil. Racemes lax, pedicels c 4cm. Uppermost sepal deeply helmet-shaped, 2.5 × 1.5cm, shortly (c 5mm) beaked; upper lateral sepals broadly obovate, c 1.7cm long and broad; lower sepals ovate, c 15 × 8mm. Petal head 10–11mm, spur strongly recurved. Carpels glabrous.

Bhutan: C – Bumthang district (Dhur Chu Valley), **N** – Upper Bumthang Chu district (Weitang and Pangothang). Clearings in moist conifer forests, and in Juniper/Rhododendron scrub, 3350–3960m. July–September.

3. A. elwesii Stapf; *A. uncinatum* sensu F.B.I. non L.

Similar to *A. bulbilliferum* but leaf axils without bulbils; uppermost sepal more narrowly helmet-shaped, c 2 × 1cm, with beak c 8mm; upper lateral sepals c 1.5 × 1.3cm; lower pair 10–13 × 4–5mm.

Chumbi : Bhutan/Tibet border; **Sikkim**: Lachen and Lachung. Climbing on shrubs in moist forests, 2440m. September.

4. A. fletcherianum Taylor

Stems 7–12cm. Leaves 2–4cm diameter, ultimate lobes 3–4mm broad, acute, glabrous; petioles 3–8cm. Flowers solitary, pedicel bearing a ternately divided bracteole 0.7–1.5cm, above the middle. Uppermost sepal boat-shaped, 2.5–3 ×

1–1.2cm; lateral sepals ovate or obovate, 2–2.5cm long and broad. Petal head 4–5mm. Carpels 6–8, sparsely pubescent.

Bhutan: N – Upper Mangde Chu district (Saga La and Khem La), Upper Bumthang Chu (Waitang and Pangothang), Upper Kuru Chu (Narim Thang) and Upper Kulong Chu (Me La). Alpine rocks and cliff ledges, grassy slopes and screes, 4110–4720m. July–September.

5. A. naviculare (Bruehl) Stapf

Stems 5–25cm. Leaves mostly basal, 1–3.5cm diameter, lobes with obtuse or subacute teeth, finely pubescent; petioles 2–8cm. Flowers 1–4. Uppermost sepal boat-shaped, 2–2.5 × 0.8cm; upper lateral sepals obovate, 1.5 × 1cm; lower sepals ovate, c 8 × 3mm. Petal head 3–4mm. Follicles 10–12mm, hairy, surrounded by persistent sepals.

Bhutan: C –Thimphu district (Kuma Thang), **N** – Upper Mangde Chu, Upper Bumthang Chu, Upper Kuru Chu and Upper Kulong Chu districts; **Sikkim**. Alpine meadows, Juniper scrub and on cliff ledges, 3960–4570m. July–October.

6. A. hookeri Stapf; *A. pulchellum* Handel-Mazzetti

Similar to *A. naviculare* but leaves more acutely toothed; sepals deciduous; petals usually glabrous, head 4–6mm, dorsally pouched.

Bhutan: N – Upper Mangde Chu district (Phage La); **Sikkim**: Jongri. Alpine meadows, open rocky slopes and amongst Juniper scrub, 4265–4570m. August.

7. A. sherriffii Lauener

Stems c 35cm, leafy. Leaves 4–5cm diameter, minutely pubescent; petioles 6–12cm. Flowers solitary. Uppermost sepal ± helmet-shaped, c 3 × 1.5cm; upper lateral sepals obovate, c 2.5 × 2cm; lower sepals elliptic, c 2 × 0.7cm. Petals hairy, especially on the claw; head S-shaped, c 13mm. Carpels sparsely pubescent.

Bhutan: N – Upper Kulong Chu district (Pang La). Rocky alpine hillsides, 4420m. August.

8. A. novoluridum Munz; *A. luridum* Hook. f. & Thomson non Salisbury. Fig. 25a,b.

Rootstock rhizomatous, surrounded at apex by withered remains of petioles. Stems 0.4–1(–2)m. Leaves (5–)7–15cm diameter, divided into narrow pubescent segments c 0.5–1.5cm broad. Flowers reddish-purple in dense yellowish-pubescent racemes; pedicels mostly up to 5mm. Uppermost sepal ovate, 1.3–1.7cm; spur rounded, 4–5mm; upper lateral sepals obovate, c 10 × 7mm; lower sepals elliptic, c 10 × 4mm. Petals T-shaped, head c 5mm, set at right angles on short (3–4mm) claw; spur bulbous. Follicles 3, 10–12mm, pubescent at first.

Bhutan: C –Thimphu and Tongsa districts, **N** – Upper Mo Chu, Upper Mangde Chu, Upper Bumthang Chu and Upper Kulong Chu districts; **Sikkim**. Margins of Fir forest, in Juniper/Rhododendron scrub and on alpine slopes and cliffs, 3800–4570m. August–September.

9. A. scaposum Franchet; *A. jucundum* Diels

Similar to *A. novoluridum* but leaves with segments 2–3cm broad; racemes ± distantly flowered, pedicels 2–2.5cm, densely pubescent; sepals purple or white, uppermost c 1.3cm tall, spur 8–10(–15)mm; upper lateral sepals obovate, c 10 × 8mm; lower sepals elliptic, 8 × 2–3mm; petals T-shaped, head 5–6mm, spur downwardly curved, claw 10–14mm.

Bhutan: C–Tashigang district (Tashiyangtse and Tobrang), **N**–Upper Kulong Chu district (Lao). Fir and Rhododendron forest, 2740–3050m. August.

10. A. orochryseum Stapf. Med: *Bongkar.*

Stems slender, 40–50(–100)cm, finely pubescent. Leaves 3–9cm diameter, divided into segments 2–3cm broad, margins of upper ones usually with oblong finger-like teeth; petioles up to 25cm. Flowers 2–10 in rather lax racemes, white tinged with yellow and blue; pedicels up to 5cm, yellowish pubescent. Uppermost sepal helmet-shaped, 2–2.8 × 1–1.4cm; upper lateral sepals broadly obovate, c 1.5cm long and broad; lower sepals elliptic, c 10 × 7mm. Petals hairy, head c 8mm; spur bulbous, upwardly curved.

Bhutan: C–Tongsa district (Thampe La), **N**–Upper Mo Chu district (Zaradingthang) and Upper Bumthang Chu district (Tolegang). Margins of Fir forests, in Juniper/Rhododendron scrub and on alpine screes, 3950–4720m. August–September.

11. A. heterophylloides (Bruehl) Stapf; *A. leucanthum* (Bruehl) Stapf

Similar to *A. orochryseum* but leaves more roundly toothed; flowers white or tinged blue.

Bhutan: N–Upper Mo Chu district (Shodug and Yale La); **Sikkim**. Alpine slopes, 4000–4500m. August–September.

The E Himalayan plants belong to var. **leucanthum** (Bruehl) Lauener.

12. A. parapolyanthum Lauener

Stems c 75cm. Leaves up to 12cm diameter, divided into narrow segments 2–5mm broad, pubescent. Pedicels 1–2cm, densely pubescent. Uppermost sepal boat-shaped, c 2 × 0.5cm; upper lateral sepals obovate, c 1.5 × 1cm; lower sepals oblanceolate, c 13 × 4mm. Petal head T-shaped, c 6mm, spur bulbous.

SE Tibet: between Me La and Cho La. Amongst Juniper scrub and on stony alpine hillsides, 3650m. August.

13. A. gammiei Stapf; *A. napellus* sensu F.B.I. non L., *A. napellus* L. var. *rigidum* Hook. f. & Thomson, *A. dissectum* D. Don non Reichenbach, *A. wallichianum* Lauener

Lax herb, stems c 60cm. Leaves c 7cm diameter, deeply dissected into narrow lobes c 3–5mm broad. Flowers 2–3 at branch ends, sometimes forming diffuse widely branched racemes. Uppermost sepal 2–2.2 × 0.8cm; upper lateral sepals broadly obovate, 1.5cm long and broad; lower sepals elliptic, c 1.5 × 0.7cm. Petal head c 7mm, spur recurved.

Sikkim: Kapup. Sandy lake shore, 3800m. September.

14. A. laciniatum (Bruehl) Stapf

Similar to *A. gammiei* but taller and more erect, stems 0.6–1.5m; leaves less deeply dissected, up to 15cm diameter; flowers widely spaced in racemes; uppermost sepal 2.2–2.5 × 1.3–1.5cm; upper lateral sepals obovate, c 1.7 × 1.5cm; lower sepals c 1.5 × 0.5cm; petal head c 12mm, spur shortly recurved; follicles 3, linear-oblong, 18–25 × 5–6mm, pubescent.

Bhutan: C –Thimphu district (Chelai La) and Tongsa district (Rinchen Chu), **N** –Upper Mo Chu district (between Chamsa and Kohina) and Upper Kulong Chu district (Pang La); **Sikkim**. Grassy alpine slopes and amongst shrubs, 3500–4570m. July–August.

Roots used medicinally and for making poison for arrows.

15. A. nakaoi Tamura; *A. parabrachypodum* Lauener

Stems erect, up to 1m. Leaves 3–10cm diameter, lobes oblong, acute. Flowers few or numerous, subpaniculate, branches bearing 2–3 flowers. Uppermost sepal c 2 × 1cm; upper lateral sepals suborbicular, c 1.5cm diameter; lower sepals ovate, c 13 × 6mm. Petals ± T-shaped, head 7–8mm, spur rather bulbous, recurved.

Bhutan: C –Bumthang district (Dhur Chu), **N** –Upper Mangde Chu (Phage La), Upper Bumthang Chu (Pangotang and Mon La Karchung La) and Upper Kulong Chu district (Pang La and Me La); **Sikkim**. Amongst Rhododendron scrub and on rocky alpine slopes and screes, 3800–4400m. August–October.

16. A. ferox Seringe

Similar to *A. nakaoi* but leaves up to 12cm diameter, segments acuminate; flowers numerous in racemes; uppermost sepal 2–2.5 × 0.8–1cm; upper lateral sepals broadly obovate, c 1.5 × 1.3cm; lower sepals ovate, c 10 × 3mm; petal head ± S-shaped, c 6mm, spur recurved.

Bhutan: C –Tongsa district (Pele La); **Sikkim**. Moist Spruce forest, amongst shrubs, 2600–3800m. September–October.

Tubers are used to poison arrows and are extensively used medicinally (13).

17. A. patulum Tamura. Med: *Tsendug*.

Stems tall with wide branches 15–25cm. Leaves 10–13cm diameter, lobes acute, subglabrous. Flowers in terminal and lateral racemes, pedicels 6–9cm, ± horizontal. Uppermost sepal helmet-shaped, c 18 × 10–13mm, scarcely beaked; upper lateral sepals suborbicular, 14–16mm diameter; lower sepals oblong-elliptic, 10–13 × 3mm. Petal head ± S-shaped, 10–12mm, spur somewhat recurved.

Bhutan: C –Thimphu district (between Barshong and Zambuthang), **N** –Upper Mo Chu district (Shukuna Mountain). Alpine slopes, 2740–3000m. July–September.

18. A. funiculare Stapf

Similar to *A. patulum* but more slender, stems 40–50cm; leaves c 6cm diameter, with few acute segments and teeth; racemes lax, 4–6-flowered, pedicels 6–10cm; uppermost sepal helmet-shaped, c 2 × 1cm, shortly rostrate; upper lateral sepals suborbicular, 13–14mm; lower sepals elliptic, c 1cm.

Bhutan: C–Thimphu district (Paro). Moist conifer forest, 3275m. August?

19. A. deinorrhizum Stapf

Stems 1.5–2m. Leaves c 15cm diameter, lobes and teeth acute, finely pubescent. Uppermost sepal boat-shaped, c 2 × 0.6cm, shortly beaked; upper lateral sepals obovate, c 1.5 × 1cm; lower sepals elliptic, 10 × 5mm. Petals densely pubescent, head S-shaped, c 8mm, spur recurved.

Bhutan: C–Tashigang district (Dib La). Alpine grassy slopes, 3350m. September.

Poison is prepared from the root of this species (Stapf).

20. A. spicatum (Bruehl) Stapf; *A. ferox* sensu F.B.I. p.p. non Seringe. Fig.25c.

Similar to *A. deinorrhizum* but sepals often white, tinged with purple; uppermost sepal helmet-shaped; c 2.5 × 1.3cm, shortly (2–3mm) beaked; upper lateral sepals orbicular, c 1.5cm; lower sepals elliptic, c 13 × 3mm; petals hairy or sometimes glabrous, head S-shaped, 10–11mm, spur recurved; follicles 5, oblong, c 10 × 4mm, hairy.

Bhutan: C–Thimphu and Tongsa districts, **N**–Upper Mo Chu, Upper Pho Chu, Upper Mangde Chu, Upper Bumthang Chu and Upper Kulong Chu districts; **Sikkim**. In Fir forest and Juniper/Rhododendron scrub, and on open alpine slopes and in ravines, 3500–4420m. July–September.

The root contains valuable alkaloids (Stapf in *Ann. RBG Calc.* 10(2), 1905; Caius).

21. A. bisma (Hamilton) Rapaics; *A. palmatum* D. Don

Stems 60–150cm. Leaves up to 15cm diameter, ± glabrous. Pedicels 4–10cm. Uppermost sepal helmet-shaped, c 2.5 × 1.2cm; upper lateral sepals obovate, c 2 × 1.5cm; lower sepals elliptic, c 15 × 7mm. Petal head S-shaped, c 5–6mm, spur bulbous, lamina very short (c 1mm). Follicles oblong, 2.5–3 × 0.5–0.6cm, glabrous.

Bhutan: C–Thimphu district (Chelai La) and Punakha district (Tang Chu), **N**–Upper Mo Chu district (Kohina); **Sikkim**. Moist Fir and Rhododendron forests, and on alpine slopes, 2750–3700m. August.

Extracts from the tubers of this non-poisonous species are used medicinally (Stapf, Caius).

22. A. hicksii Lauener

Stems 45–75cm. Leaves 5–10cm diameter, glabrous, teeth and lobes acute. Pedicels 1.5–4cm. Uppermost sepal helmet-shaped, c 2 × 1cm, minutely beaked;

upper lateral sepals obovate, c 1.5 × 1cm; lower sepals elliptic, c 10 × 5mm. Petal head ± T-shaped, 6–8mm, spur recurved.

Bhutan: N – Upper Mangde Chu district (Passu Sefu) and Upper Bumthang Chu district (Pangotang, Tolegang and Marlung). Alpine cliff ledges, rocky slopes and screes, 4260–4730m. July–September.

Records of the W Himalayan *A. balfouri* Stapf (*A. atrox* (Bruehl) Mukerjee) from Bhutan (117) possibly refer to this species.

22. PAEONIA L.

Erect or spreading shrubs. Leaves alternate, bipinnate, often ternately so. Flowers actinomorphic, solitary or few, terminal. Sepals 5, herbaceous. Petals 5 or more. Stamens numerous. Carpels 5, borne at first within a cup-like disc; styles short, thick. Fruit a cluster of leathery follicles.

1. P. suffruticosa Andrews; *P. moutan* Andrews subsp. *atava* Bruehl

Stems up to 2m. Leaves 20–45cm, leaflets ovate or 3-lobed, 4.5–9 × 2–5cm, acuminate, base rounded or cuneate, glabrous above, pubescent on veins beneath and with a ring of hairs at junction of rachis and petiole. Flowers c 16cm across. Petals 7–8 × 6–8cm, white to rose-pink, each with a dark purplish blotch at base. Stamens 2–3cm, filaments purplish, anthers yellow. Carpels green, densely pubescent; stigmas white.

Bhutan: C – Thimphu district (between Chang Na and Paro); **Chumbi**: Tuk-Chung. 2900–3300m. May–June.

Probably cultivated or naturalised in Bhutan.

Family 43. CIRCAEASTERACEAE

by A.J.C. Grierson & D.G. Long

Weak glabrous annuals with persistent cotyledons. Leaves in a rosette at apex of short stem (elongated hypocotyl), exstipulate, venation dichotomous, open. Flowers in fascicles amongst leaves, bisexual. Sepals 2–3, free. Petals absent. Stamens 1–3. Carpels 1–3, superior, free, linear-oblong; ovule 1. Fruit a cluster of achenes.

1. CIRCAEASTER Maximowicz

Description as for Circaeasteraceae

1. C. agrestis Maximowicz. Fig. 26e–g.

Hypocotyl 5–8cm. Cotyledons elliptic-oblong, 5–10 × 1.5–2mm, entire. Leaves rhombic-obovate, (0.7–)1–2 × (0.2–)0.4–1cm, obtuse or subacute, base cuneate, subpetiolate, margins with a few short teeth at apex, otherwise entire. Peduncles c 1mm at flowering, up to 10mm in fruit. Sepals oblong, c 0.5mm; stamens and carpels c 0.5mm at flowering time. Stigma linear, minute. Achenes ellipsoid, c 3mm, covered with fine hooked hairs.

Bhutan: C –Thimphu district (Pajoding) and Tongsa district (Padima Tso near Thampe La); **Sikkim**. Mossy turf in Juniper/Rhododendron scrub, 3950–4250m. July–August.

Family 44. BERBERIDACEAE

by A.J.C. Grierson

Shrubs, often spiny. Leaves alternate, simple or 1-pinnate, herbaceous or coriaceous, exstipulate. Flowers solitary or several to many in fascicles or racemes, bisexual, actinomorphic. Sepals and petals similar, usually 6 of each in whorls of 3, free, yellow or greenish, sometimes tinged with red, petals with 2 oblong glands near base. Stamens 6, slightly adnate to base of petals, anthers opening by valves. Ovary consisting of a single carpel, superior, ovules few, basal, style short or absent, stigma rounded. Fruit a berry.

1. Leaves simple; stems usually spiny ....**1. Berberis**
\+ Leaves pinnate; stems unarmed ....**2. Mahonia**

1. BERBERIS L.

Stems usually bearing 3–5-fid spines (modified leaves), rarely without spines. Leaves simple, borne in clusters on stunted short-shoots in axils of spines. Sepals 6–12 in 2–4 whorls. Petals 6.

The following local names refer to *Berberis* species in general: Dz: *Kertsa, Ketsa*; Med: *Ker Shun*; Tongsa dialect: *Tibsi*; Eng: *Barberry*.

Because of the great variability within this genus numerous varieties have been created; for the sake of simplicity these have largely been ignored in this account.

From the known distributions of the species involved, East Himalayan records of the following have been discounted: *B. chitria* Lindley, *B. collettii* Schneider, *B. edgeworthiana* Schneider, *B. everestiana* Ahrendt, *B. ludlowii* Ahrendt, *B. orthobotrys* Aitchison, *B. replicata* W. W. Smith, *B. sublevis* W. W. Smith, *B. wallichiana* DC. and *B. zayulana* Ahrendt. *B. grandibracteata* Ahrendt is only known in cultivation from seed 'believed to have come from Bhutan'; this has also been excluded.

1. Flowers solitary, mostly 1–1.5cm across, but only 0.5–0.7cm across in *B. tsarica* (flowers sometimes clustered in *B. angulosa* but then flowers 1–1.5cm across) 2
+ Flowers several or numerous in various types of inflorescence, if clustered then flowers less than 1cm across 3
2. Flowers large, 1–1.5cm across .... **Species 1–3**
+ Flowers small, 0.5–0.75cm across .... **4. B. tsarica**
3. Evergreen shrubs; flowers only in clusters 4
+ Evergreen or deciduous shrubs; flowers at least in part in racemes, often subumbellate, peduncles sometimes very short, sometimes a few fasciculate inflorescences also present 5
4. Stems unarmed or with weak spines c 5mm .... **5. B. insignis**
+ Stems spiny, with at least some spines 1–2.5cm .... **Species 6–8**
5. Leaves thickly coriaceous, evergreen, prominently reticulate above; petals longer than inner sepals .... **9. B. asiatica**
+ Leaves thin or slightly coriaceous, deciduous, not prominently reticulate above; petals shorter than inner sepals 6
6. Racemes elongate, 3–5cm, 5–15- flowered .... **Species 10–12**
+ Racemes short, 1.5–2cm, 2–3-flowered .... **Species 13 & 14**

1. B. angulosa Hook. f. & Thomson; *B. parisepala* Ahrendt. Nep: *Chutro* (34).
Shrub 1–1.5m, deciduous; stems brownish, strongly grooved, minutely pubescent; spines 3-fid, internodes usually 1.5–2cm. Leaves herbaceous, elliptic-obovate, 1–2 × 0.6–1cm, acute, base attenuate, entire or sometimes spinose-toothed. Flowers solitary or in fascicles of 2–3, pedicels c 1cm. Outer sepals ovate, c 10 × 5mm, inner broader, up to 7mm. Petals obovate, c 7 × 5mm. Fruit ellipsoid, c 10 × 7mm; style short or absent.

Bhutan: N–Upper Mo Chu district (Naha), Upper Pho Chu district (Chesha La), Upper Mangde Chu district (Wortang) and Upper Kulong Chu district (Me La); **Sikkim:** above Chowbhanjan. Mountain screes and open rocky hillsides, 3650–4100m. May–June.

Plants with flowers in clusters have been distinguished as var. *fasciculata* Ahrendt.

2. B. macrosepala Hook. f. & Thomson
Similar to *B. angulosa* but stems less grooved; leaves 2.5–3.5 × 1–1.5cm, usually spinose-toothed; pedicels 2–3cm; berries ellipsoid, c 1.5 × 1cm, red.

Bhutan: C–Tongsa, Mongar and Sakden districts, **N**–Upper Pho Chu, Upper Mangde Chu and Upper Kulong Chu districts; **Sikkim.** Rocky alpine hillsides and screes, 3800–4400m. June–July.

3. B. concinna Hook. f. & Thomson

Similar to *B. angulosa* but leaves subcoriaceous, suborbicular or obovate, 1.25–2 × 0.7–1.25cm, strongly spinose-dentate, greyish glaucous beneath; pedicels 1–3cm; berries ellipsoid, c 1.5 × 1cm, red.

Sikkim: Lachen Valley and Tonglo. Rocky mountainsides, 3350–3950m. June–July.

4. B. tsarica Ahrendt

Shrub 30–100cm, deciduous; stems angular, densely spiny, spines mostly 5-fid, internodes c 7mm. Leaves oblanceolate, 7–15 × 2.5–4.5mm, entire or with a few spinose teeth. Flowers solitary; pedicels 3–10mm. Sepals elliptic, 4–5 × 3mm. Petals obovate, c 4 × 2.5mm. Berries broadly ellipsoid, 8–10 × 5–7mm, red.

Bhutan: N – Upper Mo Chu (Lingshi), Upper Bumthang Chu (Tolegang) and Upper Kulong Chu (Me La) districts; **Sikkim.** Open hillsides, 3950–4250m. May–June.

5. B. insignis Hook. f. & Thomson; *B. incrassata* Ahrendt. Nep: *Chutro* (34).

Evergreen shrub, 1.5–2m; stems yellow or purplish, unarmed or with a few weak spines 5–7mm. Leaves elliptic-lanceolate, 9–15 × 2–6cm, acuminate, base attenuate, subpetiolate, margins spinose-dentate, veins prominent beneath. Flowers 5–numerous in axillary fascicles; pedicels 1–1.5cm. Sepals gradually increasing in size, outer broadly ovate, c 2 × 2mm, innermost elliptic, c 7 × 4mm. Petals obovate, c 6 × 3.5mm. Berries ellipsoid, c 8 × 5mm, black.

Bhutan: C – Tongsa, Mongar and Tashigang districts, **N** – Upper Mo Chu district (Gasa); **Sikkim.** Margins of broad-leaved forests, 2550–3000m. May.

6. B. griffithiana Schneider; *B. subpteroclada* Ahrendt p.p., *B. bhutanensis* Ahrendt

Evergreen shrub 1–3m, stems grooved, yellowish. Leaves elliptic, 2–4 × 0.5–1cm, acute, base attenuate, margins ± inrolled, spinose-toothed, veins ± prominent beneath. Flowers in fascicles of 3–7, pedicels 1–2.5cm. Sepals gradually increasing in size, outermost ovate, c 2 × 1.5mm, innermost obovate-elliptic, 6.5–9 × 4–6mm. Petals obovate, 5–6 × 3–4mm. Ovary cylindrical, ovules 2–4. Berries (submature) obovoid, 8–10 × 3–4mm, reddish.

Bhutan: C – Ha, Tongsa, Bumthang and Tashigang districts, **N** – Upper Bumthang Chu district. Margins of forests, 2440–3200m. April–May.

FIG. 26. **Berberidaceae, Circaeasteraceae, Podophyllaceae** and **Lardizabalaceae. Berberidaceae**. a & b, *Berberis asiatica*: a, portion of shoot showing spines, leaves and inflorescence; b, stamen, anther opening by valves. c & d, *Mahonia napaulensis*: c, portion of shoot showing leaves and terminal cluster of racemes; d, flower. **Circaeasteraceae**. e–g, *Circaeaster agrestis*: e, plant with cotyledons and flowers; f, flower; g, mature achene. **Podophyllaceae**. h & i, *Podophyllum hexandrum*: h, flower with two deflexed immature leaves; i, half of mature leaf. **Lardizabalaceae**. j–m, *Holboellia latifolia*: j, portion of shoot with inflorescence; k, leaf; l, male flower; m, female flower, sepals removed to show staminodes at base of carpels. Scale: c, ⅓; i × ½; a, h, k × ⅔; j × 1; e × 1½; l × 2; d, m × 2½; b × 5½; g × 9; f × 25. Drawn by M. Bates.

a
b
c
d
e
f
g
h
i
j
k
l
m
MB

7. B. praecipua Schneider; *B. subpteroclada* Ahrendt p.p. Dz: *Kepe Tsang.*

Similar to *B. griffithiana* but flowers smaller, in larger fascicles of 5–15; inner sepals 5–5.5 × 2–3.5mm; petals 4.5–5 × 3mm; ovules 1–2.

Bhutan: C – Ha district (Ha Dzong) and Thimphu district (common), **N** – Upper Mo Chu district (Gasa). Dry open hillsides and margins of coniferous forests, 2350–3650m. April–June.

Possibly only subspecifically distinct from *B. griffithiana.*

8. B. hookeri Lemaire; *B. wallichiana* auct. non DC.

Similar to *B. griffithiana* but leaves larger, elliptic, 3–6 (–8) × 1–2 (–3)cm, margins more strongly spinose-dentate; flowers 3–8 in fascicles; pedicels 1.5–2 (–3)cm; outer sepals ovate, c 3 × 1.5mm, innermost elliptic, c 8 × 4.5mm; petals obovate, 5–6 × 3.5–4.5mm; ovules 3–8.

Bhutan: C –Thimphu, Tongsa, Bumthang and Tashigang districts, **N** – Upper Mo Chu district (Laya); **Sikkim.** Margins of Hemlock/ Rhododendron forests, 3050–3500m. April–June.

9. B. asiatica DC. Dz: *Kepai Tsang.* Fig. 26a,b.

Evergreen shrub 1–3m, stems yellowish, spines 5–10mm. Leaves coriaceous, obovate, 3–6 × 1.2–3cm, acute or obtuse, mucronate at apex, base attenuate, margins entire or with a few spinose teeth, prominently reticulate especially above. Racemes short at flowering (c 1cm), elongating in fruit to 4cm, pedicels 1–1.5cm. Sepals increasing in size gradually, outermost broadly ovate, c 2mm long and broad, inner obovate, c 7.5 × 4mm. Petals obovate, c 8 × 6mm. Ovary ellipsoid, c 5 × 1.5mm, ovules 2–4. Berries ellipsoid, c 10 × 7mm, red becoming black, style c 1mm.

Bhutan: S – Chukka district, **C** – Ha, Thimphu, Punakha and Tongsa districts. Dry valleys and hillsides, 1400–2300m. April–May.

10. B. aristata DC.; *B. sikkimensis* (Schneider) Ahrendt, *B. ceratophylla* G. Don, *B. micrantha* Ahrendt?. Nep: *Chutro* (34).

Shrub 1–3m, deciduous, stems brownish, ± grooved, spines 1–2cm. Leaves obovate, 2–3 × 0.7–1.5cm, obtuse or acute, base attenuate, margins entire or with a few spinose teeth near apex, veins ± prominent. Racemes 5–6cm, usually not all subumbellate, pedicels 5–10mm. Sepals increasing gradually in size, outermost ovate, c 1.5 × 1mm, inner obovate, c 7 × 3–4mm. Petals obovate, 6–7 × 3–4mm, with 3 distinct unbranched veins or the middle one forked near apex. Berries immature, narrowly ellipsoid, c 7 × 1.5mm, including style c 1mm.

Bhutan: C –Tongsa district (Chendebi), **N** – Upper Kuru Chu district (Chenbi Rongang). Open hillsides and river banks, 1670–2500m. May.

11. B. virescens Hook. f.; *B. vulgaris* L. var. *brachybotrys* (Edgeworth) Hook. f. & Thomson, *B. ignorata* Schneider, *B. himalaica* Ahrendt, *B. paravirescens* Ahrendt *nom. nud.*

Similar to *B. aristata* but often purplish, ± terete; leaves generally entire; racemes slender, 3–5cm, usually subumbellate above; outer sepals ovate, 3–5 × 1.5–2mm, inner obovate, 5–6 × 3.5–4mm; petals c 4 × 2.5mm, often notched at apex, lateral and middle veins branched; berries ellipsoid, c 10 × 5mm, black.

Bhutan: C – Ha, Thimphu and Bumthang districts, **N** – Upper Mo Chu, Upper Bumthang Chu, Upper Kuru Chu and Upper Kulong Chu districts; **Sikkim.** Open hillsides and forest clearings, 2140–3800m. April–June.

12. B. thomsoniana Schneider; *B. umbellata* auct. non G. Don. Nep: *Chutro* (34).

Similar to *B. aristata* but stems yellowish and strongly grooved; leaves 3–6 × 1.25–1.75cm, usually entire, sometimes spinose-toothed; racemes c 4cm, subumbellate above; outer sepals narrowly elliptic, c 5 × 2mm, carinate, inner obovate, 5–6 × 3–4mm; petals obovate, 4–5 × 3–4mm; berries ellipsoid, 8–10 × 5–6mm.

Bhutan: N – Upper Mo Chu district (Lingshi); **Sikkim.** 3350–3960m. July.

13. B. cooperi Ahrendt; *B. lasioclema* Ahrendt, *B. hobsonii* Ahrendt

Shrub 1–2m, deciduous, stems strongly grooved, yellowish, spines 1–1.5cm. Leaves obovate, 1.5–3.5 × 0.7–1.5cm, acute or obtuse, mucronate, base attenuate, margins usually spinose-dentate. Inflorescence at least in part shortly racemose, peduncles 0.7–1.5cm, 2–3-flowered, otherwise fasciculate, pedicels 1–2cm, finely pubescent. Sepals oblanceolate or obovate up to 6–7 × 3–4mm, sometimes with 1 or 2 of the outermost half as large. Petals obovate, 5–5.5 × 3.5–4mm. Ovules 3–4. Berries red, broadly ellipsoid or subglobose, c 8 × 7–8mm, with short styles.

Bhutan: C – Ha, Thimphu and Tongsa districts, **N** – Upper Mo Chu, Upper Pho Chu and Upper Bumthang Chu districts; **Chumbi.** Open hillsides, 3300–3900m. May–June.

14. B. beesiana Ahrendt

Similar to *B. cooperi* but stems terete; leaves obovate, 1.5–2 × 0.5–0.7cm, usually entire; sepals 5–5.5 × 3–3.5mm; petals obovate, 3.5–4 × 2.5–3mm; berries globose, c 6mm.

Bhutan: N – Upper Kulong Chu district (Me La). Open valleys, 3800m. June.

2. MAHONIA Nuttall

Unarmed shrubs. Leaves unevenly 1–pinnate, rachis articulated at each leaflet node. Flowers in narrow fascicled racemes. Sepals usually in 3 series of 3, yellow. Petals 6 in 2 series of 3.

1. M. napaulensis DC.; *M. acanthifolia* G. Don, *M. griffithii* Takeda, *M. sikkimensis* Takeda, *M. hicksii* Ahrendt, *Berberis napaulensis* (DC.) Sprengel. Nep: *Chutro, Kesari* (34). Fig. 26 c,d.

Shrub 1–3m, stems leafy near apex. Leaves coriaceous, 30–45cm, leaflets opposite, 6–13 pairs, ovate to suborbicular, 3–8 × 2–4cm, acute, base rounded or truncate, coarsely spinose-dentate with 3–7 teeth on each side, glabrous. Racemes 10–25 cm in fascicles of 6–10, pedicels (2–) 5–10mm, bracts oblong-ovate, 3–6 × 1.5–3mm. Sepals ovate to obovate, 2–8 × 2–4mm. Petals obovate, 4.5–6 × 2.5–4mm. Berries ellipsoid, 8–12 × 5–7mm, blue-black, glaucous, style 0.5–2mm.

Bhutan: S–Chukka district, C–Thimphu, Punakha, Tongsa, Mongar and Tashigang districts; **Sikkim.** Forest margins, 1525–3000m. April.

The other four E Himalayan taxa, here placed in synonymy, were distinguished on the basis of leaf length, number and size of leaflets, relative sizes of inflorescence and flower parts, all of which appear to overlap and could represent the variability of a single species. Herbarium material from this region is scanty, making it difficult to assess the differentiation that the above authors maintain is present. Until the genus is better collected it seems preferable to regard *Mahonia* as being represented in this area by a single species.

The record (73) from Sikkim of *M. borealis* Takeda, a West Himalayan species, is erroneous.

Family 45. PODOPHYLLACEAE

by A.J.C. Grierson & D.G. Long

Perennial, shortly rhizomatous herbs. Stem erect, unbranched. Leaves 2 (rarely 3), alternate, borne at apex of stem, simple, deeply palmately lobed and veined, exstipulate. Flowers 1–3, borne on petiole of upper leaf, actinomorphic, bisexual, erect or pendulous. Perianth parts 9 in 3 series; outer 3 sepaloid, fugacious, inner 6 petaloid. Stamens 6; anthers opening by longitudinal slits. Ovary a solitary superior carpel, unilocular, ovules numerous, marginal, stigma sessile, peltate. Fruit a fleshy berry, rupturing irregularly; seeds numerous.

1. PODOPHYLLUM L.

Description as for Podophyllaceae.

1. P. hexandrum Royle; *P. emodi* Honigberger, *P. emodi* var. *axillare* Chatterjee & Mukerjee, *P. emodi* var. *bhootanense* Chatterjee & Mukerjee, *Sinopodophyllum emodi* (Honigberger) Ying. Med: *Yelmose, Yelmosey.* Fig. 26 h,i.

Stem in flower 15–25cm, later up to 45cm, glabrous. Leaves at first deflexed, brownish, orbicular-reniform in outline, 6–15 × 3.5–10cm, palmately divided almost to base into 3 broadly elliptic or obovate segments, acute, base cuneate, margins serrate, pubescent beneath; segments undivided or 2–3-lobed to middle;

petioles up to 20cm. Flower solitary, erect. Petals obovate, 1–3 × 1–1.5cm, white or pink. Fruit ovoid, up to 6 × 3cm, reddish, borne on pedicel 2–5cm.

Bhutan: C – Ha, Thimphu and Bumthang districts, **N** – Upper Mo Chu district; **Sikkim.** Open mountain slopes and shady coniferous forests, 3100–4400m. May–June.

Rhizome used medicinally (16).

2. **P. sikkimense** Chatterjee & Mukerjee; *P. sikkimense* var. *major* Chatterjee & Mukerjee

Similar to *P. hexandrum* but leaves peltate, orbicular in outline, up to 25cm across, divided to middle into 6–8 oblong, acute lobes; flowers 2–3, ± pendulous on pedicels 3–7cm; petals elliptic c 2 × 1cm, pink or white.

Bhutan: C –Tongsa district (Phobsikha); **Sikkim; Chumbi.** In damp scrub or forest, 3350m. April–May.

Further studies are necessary on the generic limits within Podophyllaceae. *P. hexandrum* has recently been placed in a new monotypic genus *Sinopodophyllum* Ying, and *P. sikkimense* may belong to the genus *Dysosma* Woodson, as it is closely related to *D. aurantiocaule* (Handel-Mazzetti) Hu. We are grateful to Dr P.M. Dewick and Mr J. Shaw of Nottingham University for this information.

Family 46. LARDIZABALACEAE

by A.J.C. Grierson & D.G. Long

Monoecious, twining or erect shrubs. Leaves alternate, trifoliate, digitately compound or pinnate, exstipulate. Flowers in racemes, sometimes clustered, unisexual, actinomorphic. Sepals 6 in 2 valvate series, petaloid. Petals 6, nectary-like, minute or absent. Stamens 6, filaments free or connate, anthers free, opening by lengthwise slits; staminodes 6, or absent in female flowers. Carpels 3, superior, free, erect but soon divergent; pistillodes usually present in male flowers. Mature carpels fleshy, opening by dorsal suture; seeds ovoid, numerous.

1. Erect shrubs; leaves pinnate; flowers in elongate racemes ....**1. Decaisnea**
+ Twining shrubs; leaves trifoliate or digitately compound flowers in short, usually clustered racemes ....**2. Holboellia**

1. DECAISNEA Hook. f. & Thomson

Erect shrubs. Leaves unequally pinnate, rachis jointed at each pair of leaflets. Flowers in elongate racemes. Sepals lanceolate, acuminate. Petals absent. Male flowers with connate stamens. Female flowers with minute staminodes; carpels oblong. Fruiting carpels 1–3, spreading.

1. **D. insignis** (Griff.) Hook. f. & Thomson. Dz: *Lasey Sham.*

Shrub 3 – 6m. Leaves 45 – 60cm; petioles swollen at base when fresh, becoming constricted when dry; leaflets up to 8 pairs, opposite, ovate, 5 – 8 × 1 – 4cm, acuminate, base rounded or cuneate, glabrous, margins entire; petiolules up to 10mm. Racemes terminal or in axils of upper leaves, 30 – 45cm. Sepals greenish-yellow, 3 – 3.5cm. Male flowers with stamens c 10mm, filaments united in lower half. Female flowers with carpels c 10mm. Fruiting carpels oblong, 8 × 1.5 – 2cm, cylindric, recurved; seeds ovoid, compressed, c 10 × 7mm, brown, glossy.

Bhutan: **C** – Thimphu, Punakha and Mongar districts, **N** – Upper Mo Chu district (N of Tamji); **Sikkim.** Margins of Cool broad-leaved forests, 2000 – 2600m. May.

Fruit edible.

2. HOLBOELLIA Wall.

Twining shrubs. Leaves trifoliate or digitate with up to 9 leaflets. Racemes solitary or clustered, axillary, often in axils of fallen leaves. Flowers fragrant. Sepals oblong, elliptic or ovate, obtuse. Male flowers with 6 free stamens; pistillodes minute. Female flowers with 3 oblong-ellipsoid carpels; staminodes minute. Fruiting carpels 1 – 3, spreading.

1. H. latifolia Wall. Dz: *Enterobjay;* Sha: *Khrogchang Sey, Throckchangru;* Nep: *Gufla* (34). Fig. 26 j – m.

Leaflets ovate-elliptic, 4.5 – 13 × 1.5 – 7cm, acute or acuminate to a fine tip, base rounded or cuneate, glabrous, margins entire; petiolules up to 4cm. Racemes 3 – 12cm, bearing 3 – 7 flowers. Sepals green or purplish, elliptic or ovate, usually narrower in male flowers, 10 – 15 × 3 – 4.5(– 6)mm, broader in female flowers, 10 – 16 × 5 – 7mm. Male flowers with stamens 8 – 15mm; pistillodes linear, c 1.5mm. Female flowers with oblong-ellipsoid carpels 8 – 10mm; staminodes ovoid, 1 – 1.5mm. Fruiting carpels ellipsoid, 6 – 10 × 3 – 4cm, red; seeds ovoid, blackish, ± compressed, 8 × 6mm, glossy.

Bhutan: **S** – Chukka and Deothang districts, **C** – Ha to Tashigang districts, **N** – Upper Mo Chu district; **Sikkim.** Forest margins, especially Cool broad-leaved forests, 1880 – 2900m. April – June.

Fruit edible. Leaflets vary in number and size and this has been proposed as the basis for recognising two varieties. In var. **angustifolia** (Wall.) Hook. f. & Thomson they are more numerous and tend to be narrower than in var. **latifolia** where they are fewer and broader. There appears to be little significance in this and the suggestion (73) that the plant with narrow leaflets grows at lower elevations is quite incorrect.

Family 47. MENISPERMACEAE

by A.J.C. Grierson.

Twining, rarely erect, dioecious shrubs or trees. Leaves alternate, exstipulate, simple, entire, rarely shallowly toothed or lobed, petiolate, sometimes peltate, petiole usually attached at base of lamina. Inflorescence cymose, paniculate, fasciculate or capitate, axillary or basal on older wood. Flowers small, unisexual, actinomorphic, rarely asymmetrical. Male flowers: sepals in 2–4 series of 3 or sometimes 4–5, imbricate; petals mostly smaller than innermost sepals or absent; stamens 3–6, opposite petals, free or united. Female flowers: sepals and petals usually similar to those of male flowers, staminodes sometimes present, carpels 1–6, free, style short, simple or 2–3-branched, stigma terminal. Fruit drupaceous, scar of style subterminal or near base due to excentric growth; stone often stongly curved, horseshoe-shaped; endosperm ruminate.

1. Leaves elliptic or oblanceolate, narrowly rounded or cuneate at base........2
\+ Leaves ovate or deltoid, broadly rounded, truncate or cordate at base......4

2. Leaves pinnately veined ....**9. Pycnarrhena**
\+ Leaves prominently 3-veined at base 3

3. Erect shrubs ....**4. Cocculus**
\+ Twining shrubs ....**10. Haematocarpus**

4. Leaves peltate..5
\+ Leaves with petioles attached at base of lamina 9

5. Inflorescence umbellate, sometimes condensed to disc-like capitula **6. Stephania**
\+ Inflorescence paniculate, cymose or racemiform...............................6

6. Male inflorescence consisting of dense sub-corymbose cymes; female inflorescence racemiform, flowers clustered in axils of suborbicular bracts 0.5–1.5cm long and broad.. **7. Cissampelos**
\+ Inflorescence of both sexes paniculate or racemiform; bracts, if present, linear and no longer than 2mm 7

7. Petiole inserted 6–25mm from basal margin of lamina **8. Cyclea** *(C. barbata)*
\+ Petiole inserted 1–2mm from basal margin of lamina 8

8. Panicles narrow, racemiform; stone oblong, much compressed, surrounded by a notched wing ....**1. Aspidocarya**
\+ Panicles broad; stone suborbicular, somewhat compressed, surrounded by several rows of peg-like projections ....**8. Cyclea** *(C. bicristata)*

9. Leaves deeply cordate or hastate at base, basal lobes usually acute **2. Parabaena**
+ Leaves not or only shallowly cordate and then basal lobes always rounded 10

10. Stamens connate into a column; anthers borne on margin of peltate disc; carpel solitary 11
+ Stamens with filaments free or connate but anthers free; carpels 3 12

11. Inflorescences borne in leaf axils, male ones consisting of short-stalked subcorymbose cymes; female inflorescences racemiform, flowers clustered in axils of suborbicular bracts 0.5–1.5cm long and broad .... **7. Cissampelos**
+ Inflorescences usually borne on old wood, those of both sexes paniculate with narrow racemiform branches; female inflorescences with linear bracts c 1mm **8. Cyclea** *(C. bicristata)*

12. Inflorescences cymose .... **5. Pericampylus**
+ Inflorescences racemose or narrowly racemiform-paniculate 13

13. Filaments united; stone compressed, surrounded by a toothed marginal wing **1. Aspidocarya**
+ Filaments free; stone plano-convex, surface warted but with a smooth central rib .... **3. Tinospora**

1. ASPIDOCARYA Hook. f. & Thomson

Twiner. Leaves subpeltate or petioles basally attached to lamina. Inflorescence axillary, narrowly paniculate, racemiform. Male flowers: sepals (6–) 12 in 2–4 series, outer ones minute, innermost larger than others; petals 6; stamens united into a column and bearing 6 anthers around terminal disc. Female flowers: sepals and petals similar to those of male; staminodes 6; carpels 3, stigmas subcapitate or 3-fid. Fruit oblong, compressed, stylar scar subterminal, stone keeled on the back, margin surrounded by a notched wing and with a line of weak projections near margin on ventral surface.

FIG. 27. **Menispermaceae**. a–c, *Cyclea bicristata*: a, male flower; b, the same, sepals partly removed to show staminal column; c, female flower. d–k, *Stephania glabra*: d, peltate leaf; e, portion of shoot with inflorescence; f, outer sepal; g, inner sepal; h, petal; i, staminal column; j, ovary; k, fruit, flesh removed. l–q, *Cocculus laurifolius*: l, portion of shoot with leaf and female inflorescence; m, outer sepal; n, inner sepal; o, stamen with bifid petal clasping base of filament; p, female flower, perianth fallen; q, fruit, flesh removed. r–u, *Parabaena sagittata*: r, fruit, flesh removed; s, sepal; t, petal; u, staminal column. v, *Aspidocarya uvifera*: fruit, flesh removed. w–ab, *Cissampelos pareira*: w, portion of shoot with leaves and male inflorescences; x, male flower; y, portion of shoot with leaves, bracts and fruit; z, portion of female inflorescence; aa, female flower; ab, fruit, flesh removed. Scale: e × ⅓; d, l, w, y × ⅔; z × 1; k × 2½; v × 2⅔; q, r, ab × 4; f, g, i, p, t, u × 7; n × 8; a, b, h × 9; m × 10; c, j, o, s, x × 14; aa × 17. Drawn by M. Bates.

a
b
c
d
e
f
g
h
i
j
k
l
m
n
o
p
q
r
s
t
u
v
w
x
y
z
aa
ab
M.B.

1. A. uvifera Hook. f. & Thomson. Fig. 27 v.

Leaves ovate, 9–17 × 6–14cm, acuminate, base truncate or cordate, entire, glabrous above, sparsely pilose beneath; petioles 5–12cm, attached to base of lamina or inserted 1–2mm from its basal margin. Inflorescence 10–45cm, densely brownish pubescent. Male flowers: outer sepals lanceolate or oblong, c 2.5–3 × 0.5mm, innermost obovate, c 3.5–4 × 2mm; petals obovate, cuneate, 2 × 1–1.25mm, margins thickened near base, involute; staminal column c 2mm. Fruiting inflorescence 12–22cm; drupes oblong-ovoid c 2 × 1.5cm, stones c 1.5 × 0.8cm.

Bhutan: S – Gaylegphug district (Karai Khola and below Shamkhara), **C** – Punakha and Tongsa districts; **Sikkim.** Subtropical and Warm broad-leaved forests, 500–1700m. March–May.

Fruit edible (34).

2. PARABAENA Miers

Twiner. Leaves ovate or oblong, cordate or hastate at base. Inflorescences axillary, cymes dichotomously branched. Male flowers: sepals 6, oblong-ovate; petals 6, small, obovate-cuneate; stamens 6, filaments connate into a column, anthers free, forming a densely packed head. Female flowers: sepals and petals as in male flowers, staminodes 6, oblong; carpels 3 erect, style recurved. Fruit ovoid, ± plano-convex, rostrate; stone covered with short thick spines on the back, ventrally concave, imperforate.

1. P. sagittata Hook f. & Thomson. Nep: *Karpati Lahara* (34). Fig. 27 r–u.

Leaves 12–18 × 6–10cm, acuminate, base ± deeply cordate with acute basal lobes, margins subentire or shallowly toothed, weakly hirsute beneath; petioles 3–12cm. Cymes 3–10cm, flowers cream-coloured. Male flowers: sepals c 2 × 0.75mm, pubescent; petals c 1 × 0.5mm, scarious at margins; staminal column c 1mm. Fruit c 7 × 5mm.

Bhutan: S – Gaylegphug and Sarbhang districts, **C** – Mongar district; **Sikkim.** Subtropical and Warm broad-leaved forests, 300–1200m. May–June.

3. TINOSPORA Miers

Twiners. Leaves ovate, cordate, palmately 5–7-veined at base. Inflorescence borne on leafless stems, racemiform, male more slender than female, flowers in clusters. Male flowers: sepals 6 in two whorls; petals 6, obovate, clawed at base, margins inflexed, each petal enfolding a stamen. Perianth of female flowers similar to male; staminodes 6, shorter than petals; carpels 3, ellipsoid, erect, style short, thick, stigma peltate. Drupes ovoid or ellipsoid; stone plano-convex, ridged dorsally, imperforate but deeply pitted ventrally.

1. T. cordifolia (Willdenow) Hook. f. & Thomson. Nep: *Gurjo* (34).

Leaves broadly ovate, 6–15 × 6–13cm, abruptly acuminate, base cordate, usually with glandular domatia in vein axils on lower surface, otherwise glabrous. Inflorescences 5–12cm, males in few-flowered clusters, female flowers usually borne singly along axis. Male flowers on pedicels 3–4mm; outer sepals ovate, 1–1.5mm, inner elliptic, 3–4 × 1.5–2mm; petals obovate, c 2.5 × 1mm; stamens club-shaped, 3mm. Female flowers with sepals and petals similar to male; staminodes linear, c 1.5mm; carpels ellipsoid, 1.75mm. Drupes red; stone broadly ellipsoid, 6–7mm, rounded at both ends, papillose.

Sikkim: Darjeeling Terai (Punkabari). Subtropical and terai forests, 150m. February.

Foliage eaten by elephants (34).

2. T. sinensis (Loureiro) Merrill; *T. malabarica* (Lamarck) Hook. f. & Thomson

Similar to *T. cordifolia* but leaves pubescent, especially beneath, sometimes lobed or sinuate; stone ± keeled at apex, pointed at base, surface rough and warted.

Sikkim: Mungpoo (34). Subtropical forests. February–May.

4. COCCULUS DC.

Erect shrub or tree. Leaves elliptic, entire. Inflorescence axillary, cymose or paniculate. Male flowers: sepals 6 in 2 series, the outer smaller; petals 6, bifid at apex, base minutely auriculate, clasping filament; stamens 6, free. Female flowers: sepals and petals as in male but caducous; staminodes 6, minute; carpels 3–6, styles cylindric. Fruit subglobose, somewhat compressed, stylar scar near base; stone tightly coiled, deeply pitted on either side, ridged around the edge.

1. C. laurifolius DC. Fig. 27 l–q.

Shrub or tree 2–7m. Leaves elliptic or oblanceolate, 8–12 × 2–3.5cm, acuminate, base cuneate, prominently 3-veined, glabrous. Inflorescences 2–6cm, female usually shorter and fewer-flowered. Male flowers: sepals greenish-yellow, elliptic, outer ones c 1 × 0.5mm, inner c 2 × 1mm; petals c 0.5mm, bifid to middle; stamens c 1.25mm. Female flowers: carpels elliptic-ovoid, c 1mm, style ± reflexed. Fruit 4–5mm diameter.

Bhutan: C –Tongsa district (near Yoormu). Warm riverine broad-leaved forest, 1250m. May.

5. PERICAMPYLUS Miers

Twiner. Leaves not peltate. Inflorescence axillary, cymose-paniculate. Male flowers: sepals 6–9, outermost minute, innermost 3 larger and broader than others; petals 6, smaller than innermost sepals; stamens 6, free. Female flowers: sepals and petals as in male; staminodes 6; carpels 3, style bifid, recurved. Fruit suborbicular, somewhat compressed; stone concave, imperforate, bearing a double row of projections around the edge and a double row on either side near margin.

1. P. glaucus (Lamarck) Merrill; *P. incanus* (Colebrooke) Hook. f. & Thomson, *P. aduncus* Miers. Nep: *Tschethangru* (38), *Pipal-pati Lahara* (34).

Leaves broadly ovate, 3.5–10cm long and broad, acute, base rounded, truncate or cordate, margins entire or slightly undulate, finely pilose above, pubescent beneath. Inflorescences 2–4cm, brownish pubescent. Male flowers: sepals greenish yellow, pubescent, outermost lanceolate, c 0.75mm, innermost obovate, c 1.5 × 1mm; petals oblanceolate, c 1 × 0.5mm; stamens c 1.5mm. Female flowers: staminodes linear, c 0.75mm, ± adnate to petals; carpels ovoid, c 0.5mm, style almost as long. Fruit purple or blackish, 5–7mm diameter.

Bhutan: S–Chukka and Gaylegphug districts, **C**–Punakha and Tongsa districts; **Sikkim.** Warm broad-leaved forests, 980–1600m. March–May.

6. STEPHANIA Loureiro

Twiners, sometimes tuberous rooted. Leaves peltate, ovate or deltoid, palmately veined. Inflorescences umbellate or subcapitate, axillary or borne on old wood. Male flowers: sepals usually 6 in 2 series; petals usually 3; stamens connate into column, anthers borne at margin of peltate head, dehiscing transversely. Female flowers sometimes asymmetrical: sepals and petals similar to those of male flowers but sometimes fewer; carpel 1, style short, stigma stellately divided. Fruit obovoid, somewhat compressed, stylar scar near base; stone horseshoe-shaped, usually bearing 4 rows of projections around margin, perforate or imperforate.

1. Leaves broadly ovate or suborbicular, often broadly and obscurely toothed **Species 1 & 2**
+ Leaves narrowly deltoid, always entire .... **Species 3 & 4**

1. S. glabra (Roxb.) Miers; *S. rotunda* sensu F.B.I. p.p. non Loureiro. Nep: *Tamarke Pailo, Tamarke Lahara, Tambarki* (34), *Nimi Lahara* (34), *Barkuli Lahara* (34). Fig. 27 d–k.

Rootstock often tuberous. Leaves broadly ovate or suborbicular, 5–15 × 3–15cm, acute or subacute, often mucronate, base rounded, glabrous; petioles 3–11cm, inserted 2–3cm from base of lamina. Inflorescence usually axillary; peduncles 2–7cm, umbels slender in male, generally stouter in female, 6–12-rayed; pedicels up to 4mm. Male flowers: sepals oblong or obovate, c 2.5 × 0.75–1mm; petals obovate-cuneate, c 1.5 × 1mm, with two marginal glands near middle; staminal column c 2mm. Female flowers: sepals and petals similar to those of male flowers but often fewer; carpel ovoid, c 1mm. Fruit suborbicular, 5–8mm diameter, reddish at maturity; stone with 2 rows of (c 18) peg-like processes around the edge and 1 row on either side near the margin, perforate.

Bhutan: S – Phuntsholing, Chukka and Deothang districts, **C** –Tongsa, Mongar and Tashigang districts; **Sikkim.** Climbing on shrubs, often near streams, in Subtropical and Warm broad-leaved forests, 600–1600m. May–June.

The stem yields a fibre that is used to make fishing-lines (34).

2. S. glandulifera Miers; *S. rotunda* sensu F.B.I. p.p. non Loureiro. Dz: *Chechu Robji;* Sha: *Zala Mathang Ru;* Nep: *Tambarki* (34).

Similar to *S. glabra* but umbels often borne on mature leafless branches; bark thick, warted; sepals in male flowers glandular-papillose near apex; petals not glandular; female flowers apparently unknown; fruit similar to *S. glabra.*

Bhutan: S – Chukka district (Chukka Dzong), **C** – Punakha district (Mishichen and Ngawang); **Sikkim.** Warm broad-leaved forests, 1700–1900m. February–May.

3. S. elegans Hook. f. & Thomson

Slender twiner. Leaves deltoid, 5–9 × 3–5cm, acuminate, base truncate or somewhat rounded, margins entire, glabrous; petioles 2–4cm. Umbels usually axillary, 3–4-rayed, peduncles 3–6cm, pedicels short (c 1mm). Male flowers: sepals obovate, c 2 × 1mm; petals broadly obovate, c 0.75 × 0.5mm, fleshy; staminal column c 1mm. Female flowers similar. Fruit reddish, obovoid, c 5mm diameter; stone similar to that of *S. glabra* but projections (c 10) ridge-like and less prominent.

Bhutan: S – Chukka district, **C** –Thimphu district, **N** – Upper Mo Chu district; **Sikkim.** Warm broad-leaved forests, 1220–2100m. July.

4. S. japonica (Thunberg) Miers; *S. hernandiifolia* (Willdenow) Walpers. Nep: *Charcharey Lahara* (38), *Tambarki* (34).

Similar to *S. elegans* but leaves more rounded at base, pale and sparsely pubescent beneath; umbels 6–7-rayed, peduncles 2–6cm; male flowers ± sessile in dense capitate clusters, sepals 6–8, oblanceolate, c 1.5 × 0.5mm, petals 3–4, obovate, c 1 × 0.75mm; female flowers similar but sepals often fewer; fruit suborbicular, c 5mm diameter, stone bearing a double row of 10–12 peg-like projections around edge with as many marginal ridges on either side.

Bhutan: S – Samchi, Gaylegphug and Deothang districts; **Sikkim.** Subtropical forests, 300–900m. June.

The above description refers to var. **discolor** (Miquel) Forman.

7. CISSAMPELOS L.

Twiners. Leaves peltate or lamina basally attached to petiole. Male inflorescence a dense axillary cyme with slender peduncle. Male flowers: sepals 4, obovate, free; petals connate into a shallow cup; stamens united into a column bearing 4 anthers at margin of a peltate disc. Female inflorescences racemiform; flowers borne in clusters in axils of accrescent bracts. Female flowers asymmetric; sepal 1; petal 1, borne in axil of sepal; carpel 1, style 3-lobed. Fruit suborbicular, somewhat compressed, stone bearing 2 rows of weak transverse ridges around the edge, imperforate.

1. C. pareira L. Sha: *Jingru;* Nep: *Thamarke, Batulpati* (34). Fig. 27 w–ab.

Leaves broadly ovate, 2.5–8cm long and broad, obtuse or acute, mucronate, base cordate, rounded or truncate, entire, pubescent; petioles 2–9cm, inserted at base of lamina or up to 10 (–18)mm from margin. Male flowers: greenish yellow in subcorymbose cymes on slender, pubescent peduncles 2–4cm; sepals oblanceolate, c 1.25 × 0.5mm; petals connate into a shallow cup c 0.25mm; staminal column c 0.5mm. Female flowers greenish in false racemes up to 18cm, each fascicle of flowers in axil of a suborbicular bract up to 1.5cm; sepal obovate, c 1.25 × 0.75mm; petal broadly obovate, c 0.75mm long and broad; carpel narrowly ovoid c 1mm, hispidulous. Fruit reddish, c 5mm diameter.

Bhutan: S – Chukka to Deothang districts, common, **C** – Punakha to Tashigang districts; **Sikkim.** Subtropical and Warm broad-leaved forests, 500–1700m. March–May.

This description relates to var. **hirsuta** (DC.) Forman. Stems yield a strong, useful fibre (34).

8. CYCLEA Wight

Twiner. Leaves peltate or not, entire. Inflorescences axillary or borne on old wood, paniculate. Male flowers: sepals 4–5, usually connate; petals 4–5, connate or sometimes absent; stamens connate into a column and bearing 4–5 anthers at margin of a peltate disc. Female flowers: sepals and petals usually 2 (sometimes petals absent), connate; carpel solitary, style short, 3-lobed. Fruit obovate or suborbicular, somewhat compressed; stone bearing a double row of peg-like projections around the edge and 2 similar marginal rows on either side.

1. C. bicristata (Griffith) Diels; *Lophophyllum bicristatum* Griffith. Fig. 27 a–c.

Leaves subpeltate, petiole inserted 1–2mm from basal margin of lamina, or petiole basally attached to lamina, ovate, (7–)10–18 × (5–)8–14cm, acute, base cordate, pubescent beneath; petioles 5–7cm. Inflorescence usually borne on old wood. Male flowers: sepals connate into a cup-shaped calyx c 1.5mm, ± fleshy; petals absent; staminal column c 1.5mm. Female flowers: sepals 2 in an opposite pair, c 1mm, free, pouched at base; petals absent; carpel ovoid ± as long as sepals. Fruit c 4mm diameter; stone imperforate with c 15 peg-like projections in each row, marginal ones ± hooked.

Bhutan: S – Samchi (Buduni and Deo Pani Khola) and Deothang districts (near Deothang), **C** – Punakha district (Rinchu); **Sikkim.** Subtropical forests, 350–1500m. March.

2. C. barbata Miers; *C. wallichii* Diels

Similar to *C. bicristata* but ± hispid throughout; leaves ovate, peltate, 6–17 × 4–12cm, acuminate, base truncate or broadly rounded; petioles 3–6cm, inserted 6–25mm from base of lamina; panicles 7–20cm, axillary or borne on old wood, flowers in dense ovoid clusters; male flowers greenish, sepals 2–3mm, united to middle, ± hispid, petals obovate, c 1.5mm, connate near base, staminal column c 3mm; female flowers with 2 obovate sepals c 0.5mm, petals 2, reniform, 0.5 × 0.75–1mm, opposite to and broader than sepals, carpel ovoid, c 1mm, pilose, style branches c 0.5mm; fruit c 5mm diameter, stone with projections less prominent than in *C. bicristata.*

Sikkim: Tista valley. Subtropical forests.

9. PYCNARRHENA Miers.

Twiner. Leaves elliptic or oblanceolate, pinnately veined; petioles swollen at base and apex. Inflorescences axillary, fasciculate. Male flowers: sepals 6–9, imbricate, outer ones minute, the inner larger and concave; petals 3–5, smaller than inner sepals; stamens 6–10, filaments ± united, anthers densely clustered. Female flowers: sepals and petals as in male, carpels 3–6 with recurved styles. Fruit ellipsoid, stone thin-walled, smooth.

1. P. pleniflora Hook. f. & Thomson

Leaves elliptic or narrowly obovate, 10–18 × 4–7cm, shortly and bluntly acuminate, base cuneate or rounded, midrib pubescent at first, otherwise glabrous; petioles 1–2cm, ± pubescent. Male flowers: peduncles 3–4mm; sepals c 2mm, outer ciliate, inner glabrous; anthers 6–10 in a cluster c 1mm diameter, ± sessile. Female flowers unknown. Fruit (immature) ellipsoid, c 1cm, on peduncle 2–4mm.

Bhutan: S – Gaylegphug district (Birti and Gaylegphug). Subtropical forests, 270–900m. April.

10. HAEMATOCARPUS Miers

Twiner. Leaves elliptic, 3-veined at base. Inflorescence axillary, terminal or on old wood, paniculate or racemiform. Male flowers: sepals 12–15 in whorls of 3, outer minute, inner larger; petals 6, inner 3 minutely auriculate at base and clasping the adjacent stamen; stamens 6, free, connective enlarged and projecting inwards. Female flowers: sepals and petals as in male flowers; staminodes 6, minute; carpels 6, styles reflexed. Fruit ellipsoid; stone smooth with inner longitudinal septum.

1. H. validus (Miers) Forman; *H. comptus* Miers, *H. thomsonii* Miers, *H. incusus* Miers

Leaves elliptic, 7–16 × 3–8.5cm, acuminate, base rounded, prominently 3-veined especially beneath, glabrous, petioles 1–3.5cm. Inflorescences up to 35cm. Male flowers: pedicels 2–4mm; outer sepals minute, inner elliptic, c 3mm, with dark red streaks; petals elliptic, 1.5–2mm; stamens c 1mm. Fruit red, c 4.5 × 2.5cm.

Sikkim: locality unknown.

Family 48. NYMPHAEACEAE

by E. Campbell

Perennial aquatic herbs with stout rhizomes. Leaves alternate, peltate, floating or emersed, palmately veined, minutely stipulate. Flowers solitary on long simple axillary pedicels, actinomorphic, bisexual, emersed or at water surface, often showy. Sepals 2–6, free, sometimes petal-like. Petals 3–many, free. Stamens many, sometimes petaloid. Ovaries superior or semi-inferior; carpels numerous, free or united, embedded in large receptacle; stigmas sessile on carpels or borne on short style; ovules 1–2 or many per carpel. Fruit a 1-seeded indehiscent nut or a many-seeded spongy berry. Seeds sometimes arillate.

1. Leaves not cordate; sepals and petals similar; stamens not petaloid; carpels free but embedded in receptacle, with 1–2 ovules, short style and capitate stigma; fruit a 1-seeded nut .... **1. Nelumbo**
\+ Leaves deeply cordate; sepals and petals dissimilar; stamens petaloid; carpels united within receptacle, with many ovules, stigma sessile on carpel but bearing inwardly curving appendage; fruit a many-seeded berry .... **2. Nymphaea**

1. NELUMBO Adanson

Leaves large, peltate, floating or emersed. Flowers fragrant, on stout pedicel, usually borne above water surface. Sepals 2–5, petal-like. Petals numerous. Stamens numerous, not petaloid, each with a fleshy terminal appendage. Ovaries superior; carpels 20–30, free, sunk into a large flat-topped receptacle; style short, exserted, stigma capitate; ovules 1–2 per carpel. Fruits 1-seeded, nut-like, indehiscent, set in large pits in enlarged obconic pithy receptacle. Seeds without aril.

1. N. nucifera Gaertner; *Nelumbium speciosum* Willdenow. Eng: *Sacred Bean, Egyptian Lotus.* Fig. 31 q–s.

Leaves orbicular, 30–90cm across, sometimes shallowly bilobed, margins entire, wavy; petioles 60–180cm, often minutely prickly. Pedicels up to 180cm, minutely prickly. Flowers rose-pink or sometimes white, 10–25cm across. Sepals and petals elliptic-obovate, 8–11 × 3.5–5cm, obutse or subacute. Receptacle 2–3cm across, 9–11cm in fruit. Ripe carpels ellipsoid, c 1.5 × 1cm.

Bhutan: C – Punakha district (Punakha). In ponds and backwaters of rivers, 1400m. May – September.

Cultivated as an ornamental and as a sacred plant. Tubers, leaves and fruits edible.

2. NYMPHAEA L.

Similar to *Nelumbo* but leaves always floating, deeply cordate and shallowly peltate, margins dentate; flowers borne at water surface; sepals and petals dissimilar; stamens petaloid, without terminal appendage; ovary semi-inferior, composed of numerous radiating united carpels sunk in a flat-topped receptacle, upper surface of each carpel with a stigmatic ridge and inwardly-curved linear appendage; ovules many in each carpel; fruit a spongy many-seeded berry; seeds arillate.

1. N. lotus (L.) Willdenow

Leaves ovate to orbicular, 10 – 30 × 10 – 30cm, obtuse, base deeply cordate, peltate, margins undulate, sharply dentate; petioles 60 – 160cm, usually pubescent. Flowers slightly fragrant, borne on pubescent pedicels up to 160cm. Sepals 4, ovate-elliptic, 4 – 8 × 1 – 2.5cm, subacute, green outside, pink or white within. Petals linear-elliptic, 3 – 7.5 × 1.5 – 2.5cm, white or pink. Receptacle 2 – 3cm across, 6 – 9cm across in fruit. Seeds spherical, c 1.5mm.

Sikkim: Jalpaiguri terai. August.

A second, unidentified, species of *Nymphaea* is cultivated at Punakha Dzong, with entire leaves and red, pink or white flowers.

Family 49. SAURURACEAE

by A.J.C. Grierson & D.G. Long

Perennial stoloniferous aromatic herbs. Leaves alternate, palmately veined, entire, stipulate. Flowers in dense terminal and leaf-opposed spikes, subtended by a whorl of 4 showy bracts; actinomorphic, bisexual. Perianth absent. Stamens usually 3 – 6, filaments adnate to base of ovary. Ovary superior, 1-locular, of 3 partially connate carpels; ovules several, parietal; styles 3, recurved. Fruit a capsule, somewhat fleshy, dehiscing at apex.

1. HOUTTUYNIA Thunberg

Description as for Saururaceae

1. H. cordata Thunberg. Sha: *Mombering, Nombaring.* Fig. 28 a,b.

Stems 15 – 45cm. Leaves broadly ovate, 3.5 – 9 × 3 – 8cm, acute or shortly acuminate, base cordate; petioles 2–4cm; stipules oblong, 1.5–2.5cm, adnate to petiole in lower half. Flower spikes 1–2.5cm; basal bracts white, oblong or obovate, 1–1.5 × 0.7–1cm, obtuse.

Bhutan: S–Chukka district, **C**–Punakha, Tongsa, Mongar and Tashigang districts, **N**–Upper Mo Chu and Upper Kulong Chu districts; **Sikkim.** Weed of cultivation and on roadsides in Cool broad-leaved forest, 1000–2400m. June–August.

Leaves eaten as a green vegetable.

Family 50. PIPERACEAE

by D.G. Long

Small herbs or shrubs, often climbing, nodes often swollen. Leaves simple, alternate, opposite or whorled, entire, usually palmately veined at base, rarely pinnately veined; stipules present or absent, often adnate to petiole, caducous. Flowers in terminal or leaf-opposed spikes, unisexual or bisexual, minute, much reduced, borne in axils of peltate or basifixed bracts. Perianth absent. Male flowers with 2–4 stamens, anther-cells 2, sometimes becoming confluent by dehiscence, sessile or on short filament. Female flowers with superior 1-celled ovary, stigma sessile, brush-like or with 3–5, simple, recurved branches, ovule 1. Fruit a drupe, free or sometimes becoming coherent in a fleshy spike.

FIG. 28. **Saururaceae, Aristolochiaceae, Chloranthaceae** and **Piperaceae. Saururaceae**. a & b, *Houttuynia cordata*: a, upper portion of stem with leaves and flowers; b, ovary with adnate stamens. **Aristolochiaceae**. c–f, *Asarum himalaicum*: c, portion of rhizome with leaf and flower; d, flower, one perianth lobe removed to show upper part of ovary with adnate stamens; e, stamen; f, style. g & h, *Aristolochia griffithii*: g, portion of shoot with leaves and flowers; h, ovary with adnate bundles of anthers. i, *Aristolochia tagala*: dehisced capsule. **Chloranthaceae**. j–m, *Chloranthus elatior*: j, upper portion of stem with leaves and inflorescence; k, node of inflorescence with two opposite bracts each bearing a staminal scale and ovary; l, ovary with staminal scale partially removed; m, staminal scale from within. **Piperaceae**. n–q, *Peperomia tetraphylla*: n, portion of rhizome with erect stem bearing leaves and terminal spike; o, portion of spike showing flowers; p, single flower showing ovary with a stamen on either side protruding from peltate scale; q, diagrammatic section of part of spike to show side view of flower. r–t, *Piper attenuatum*: r, portion of shoot with leaf and infructescence; s, portion of male spike showing clusters of stamens emerging from ends of adnate bracts; t, portion of female spike bearing an ovary at the end of each adnate bract. u–ab, *Piper longum*: u, portion of shoot bearing leaf and male spike; v, portion of male spike; w, diagrammatic section of male spike to show arrangement of peltate scales around rachis with stamens at their base; x, immature stamen; y, mature stamen with dehisced anther; z, portion of shoot bearing leaf and female spike; aa, portion of female spike showing peltate scales with styles emerging between them; ab, ovary. Scale: g, j × ⅓; a, c, i, r, u, z × ⅔; d, n × 1; aa × 3; f, h × 3½; e × 4½; k, l, m, s, t, v, ab × 7; b, w × 8; o × 10; x, y × 20; p, q × 30. Drawn by M. Bates.

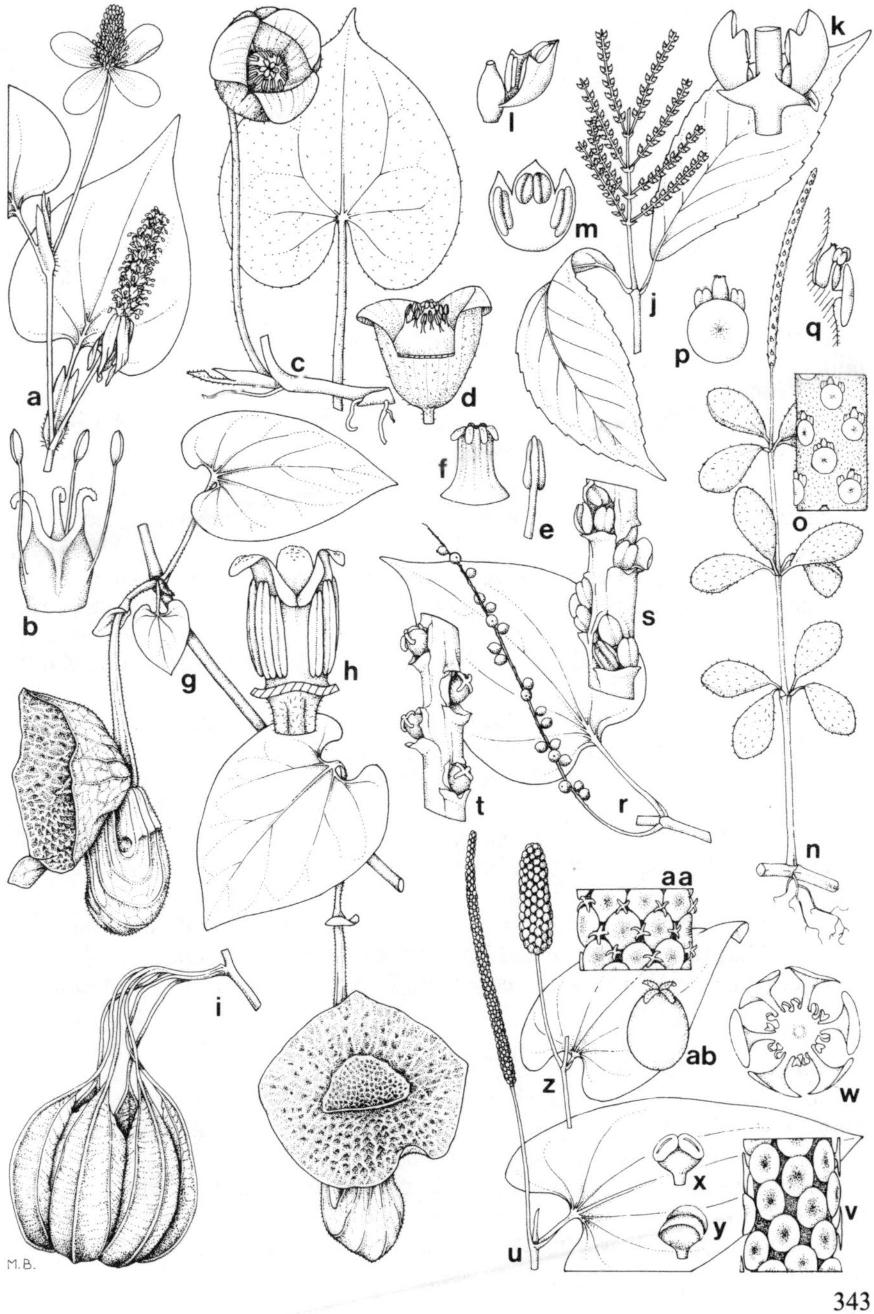
a
b
c
d
e
f
g
h
i
j
k
l
m
n
o
p
q
r
s
t
u
v
w
x
y
z
aa
ab
M.B.

1. Small herbs; leaves alternate, opposite or whorled; flowers bisexual; stigma 1, brush-like .... **1. Peperomia**

\+ Free-standing or scandent shrubs; leaves alternate, flowers usually unisexual; stigmas 3–5, filiform .... **2. Piper**

1. PEPEROMIA Ruiz & Pavon

Small succulent herbs. Leaves alternate, opposite or whorled, palmately 3–5-veined from base or pinnately veined, pellucid-dotted, exstipulate. Flowers bisexual, in axil of peltate bract. Stamens 2, very short, anther cells horizontal, reniform, placed side-by-side. Ovary minute with 1 brush-like stigma. Drupes minute, sessile.

1. Leaves alternate, cordate at base .... **3. P. pellucida**

\+ Leaves opposite or whorled, rounded, cuneate or attenuate at base 2

2. Leaves broadly obovate, broadly elliptic or suborbicular, often rounded at base, equally but faintly 3-veined from base; petioles 1–2mm; spikes finely pubescent **1. P. tetraphylla**

\+ Leaves oblanceolate or narrowly obovate, attenuate at base, pinnately veined with distinct midrib; petioles 2–5mm; spikes glabrous ....**2. P. heyneana**

1. P. tetraphylla (Forster) Hooker & Arnott; *P. reflexa* (L.f.) Dietrich non Kunth. Fig. 28 n–q.

Bushy succulent herb 6–18(–25cm), much-branched at base and often rooting at nodes, stems ribbed, glabrous or sparsely pubescent. Leaves opposite or in whorls of 4, fleshy, obovate, broadly elliptic or suborbicular, 7–12 × 6–10mm, obtuse, base rounded or broadly cuneate, 3-veined from base, veins pale when fresh, almost invisible when dry, glabrous or sparsely pubescent; petioles 1–2mm, pubescent. Spikes terminal, erect, 15–40 × 1.5mm, finely pubescent. Drupes ellipsoid, c 1mm, tapering to apex.

Bhutan: S – Samchi, Chukka, Gaylegphug and Deothang districts, **C** – Punakha, Tongsa, Mongar and Tashigang districts; **Sikkim.** On mossy boulders and tree trunks in Warm broad-leaved and Evergreen oak forests, 1150–2100m. February–July.

2. P. heyneana Miquel; *P. kurzii* C. DC.

Similar to *P. tetraphylla* but leaves more thinly fleshy, oblanceolate or narrowly obovate, 8–18 × 3–9mm, base attenuate, pinnately veined with distinct midrib; petioles 2–5mm, glabrous; spikes glabrous.

Bhutan: S – Chukka, Sarbhang and Deothang districts, **C** – Punakha district (Tinlegang), **N** – Upper Mo Chu district (Kencho); **Sikkim.** Epiphytic and on mossy rocks in Subtropical, Warm broad-leaved and Evergreen oak forests, 1030–2200m. May–August.

3. P. pellucida (L.) Kunth; *P. exigua* (Blume) Miquel

Succulent creeping or erect herb 15–40cm, glabrous throughout. Leaves thin, broadly ovate, 1.5–3 × 1–3cm, acute, base cordate, palmately 5-veined from base. Spikes 2.5–5cm, terminal and leaf-opposed. Drupes c 0.5mm.

Bhutan: S – Phuntsholing district (Phuntsholing) and Gaylegphug district (Tatapani); **Sikkim.** On wet rocks and stones in Subtropical forests, 500–1600m. May–August.

Native of tropical America, now a pantropical weed.

2. PIPER L.

Free-standing or climbing shrubs; stems with swollen nodes, and when climbing with prostrate, creeping stems and spreading aerial branches. Leaves alternate, usually palmately, sometimes pinnately veined; stipules present or not. Flowers unisexual, each in axil of peltate or basifixed bract. Male flowers with 2–4 stamens, anther cells placed side-by-side or end-to-end, sometimes confluent. Female flowers with ovoid ovary bearing 3–5 filiform stigmas. Drupes ovoid or globose, sessile or shortly stalked.

Records of *P. hymenophyllum* Miquel and *P. pedicellosum* C. DC. from Bhutan (117) may be erroneous as these species are unknown elsewhere in N India and the Himalaya.

The following *Piper* species have been described from Sikkim and are probably synonyms of other species, but it has not been possible to examine the type specimens: *P. calvilimbum* C. DC., *P. filipedunculum* C. DC., *P. isopleurum* C. DC., *P. laxivenum* C. DC., *P. mungpooanum* C. DC., *P. ovatistigmum* C. DC., *P. sikkimense* C. DC., *P. sonadense* C. DC. and *P. tenuibracteatum* C. DC.

1. Bracts of male and female spikes basifixed, with decurrent wings, never peltate or medifixed 2
+ Bracts of male and female spikes peltate, stalked or sometimes sessile and medifixed 3

2. Leaves membranous, broadly ovate, base truncate or subcordate **1. P. attenuatum**
+ Leaves subcoriaceous, ovate, base rounded .... **2. P. rhytidocarpum**

3. Female flowering and fruiting spikes borne on short peduncles 2–8mm (rarely to 12mm in *P. sylvaticum*); male spikes with anther cells horizontal, reniform, placed side-by-side with low lamella between 4
+ Female flowering and fruiting spikes borne on longer peduncles 10–40mm (shorter in *P. khasianum* but then spikes interrupted); male spikes with anther cells inclined to erect, ellipsoid, placed end-to-end on conical connective... 6

4. Older stems deeply striate, not warted; leaves membranous, finely reticulate and gland-dotted beneath, especially on veins, veins puberulous beneath, not very prominent; petioles 10–25mm; fruiting spikes cylindric, 1.5–3cm **5. P. sylvaticum**
\+ Older stems weakly striate, often warted; leaves subcoriaceous, not reticulate or glandular beneath, veins glabrous, prominent beneath; petioles 3–14mm; fruiting spikes globose or shortly cylindric, 0.6–1.5cm 5

5. Leaves on aerial shoots with base almost symmetric, cuneate or rounded but shortly decurrent on petiole (leaves on creeping stems cordate); axillary buds numerous, persistent and conspicuous; petioles 7–14mm; fruiting spikes globose or shortly cylindric ....**3. P. mullesua**
\+ Leaves on aerial shoots with base slightly oblique, rounded and shallowly cordate; axillary buds few, minute and inconspicuous; petioles 3–10mm; fruiting spikes shortly cylindric, rarely subglobose ....**4. P. peepuloides**

6. Leaves deeply and symmetrically cordate, 7-veined from base without additional lateral veins above; veins minutely puberulous beneath ....**6. P. longum**
\+ Leaf base cuneate, rounded or obliquely cordate, some veins arising suprabasally (in *P. hamiltonii* veins all basal but leaf base cuneate or rounded); veins glabrous (or sparsely puberulous in *P. betleoides* and *P. chuvya)* 7

7. Leaf base strongly asymmetric, usually cordate on one or both sides (sometimes rounded-asymmetric in *P. pedicellatum,* but then male bracts long-stalked and female spikes not interrupted)...8
\+ Leaf base asymmetric, usually cuneate, never cordate (or rounded and symmetric in *P. hamiltonii;* sometimes rounded-asymmetric in *P. suipigua* but then male bracts subsessile and female spikes interrupted) 10

8. Free-standing shrub; leaves weakly cordate or rounded on one side; exstipulate; petioles 4–9mm; male spikes 10–22cm; male and female peduncles 1–2cm; bracts long-stalked ....**7. P. pedicellatum**
\+ Climbing shrubs; leaves strongly cordate on one or both sides; stipules 1–1.4cm, adnate to petiole; petioles 7–20mm; male spikes 4.5–8cm on peduncles 2–3.5cm; female peduncles 1.1–4cm; bracts subsessile 9

9. Leaves membranous, ovate or ovate-lanceolate; petioles 14–20mm; male spikes 4.5–6cm on peduncles 2.5–3.5cm; female spikes 2.5–3.5cm, not interrupted in fruit, on peduncles 2–4cm ....**8. P. betleoides**
\+ Leaves thinly coriaceous, ovate or usually broadly ovate; petioles 7–15mm; male spikes 7–8cm on peduncles 2–2.5cm; female spikes 9–10cm, interrupted in fruit, on peduncles 1.1–1.5cm ....**9. P. chuvya**

10. Leaves elliptic, with short blunt point..........................**12. P. hamiltonii**
\+ Leaves ovate to lanceolate, acuminate 11

11. Leaves thinly coriaceous, ovate, 8–15cm, shortly acuminate; petioles 8–14mm; fruiting spikes 9–13cm; drupes ovoid, 5–6mm, sessile........**10. P. suipigua**
\+ Leaves membranous, narrowly ovate to lanceolate, 6–10cm, long-acuminate; petioles 4–8mm; fruiting spikes 2–3cm; drupes globose, 3–4mm diameter on short stout stalk 0.5–1mm..**11. P. khasianum**

1. P. attenuatum Miquel. Nep: *Chabo* (117). Fig. 28 r–t.

Climbing shrub, older stems distinctly striate when dry. Leaves membranous, green or yellow-green when dry, broadly ovate, 7–9.5 × 4–7cm, shortly acuminate, base truncate or weakly cordate, 7–9-veined in basal 7–10mm, with no main veins arising above, glabrous or sparsely pubescent beneath; petioles slender, 2–6cm, sometimes pubescent; stipules narrow, adnate to petiole, 5–8mm; axillary buds 5–8mm. Flowering spikes slender, 6–12cm, glabrous, on short peduncles 4–15mm. Male spikes with oblong, basifixed bracts with wings decurrent on rachis; stamens usually 3, anther cells horizontal, end-to-end, confluent, almost sessile. Female spikes with scales as in males, ovary globose, surrounded by rim composed of bracteoles and decurrent wings of bracts, stigmas 4. Fruiting spike interrupted, up to 12cm; drupes globose, c 4mm, sessile, loosely aggregated.

Bhutan: S – Gaylegphug district (Gaylegphug) and Deothang district (Chenari and Deothang); **Sikkim:** terai, Rungit Valley, etc. Subtropical forests, 200–460m. August–October.

2. P. rhytidocarpum Hook. f. Nep: *Chabo, Tsabo.*

Similar to *P. attenuatum* but stems almost smooth when dry; leaves more coriaceous, ovate, 9–12 × 5–7cm, more finely acuminate, base rounded, 5-veined in basal 2–5mm, with 2 lateral veins arising 2–4cm above base, glabrous; petioles 1.5–3cm; axillary buds c 15mm; spikes stouter, 7–30cm; anther cells not confluent, lateral and ± vertical, borne on filament c 0.5mm; stigmas 3–4; drupes ellipsoid, c 7 × 4mm.

Bhutan: S – Sarbhang district (Singi Khola and near Phipsoo). Subtropical forest, in ravines and on river banks, 280–390m. March.

3. P. mullesua D. Don; *P. guigual* D. Don, *P. brachystachyum* Hook. f. *nom. superfl., Chavica sphaerostachya* Miquel. Med: *Pipiling;* Nep: *Dala Charbo* (34), *Peepla.*

Terrestrial or epiphytic shrub 0.5–1m, sometimes climbing, older stems warted, creeping stems pubescent. Leaves dark green when dry, those of creeping stems ovate-cordate; those on aerial branches elliptic to ovate, 7–15 × 2.5–8cm, acuminate, base ± symmetric, cuneate or rounded but then shortly decurrent on petiole, 3–5-veined in basal 3–10mm, with 1–2 pairs of prominent lateral veins borne 1.5–3(–5)cm above base, glabrous; petioles 7–14mm; stipules absent; axillary buds 5–11mm, persistent and conspicuous. Male spikes slender, erect, 3–7cm, on peduncles 3–4mm; bracts peltate, subsessile, stamens 2, anthers slightly exserted beyond bracts, cells 2, ± horizontal, reniform, placed side-by-side, with low lamella between, not confluent. Female spikes globose or shortly cylindric on short peduncle 2–8mm, stigmas 3, minute. Fruiting spikes 6–12 × 6–10mm, drupes c 2mm, densely clustered.

Bhutan: **S** – Samchi, Phuntsholing, Chukka, Sarbhang and Gaylegphug districts, **C** – Tongsa and Tashigang districts, **N** – Upper Mo Chu district (Kencho); **Sikkim.** Subtropical, Warm broad-leaved and Evergreen oak forests, 200–2200m. February–June.

The records of *P. diffusum* Vahl from Bhutan (117) probably refer to this species, and records of *P. thomsonii* (C. DC.) Hook. f. from Bhutan and Sikkim (80, 117) refer to *P. sylvaticum, P. suipigua* and *P. mullesua.* The record of *P. wallichii* (Miquel) Handel-Mazzetti (*P. aurantiacum* C. DC.) from Bhutan (135) is based on a Griffith collection of *P. mullesua.*

4. P. peepuloides Roxb.; *Chavica peepuloides* (Roxb.) Miquel, *Chavica neesiana* Miquel. Nep: *Ruk Peepla.*

Similar to *P. mullesua* but leaves on aerial branches tapering to a round, shallowly cordate, usually oblique base, prominently 5-veined in basal 5–10mm, but without lateral veins above; petioles 3–10mm; axillary buds inconspicuous, minute; female spikes shortly cylindric, rarely subglobose, in fruit 6–15 × 5–6mm, on very short peduncles 1–4mm; drupes c 1mm.

Bhutan: **S** – Samchi district (Tamangdhanra forest), Chukka district (Marichong) and Sarbhang district (Burborte Khola), **C** – Mongar district (Ngasam); **Sikkim:** Darjeeling foothills. Subtropical forests, 280–1220m. November–December.

5. P. sylvaticum Roxb.; *P. dekkoanum* C. DC., *P. thomsonii* (C. DC.) Hook. f.

Climbing shrub, stems deeply striate when dry. Leaves on creeping stems broadly ovate-cordate; those on aerial branches membranous, ovate to lanceolate, 7–13 × 3–7cm, acuminate, base cuneate or rounded, symmetric or slightly oblique, 5-veined at base, with 2 lateral veins arising 1.5–2cm above base, main veins densely puberulous, not very prominent, minor veins reticulate and gland-dotted beneath; petioles 1–2.5cm; stipules narrow, inconspicuous, adnate to petiole; axillary buds large 10–14mm. Male spikes and anthers as in *P. mullesua,* on peduncles 4–10mm. Female spikes cylindric, on short peduncle 4–8(–12)mm, stigmas 3–4. Fruiting spikes 15–30 × 7–10mm; drupes globose, 3mm, densely arranged.

Bhutan: **C** – Mongar district (Shongar); **Sikkim:** Kalimpong, Birick, Jalpaiguri etc. Subtropical and terai forests, 380–965m. March–May.

6. P. longum L. Nep: *Pipla* (34). Fig. 28 u–ab.

Slender climbing shrub, stems puberulous, striate when dry. Leaves membranous, broadly ovate on creeping stems, 7–12 × 4–9cm, subacute or shortly and bluntly acuminate, base deeply cordate-auriculate, symmetric, palmately 7-veined from base, veins minutely puberulous beneath, faintly prominent, eglandular; petioles long, 3–6cm; stipules narrow, adnate to petiole, 1–2cm; axillary buds slender, 2–5mm; leaves on flowering shoots narrower, subsessile or on short petioles up to 1.5cm. Male spikes slender, 4–9cm, on peduncles 2–4cm; bracts peltate, distinctly but shortly stalked, stamens 2, shorter than bracts, anther cells inclined, end-to-end on short filament. Female spikes cylindric, 15–20 × 3–4mm, on peduncle 1–2cm, stigmas 3–4, large. Fruiting spikes swollen, 2–2.5 × 0.6–0.8cm, drupes densely arranged, coherent.

Bhutan: **S**–Phuntsholing district (Phuntsholing); **Sikkim:** Darjeeling and Sookna. Subtropical forests, 150–800m. April–August.

Cultivated and possibly naturalised. Grown for its fruit, used as a pepper, and for its roots which are used medicinally. The true *Black Pepper, P. nigrum* L., is sold in markets in Bhutan but is imported from India.

7. P. pedicellatum C. DC.; *P. boehmeriifolium* sensu F.B.I. p.p. non (Miquel) C. DC., *P. clarkei* C. DC., *P. nigramentum* C. DC. Nep: *Bale Chabo, Long Pipla.*

Erect shrub 1–2.5m, blackish when dry, stems shallowly striate, glabrous. Leaves membranous, lower broadly ovate, 8–13 × 6–8cm, shortly acuminate, base distinctly oblique, cuneate on one side, cordate or rounded on other side, 5–7-veined at base, with 2 further lateral veins arising 2–3cm above base, glabrous and eglandular but often reticulate beneath; upper leaves ovate-elliptic, 3–6cm broad; petioles 4–9mm; stipules absent; axillary buds few, slender c 5mm. Male spikes slender, flexuous, 10–22 × 0.2–0.3cm, on peduncles 1–2cm; bracts peltate, minute, long-exserted on slender stalks, stamens 2, much shorter than bracts, anther cells almost erect on very short filaments. Female spikes slender, 4–9cm, on peduncles 1–2cm, stigmas 3. Fruiting spikes swollen, 8–10mm broad, drupes densely aggregated.

Bhutan: **S**–Samchi, Phuntsholing, Chukka, Sarbhang and Gaylegphug districts, **C**–Tongsa, Mongar and Tashigang districts; **Sikkim:** Tista valley, Darjeeling etc. Subtropical and Warm broad-leaved forests, 290–1800m. March–July.

Leaves sometimes use for chewing with betel-nut. In the past confused with *P. boehmeriifolium* (Miquel) C. DC., from Assam and Burma, which differs in its larger, oblong leaves which are distinctly gland-dotted beneath.

8. P. betleoides C. DC.; *P. betle* sensu F.B.I. p.p. non L., *P. betleoides* C. DC. var. *glabrifolium* C. DC. Nep: *Pan.*

Climbing shrub, stems weakly striate when dry, glabrous or minutely puberulous. Leaves membranous, lower ones broadly ovate-cordate, upper ones ovate to ovate-lanceolate, 9–17 × 4–7.5cm, acuminate, base strongly asymmetric, unequally cordate on both sides, 5-veined at base, with 2–3 further veins arising 1.5–2.5cm above base, veins prominent and sometimes sparsely puberulous beneath, lower surface minutely gland-dotted; petioles 1.4–2.0cm; stipules conspicuous, 1–1.4cm, subpersistent, adnate to petiole; axillary buds absent. Male spikes 4.5–6 × 0.2–0.3cm, on slender peduncles 2.5–3.5cm, bracts peltate, short-stalked, stamens just exserted beyond bracts, anther cells erect on short stout filaments. Female spikes

2.5 – 3.5cm, on slender peduncles 2 – 4cm, stigmas 4(– 5). Fruiting spikes swollen, 8 – 10mm broad, drupes 2mm, densely aggregated.

Bhutan: **S** – Phuntsholing district (above Phuntsholing), and Sarbhang district (Singi Khola and Phipsoo), **C** – Mongar district (Shongar Chu); **Sikkim:** Rungit Valley, Darjeeling etc. Subtropical forests, 280 – 1300m. April – July.

Leaves widely used for chewing with betel-nut as a stimulant. Sold in markets in S Bhutan, but apparently not cultivated. The true *Piper betle* L. does not occur in the E Himalaya.

9. P. chuvya (Miquel) C. DC.; *P. betle* sensu F.B.I. p.p. non L. Nep: *Chaba.*

Similar to *P. betleoides* but a stouter climber, leaves thinly coriaceous, ovate, often broadly ovate, 10 – 20 × 6 – 12cm, 5 – 7-veined at base with 2 further veins arising 2 – 5cm above base, puberulous and gland-dotted beneath; petioles 0.7 – 1.5cm; stipules caducous; male spikes 7 – 8cm, on peduncles 2 – 2.5cm; female spikes elongate, 9 – 10cm, on peduncles 1.1 – 1.5cm; fruiting spikes interrupted with only some drupes ripening, drupes c 4mm.

Bhutan: **S** – Samchi district (Khagra valley near Gokti); **Sikkim:** Tista valley and Badamtam. Subtropical forests, 500m. April.

10. P. suipigua D. Don; *P. nepalense* Miquel, *P. gamblei* C. DC., *P. glabramentum* C. DC., *P. subrigidilimbum* C. DC., *P. japvonum* C. DC. Dz: *De Pani;* Nep: *Dakley Chabo* (117), *Tsabo Lahara.*

Large climbing shrub, stems faintly striate when dry, glabrous. Leaves thinly coriaceous, dark green when dry, ovate, 8 – 15 × 3.5 – 6cm, shortly acuminate, base usually slightly oblique, cuneate to rounded, 3 – 5-veined at base, with 2 further lateral veins arising 1 – 2 cm above base, veins prominent on both surfaces, glabrous; petioles 8 – 14mm; stipules narrow, adnate to petiole, early caducous; axillary buds few, subulate, 1 – 2cm. Male spikes erect, 7 – 14cm, on peduncle 1 – 2cm, bracts peltate, subsessile, stamens 3, shortly exserted beyond bracts, anther cells erect, on short stout filaments. Female spikes 3 – 6cm, slender, on peduncle 1 – 1.5cm; stigmas 4. Fruiting spikes much elongated, 9 – 13cm, whitish hairy, interrupted, with scattered ovoid sessile drupes 5 – 6 × 4mm.

Bhutan: **S** – Samchi, Phuntsholing, Chukka, Gaylegphug and Deothang districts, **C** – Punakha, Tongsa and Mongar districts; **Sikkim:** Tonglu, Darjeeling etc. Warm and Cool broad-leaved forests, 900 – 3000m. May – August.

11. P. khasianum C. DC.; *P. curtistipes* C. DC., *P. mannii* C. DC. Nep: *Chabo.*

Similar to *P. suipigua* but leaves membranous, lanceolate or narrowly ovate, 6 – 10 × 2 – 4cm, long-acuminate, base cuneate, 3-veined at base, with 2 further lateral veins arising 1 – 1.5cm above base; petioles 4 – 8mm; axillary buds more numerous, 3 – 4mm; spikes shorter, 2 – 3cm, on peduncles 4 – 8mm; drupes globose, smaller, 3 – 4mm diameter, borne on short, stout stalk 0.5 – 1mm.

Bhutan: **S** – Gaylegphug district (Gaylegphug) (117); **Sikkim:** Riang, Sivoke and Brip. Subtropical forests, 300 – 600m. February – June.

12. P. hamiltonii C. DC. Nep: *Jangli Pan* (34).

Similar to *P. suipigua* but stems deeply striate when dry; leaves thinly coriaceous, pale when dry, elliptic or elliptic-obovate, 7–11 × 4–7cm, with a short obtuse point, base rounded, 5-veined in lowermost 5mm of leaf, without additional lateral veins; petioles 1–2.2cm; fruiting spikes interrupted, 11–15cm, on peduncle 1–2cm; drupes ovoid at first becoming subglobose c 2.5mm.

Sikkim: Jalpaiguri terai. Subtropical terai forest. February.

Family 51. CHLORANTHACEAE

by A.J.C. Grierson & D.G. Long

Evergreen glabrous, aromatic subshrubs; stems with swollen nodes. Leaves opposite, simple, pinnately veined; base of petioles connate forming a rim bearing minute stipules. Flowers in terminal simple or paniculately-branched, bracteate spikes; bisexual, zygomorphic. Perianth absent. Stamens 3, connate into a staminal scale bearing 3 anthers, attached to base of ovary on one side and partly enveloping it. Ovary superior, 1-locular; ovule 1; style simple, short. Fruit a drupe.

1. CHLORANTHUS Swartz

Description as for Chloranthaceae

1. C. elatior Link; *C. officinalis* Blume. Fig. 28 j–m.

Stems erect, 30–100cm. Leaves elliptic, 8–20 × 4–9cm, shortly acuminate, base cuneate, margins finely serrate; petioles 4–8mm. Spikes often paniculately branched, 3–5cm. Bracts in opposite, distant pairs, ovate, c 1.5mm, each subtending an obliquely ovoid ovary 1–1.5mm. Staminal scale ovate, c 2mm, 3-lobed, bearing a 2-locular (4-thecate) anther below the middle lobe and a 1-locular (2-thecate) anther below each outer lobe. Drupe ovoid, c 5mm, white.

Bhutan: S – Chukka district (Marichong) and Gaylegphug district (Gaylegphug); **C** – Punakha district (Rinchu); **Sikkim.** Undershrub in Subtropical and Warm broad-leaved forests, 270–1600m. May–October.

Family 52. ARISTOLOCHIACEAE

by A.J.C. Grierson & D.G. Long

Perennial rhizomatous herbs or woody climbers. Leaves simple, alternate, entire, rarely palmatifid or coarsely toothed, palmately veined, exstipulate. Flowers axillary, solitary or in short racemes, actinomorphic or zygomorphic, bisexual, often foetid; perianth shortly globose-campanulate, 3-lobed or tubular and S-shaped or straight. Stamens 6 or 12, free or often adnate to style forming a column. Ovary inferior, 6-locular; style short, columnar, with 3 or 6 stigmatic lobes; ovules numerous, axile. Fruit a dry or leathery many-seeded capsule.

1. Creeping rhizomatous herbs; flowers actinomorphic; perianth shortly globose-campanulate, 3-lobed; stamens 12, free .... **1. Asarum**
\+ Woody climbers; flowers zygomorphic; perianth with long S-shaped or straight tube; stamens 6 or 12, adnate to style and forming a column **2. Aristolochia**

1. ASARUM L.

Perennial creeping rhizomatous herbs. Leaves few, basal. Flowers solitary, axillary, actinomorphic. Perianth tube adnate to ovary, limb globose-campanulate, 3-lobed, persistent. Stamens 12, free, in 2 whorls on top of ovary. Style stout, with 6 short lobes. Fruit a coriaceous capsule enclosed in persistent perianth, bursting irregularly.

1. A. himalaicum Klotzsch; *A. himalaicum* var. *bhutanicum* W. W. Smith. Fig. 28 c–f.

Rhizomes slender, woody. Leaves broadly ovate, 4–12 × 3–10cm, shortly acuminate, deeply cordate at base, margins entire but minutely ciliate, finely pubescent on both surfaces; petioles 4–18cm. Pedicels 3–5cm. Perianth brownish- or greenish-purple, c 15mm across, tube c 12mm, pubescent, lobes triangular, 6–7mm, acute, strongly reflexed, minutely papillose within. Capsule c 15mm; seeds ovoid, dorsally rounded, ventrally grooved and arillate.

Bhutan: C–Thimphu, Punakha, Bumthang and Mongar districts; **Sikkim.** Amongst herbs and shrubs in Fir, Hemlock and Cool broad-leaved forests, 2750–3500m. May–July.

2. ARISTOLOCHIA L.

Woody twiners or climbers. Leaves entire or deeply palmatifid, rarely coarsely toothed at apex. Flowers axillary, solitary or in racemes or cymes, zygomorphic. Perianth tubular, S-shaped or almost straight, with narrow or broad limb at mouth. Column with 3 or 6 stigmatic lobes; stamens 6 or 12, solitary or in bundles of 2 or 4 below each lobe. Fruit a 6-valved capsule.

1. Leaves deeply palmatifid or with a few coarse teeth at apex **5. A. platanifolia**
\+ Leaves undivided, entire 2

2. Perianth tube narrow, almost straight, abruptly globose-swollen near base; column 6-lobed .... **6. A. tagala**
\+ Perianth tube S-shaped, inflated throughout; column 3-lobed 3
3. Limb of perianth broad and flat, 3.5–9cm across; throat and limb brownish-purple...4
\+ Limb of perianth narrow, reflexed, 2–2.5cm across; throat yellow, limb purple 5
4. Perianth tube strongly inflated at bend, 3–4cm broad, sparsely to moderately pubescent; limb 6–9cm across; anthers 12 .... **1. A. griffithii**
\+ Perianth tube weakly inflated at bend, 1–1.5cm broad, densely hirsute; limb 3.5–5cm across; anthers 6.. **2. A. nakaoi**
5. Leaves ovate, often broadly so, 10–26cm broad, densely appressed sericeous beneath .... **3. A. cathcartii**
\+ Leaves ovate-lanceolate or oblong, 5–15cm broad, pubescent beneath **4. A. saccata**

1. A. griffithii Duchartre; *Isotrema griffithii* (Duchartre) Fischer. Fig. 28 g,h.

Leaves ovate, 6–15 × 5–9cm, acute or shortly acuminate, base strongly auriculate-cordate, entire, sparsely pubescent on both surfaces, 7–9-veined at base; petioles 2.5–5cm. Flowers solitary, axillary, pendent, on pubescent pedicel 2–4cm bearing a small leaf-like bracteole. Perianth S-shaped; folded tube 4–6.5cm, green with purple veins, strongly inflated at bend, 3–4cm broad, sparsely to moderately pubescent; limb funnel-shaped, rhombic-orbicular, 6–9cm across, yellowish with dark purple veins outside, coarsely warted and brownish-purple within. Column 3-lobed, bearing 12 anthers; ovary cylindrical, 3–5cm. Fruit oblong, 12–18 × 1.5–2.5cm, 6-ribbed, dehiscing into 6 longitudinal valves. Seeds ovate, 4–6 × 3mm.

Bhutan: S–Chukka district, **C**–Thimphu, Punakha, Tongsa, Mongar and Tashigang districts, **N**–Upper Mo Chu district; **Sikkim.** Cool broad-leaved, Evergreen oak and Fir forests, 1800–3000m. April–May.

2. A. nakaoi Maekawa; *Isotrema nakaoi* (Maekawa) A. R. Smith

Similar to *A. griffithii* but leaves densely pubescent beneath; flowers smaller; folded perianth tube c 3cm, weakly inflated at bend, 1–1.5cm broad, densely brown hirsute; limb subtriangular 3.5–5cm across; anthers 6; fruit (immature) narrowly obovoid, c 4 × 2cm.

Bhutan: C–Punakha district (Wangdu Phodrang), Tongsa district (between Tongsa and Tashiling) and Tashigang district (Kori La); **Sikkim:** Gyreebans. Evergreen oak forests, 2000-2700m. March–June.

3. A. cathcartii Hook. f.

Leaves ovate, often broadly ovate, 17–35 × 10–26cm, shortly acuminate, base cordate, densely appressed sericeous beneath; petioles 6–13cm. Flowers 2–3 in short densely villous racemes on lower leafless stems, or in axils of lowermost leaves. Perianth S-shaped; folded tube 4–4.5cm, strongly inflated, white with purple veins and reticulations, pilose; mouth much broader than tube, squarish, c 2.5cm across, yellow with purple dots within; limb narrow, rim-like, reflexed, dark purple with bristle-like papillae. Column 3-lobed, bearing 6 or 12 anthers. Fruit obovoid, 2.5–5cm, 6-angled.

Bhutan: S – Sarbhang district (Singi Khola); **Sikkim:** Mungpoo, Rongsong, etc. Subtropical forests, 400–900m. March.

4. A. saccata Wall.

Similar to *A. cathcartii* but leaves narrower, ovate-lanceolate or oblong, 10–35 × 5–15cm, pubescent beneath; perianth tube yellowish-white or purple, mouth rounded, scarcely broader than tube, uniformly yellow within.

Sikkim: foothills (34). Subtropical forests, up to 1800m. March.

5. A. platanifolia (Klotzsch) Duchartre

Leaves broadly ovate or orbicular, 17–25 × 20–30cm, deeply palmatifid with oblong, toothed lobes 1–3cm broad, sometimes unlobed with 3–7 coarse teeth at apex; prominently reticulate and pubescent beneath; petioles 4–8cm. Flowers similar to those of *A. cathcartii* and *A. saccata;* perianth tube yellowish or red-brown. Capsule linear-oblong, 10–14 × 2.5–3.5cm, 6-ribbed.

Bhutan: S – Chukka district (Sinchu La, Raidak valley), **C** –Tongsa district (below Shamgong); **Sikkim.** Warm broad-leaved forests, 1500–1600m. September–October.

6. A. tagala Chamisso; *A. roxburghiana* Klotzsch. Fig. 28 i.

Leaves ovate, 8–10 × 4.5–10cm, acuminate, base deeply cordate-auriculate, glabrous or minutely pubescent on veins beneath; petioles 3–5cm. Flowers 5–15 in lax axillary racemes or cymes 5–9cm. Perianth tube narrowly tubular, almost straight, 1.5–2cm, whitish, with globose swelling 4–5mm broad near base; mouth very oblique, lower lip elongate, linear, c 2cm, dark purple. Column 6-lobed, bearing 6 anthers. Capsule obovoid, 7–9 × 3–4cm, membranous.

Bhutan: S – Samchi district (near Gokti) and Sarbhang district (Phipsoo); **Sikkim:** Tista valley, etc. Subtropical forests, 300–900m. May–June.

Family 53. DILLENIACEAE

by A.J.C. Grierson & D.G. Long

Trees or shrubs. Leaves simple, alternate, pinnately veined usually with numerous prominent lateral veins, exstipulate. Flowers solitary, clustered or in panicles, actinomorphic, bisexual. Sepals 5, imbricate, persistent. Petals 3–5, free, imbricate, deciduous. Stamens numerous, free or united in bundles at base; anthers 2-locular, opening lengthwise by slits or by apical pores. Carpels 1 – many, superior, free or somewhat connate at centre of receptacle; styles free. Fruiting carpels dehiscent and naked, or indehiscent and berry-like, enclosed within a subglobose pseudocarp composed of the enlarged, somewhat fleshy, imbricate sepals; seeds 1 or few, sometimes arillate.

1. Trees; flowers solitary or in clusters .... **1. Dillenia**
+ Shrubs, often scandent; flowers in terminal panicles .... **2. Tetracera**

1. DILLENIA L.

Trees. Leaves large. Flowers solitary and terminal or in clusters on older branches. Sepals spreading and herbaceous at first, becoming enlarged and fleshy in fruit. Petals 5, larger than sepals. Stamens almost free. Carpels 5 – 20 borne on a conical receptacle; styles radiating. Fruiting carpels enclosed within globose pseudocarp; seeds exarillate.

1. D. pentagyna Roxb. Nep: *Tantri* (34), *Tartarey, Tantare.*

Deciduous tree up to 25m. Leaves obovate, 20 – 50 × 10 – 20cm, obtuse or shortly acute, base cuneate, margins nearly entire to distinctly serrate, veins 25 – 50 pairs, pubescent on veins beneath and forming small tufts of hairs at ends of teeth; petioles 1 – 5cm. Flowers precocious, 2.5 – 3cm across, 2 – 7 in clusters on rounded projections on older branches, pedicels up to 6cm. Sepals ovate, 8 – 12 × 5 – 9mm, reddish, acute or obtuse. Petals obovate, 15 – 20 × 5 – 10mm, yellow, rounded at apex. Stamens in 2 series, the outer numerous, inwardly curved, the inner series (about 10) outwardly reflexed; anthers opening by longitudinal slits. Carpels 5, oblong, 3 – 4 × 1.2 – 1.5mm; styles c 4mm. Pseudocarps orange, c 1.5cm diameter; fruiting carpels c 8 × 5mm, each 1 – 2-seeded, seeds ovoid, 5 × 3.5mm, black, glabrous.

Bhutan: S – Phuntsholing, Sarbhang, Gaylegphug and Deothang districts; **Sikkim.** Subtropical forest slopes, 200 – 650m. March – April.

Timber reddish brown, valuable for construction work if well-seasoned (34,48).

2. D. indica L. Nep: *Panchphal, Panchkul* (34), *Panchphale.* Fig. 29 w,x.

Similar to *D. pentagyna* but evergreen; leaves elliptic to oblanceolate, 15 – 40 × 5 – 12cm, acute, attenuate at base, margins serrate, veins 30 – 50 pairs, appressed hirsute on veins especially beneath; petioles 2 – 7cm; flowers solitary, terminal, 15 – 20cm across, pedicels 4 – 8cm; sepals obovate, 4 – 6 × 3 – 5cm; petals white, obovate, 7 – 9 × 5 – 6.5cm, rounded at apex; anthers opening by apical pores; carpels 14 – 20, styles white, narrowly oblanceolate, c 18mm, flattened; pseudocarps yellowish, 10 – 12cm diameter; fruiting carpels 3.5 × 1.5cm, each with 5 seeds embedded in pulp, seeds reniform, 6 × 4mm, reddish, hirsute along outer margin with hairs 2mm.

Bhutan: S – Sankosh district (near Pinkhua) and Sarbhang district (Singi Khola); **Sikkim:** terai. Riverine subtropical forest, 340–390m. August.

The record of *D. aurea* Smith from Bhutan (80) probably refers to this species. Timber little used (34), pseudocarps edible and used as cattle-fodder, bark leaves and fruit used medicinally (13).

2. TETRACERA L.

Shrubs, often scandent with flexuous branches. Panicles few- or many-flowered, terminal. Sepals 5, reflexed in flower and fruit. Petals 3, caducous. Stamens numerous, filaments free. Carpel solitary (or rarely 2) bearing a straight style. Fruiting carpel coriaceous, 1–2-seeded, dehiscing longitudinally into 2 halves; seeds arillate.

1. T. sarmentosa (L.) Vahl; *Delima sarmentosa* L.

Erect shrub up to 3m, or scandent up to 6m, much-branched, branches scabrid, sparsely hirsute. Leaves elliptic or obovate, 5–12 × 4–7cm, obtuse or shortly acute, base rounded or cuneate, margins subentire or shallowly serrate especially in the upper part, veins 10–14 pairs, scabrid with minute papillate scales on both surfaces, sparsely hirsute between veins above and on veins beneath; petioles 1–2cm. Flowers 7–10mm across, pedicels 2–7mm. Sepals broadly elliptic, 3–4 × 2–3mm, ciliate. Petals obovate, 3–4 × 2–3mm, whitish. Carpel ovoid, c 2 × 1.5mm; style c 3mm. Fruiting carpels ovoid, 6–10 × 5–6mm, purplish, glossy; seeds ovoid, 4 × 3mm, black, ± surrounded by a pale aril which is laciniate above.

West Bengal Duars: Buxa; **Sikkim:** Darjeeling and Sivoke. Terai forests. May–October.

The above description refers to subsp. **andamanica** (Hoogland) Hoogland, the only subspecies in the area.

FIG. 29. **Theaceae, Dipterocarpaceae, Actinidiaceae** and **Dilleniaceae. Theaceae**. a & b, *Eurya acuminata*: a, portion of shoot with leaves and inflorescences; b, flower. c–e, *Schima wallichii*: c, shoot apex with leaves and flowers; d, dehisced capsule; e, seed. **Dipterocarpaceae**. f–j, *Shorea robusta*: f, leaf; g, portion of inflorescence; h, flower; i, stamen; j, capsule enclosed by accrescent calyx. k, *Vatica lanceifolia*: capsule with persistent calyx. **Actinidiaceae**. l–p, *Actinidia callosa*: l, portion of shoot with leaf and inflorescence; m, female flower; n, male flower; o, anther; p, fruit. q–s, *Saurauja armata*: q, upper portion of shoot with leaves and inflorescences; r, flower; s, stamen. t–v, *Saurauja napaulensis*: t, portion of shoot with leaf and inflorescence; u, flower; v, stamen. **Dilleniaceae**. w & x, *Dillenia indica*: w, upper portion of shoot with leaf and flower; x, pseudocarp consisting of enlarged persistent sepals enclosing fruit. Scale: x × ¼; a, c, f, l, q, t, w × ⅓; d, g × ½; j, p, r × ⅔; k × 4/5; h × 1; m, n, u × 1¼; e × 2; b × 3½; o, s × 4½; i × 5; v × 6. Drawn by M. Bates.

a
b
c
d
e
f
g
h
i
j
k
l
m
n
o
p
q
r
s
t
u
v
w
x
M.B.

Family 54. ACTINIDIACEAE

by A.J.C. Grierson & D.G. Long

Trees or shrubs, sometimes climbing. Pith of branches solid, hollow or chambered. Leaves alternate, simple, pinnately veined, glabrous or with simple hairs or scales, exstipulate. Flowers bisexual or unisexual, solitary or few to many in cymes or panicles. Sepals 5, imbricate. Petals 5, imbricate, free or shortly connate. Stamens numerous, anthers opening by longitudinal slits or apical pores. Ovary 5 – many-celled with an equal number of styles. Fruit a berry.

1. Scrambling shrubs; petals free; anthers opening by longitudinal slits; styles numerous .... **1. Actinidia**
\+ Trees or erect shrubs; petals shortly united at base; anthers opening by apical pores; styles 5 .... **2. Saurauja**

1. ACTINIDIA Lindley

Dioecious scrambling shrubs with simple hairs, pith often chambered. Flowers solitary or few in subumbellate cymes. Male flowers with well-developed stamens (two-thirds as long as petals) and rudimentary ovaries with minute styles. Female flowers with well-developed ovary and 15 – 20 styles, stamens only one-third as long as petals or apparently well-developed but with anthers empty or containing only sterile pollen. Berry oblong, often spotted with lenticels.

1. A. callosa Lindley. Nep: *Tekiphal* (34). Fig. 29 l – p.

Scandent up to 7m, stems glabrous or with a few soft simple hairs and sometimes with conspicuous lenticels. Leaves ovate-elliptic, 5 – 15 × 2 – 7cm, acuminate, base rounded or cuneate, margins finely serrate, usually with hair-like teeth, glabrous or finely pubescent beneath, lateral veins 7 – 10 pairs, sometimes with prominent parallel minor veins between them; petioles 2 – 4.5cm. Flowers solitary or in 2 – 5(– 12)-flowered cymes. Sepals ovate, 5 – 6 × 3.5 – 5mm, obtuse, ciliate. Petals white, obovate, 10 – 12 × 7 – 8mm, rounded. Male flowers with filaments c 6mm; rudimentary ovary ovoid, c 1.5mm. Female flowers sometimes with filaments 2 – 3mm; ovary oblong, c 3.5mm, densely pubescent, styles 4mm, linear. Fruit oblong, cylindrical, up to 6 × 2.5cm, bearing reflexed persistent sepals at base, usually brown-pubescent at first but later purplish and covered with circular brown lenticels c 1mm diameter.

Bhutan: **S** – Chukka, Gaylegphug and Deothang districts, **C** – Thimphu, Punakha, Tongsa and Tashigang districts, **N** – Upper Mo Chu and Upper Bumthang Chu districts; **Sikkim.** Margins of Cool broad-leaved forests, 1400 – 2450m. May – July.

A variable species so far as chambering of pith, prominence of minor veins and pubescence of fruit are concerned. Fruit edible (34).

2. A. strigosa Hook f. & Thomson. Nep: *Tekiphal* (34).

Similar to *A. callosa* but young stems, petioles and midribs beneath usually densely covered with spreading brown bristles up to 3mm.

Bhutan: C–Tongsa district (Chendebi) and Mongar district (Namning and Zimgang); **Sikkim.** Cool broad-leaved forests, 2550–2750m. May–June.

Possibly not specifically distinct from *A. callosa* as intermediate specimens occur. Fruit edible (34).

2. SAURAUJA Willdenow

Trees or shrubs. Branches, petioles and undersurface of midribs usually bearing scales; pith often hollow. Leaves often with numerous lateral veins. Inflorescence axillary, often in axils of fallen leaves; flowers solitary, cymose, paniculate or in few-flowered clusters; bisexual. Sepals 5, imbricate. Petals 5, connate at base. Stamens numerous, adnate to petals, anthers opening by apical pores. Ovary 5-celled, styles 5. Berry often globose.

1. Flowers solitary or clustered, ± sessile .... **1. S. armata**
\+ Flowers in cymes or panicles 2
2. Inflorescence of elongated (15–35cm) panicles .... **Species 2 & 3**
\+ Inflorescence of clustered, rarely solitary, short (1–10cm) cymes **Species 4–6**

1. S. armata Kurz; *S. cerea* Dyer. Fig. 29 q–s.

Shrub or tree to 7m, bearing sharp-pointed lanceolate scales c 2mm. Leaves obovate, 20–30 × 10–15cm, rounded and shortly acute, base cuneate, margins serrate with fine sharp-pointed teeth, veins 12–20 pairs, glabrous; petioles 1–4cm. Sepals broadly elliptic or suborbicular, 7–12 × 6–12cm, densely pubescent and scaly. Petals white, crimson at base, oblong or obovate, c 15 × 8–10mm, rounded. Ovary globose, c 5mm, brownish tomentose; styles 5. Fruit unknown.

Bhutan: S–Chukka district (Marichong) and Sarbhang district (between Sarbhang and Chirang). Subtropical mixed forest, 900–1200m. April–July.

2. S. napaulensis DC. Sha: *Mingdormu Shing;* Nep: *Gogun* (34). Fig. 29 t–v.

Tree to 6m bearing appressed ovate scales 1–1.5mm. Leaves elliptic-oblanceolate, (15–) 20–35 × (5–) 6.5–12cm, acuminate, base rounded, margins finely serrate, veins 30–35 pairs, glabrous or more often brownish tomentose beneath; petioles 2–5cm. Panicles 15–20cm. Sepals ovate, 5–6 × 4mm, glabrous. Petals pink, obovate, c 10 × 6mm, rounded. Fruit globose, c 8mm diameter.

Bhutan: **S** – Phuntsholing to Deothang districts, **C** – Punakha to Tashigang districts; **Sikkim.** Warm broad-leaved forests, 1300 – 2100m. May – August.

3. S. griffithii Dyer. Nep: *Gogun* (34).

Similar to *S. napaulensis* but young shoots, petioles, peduncles and leaf-undersides densely brown tomentose, scales linear-lanceolate c 1.5mm, few, weak; leaves broadly elliptic or obovate, 25 – 40 × 13 – 18cm, obtuse but with a short (c 5mm) thick point, petioles 5 – 7cm; panicles up to 40cm; sepals elliptic, c 6 × 4mm, densely pubescent; petals pale purple, obovate, c 8 × 4mm; fruit unknown.

Bhutan: **S** – Gaylegphug district (near Surey); **Sikkim:** Ryang and Sittong. Subtropical forests, 900 – 1500m. June.

4. S. fasciculata Wall. Nep: *Gogun* (34).

Shrub or tree to 5m bearing few oblong scales, c 1mm, acute or obtuse and minutely toothed. Leaves elliptic, lanceolate or ovate, 12 – 25 × 3 – 9cm, acuminate, base rounded or cuneate, margins finely serrate, veins 25 – 30 pairs, often bearing small rough scales c 0.3mm, thinly brownish-tomentose beneath. Cymes glabrous, 4 – 10cm, in clusters of 1 – 6, each 3 – 11-flowered (rarely 1-flowered); bracts narrow lanceolate, 1 – 2mm. Sepals suborbicular, c 5mm long and broad. Petals white, obovate, c 9 × 5mm, rounded. Fruit globose, 7 – 8mm.

Bhutan: **S** – Sarbhang and Deothang districts, **C** – Tongsa and Mongar districts; **Sikkim**. Warm broad-leaved and Subtropical forests, 1050 – 1550m. May – July.

5. S. punduana Wall. Dz?: *Mang, Mang-Ma* (117); Nep: *Rate Gogun* (34).

Similar to *S. fasciculata* but scales ovate, 1.5 – 3mm, acute; cymes scaly and pubescent, 3- to many-flowered; bracts elliptic, 4 – 6 × 3mm; sepals elliptic, c 10 × 6mm; petals pink, c 15 × 8mm.

Bhutan: **S** – Gaylegphug district (Tama) and ?Deothang district; **Sikkim.** Subtropical forests, 1050 – 1200m. April – August.

6. S. roxburghii Wall. Nep: *Gogun, Aule Gogun* (34).

Similar to *S. fasciculata* but scales few, ovate, acuminate; leaves with 10 – 16 pairs of veins, thinly tomentose beneath; cymes pubescent, 1 – 3 (– 4)cm; bracts ovate, 1.5 – 2cm; sepals elliptic, 4 – 5 × 3mm; petals pink c 6 × 2.5mm; fruit globose, 5 – 6mm diameter.

Bhutan: **S** – Sankosh district (near Pinkhua) and Gaylegphug district (near Gaylegphug); **Sikkim.** Subtropical forest, 270 – 1650m. May – August.

Family 55. DIPTEROCARPACEAE

by A.J.C. Grierson & D.G. Long

Evergreen resinous trees with indumentum of stellate hairs or peltate scales. Leaves alternate, simple, entire, pinnately veined; stipules small, deciduous. Flowers bisexual, actinomorphic, in axillary or terminal panicles, fragrant. Calyx tube short, bearing 5 imbricate or valvate lobes, becoming enlarged and wing-like in fruit. Petals 5, twisted in bud, free. Stamens 15–numerous, anthers opening by longitudinal slits, connective elongated above anther cells. Ovary 3-celled, superior. Fruit indehiscent, usually 1-seeded.

1. Leaves ovate-oblong, cordate at base; stamens numerous; fruit usually with 3 large and 2 smaller oblanceolate obtuse calyx lobes ....**1. Shorea**
\+ Leaves elliptic-lanceolate, cuneate at base; stamens 15; fruit with 5 equal-sized ovate-acute calyx lobes ....**2. Vatica**

1. SHOREA Roxb.

Panicles axillary or terminal. Calyx-tube short, lobes imbricate, ovate at flowering time. Petals elongate, ovate-lanceolate. Stamens numerous (20 or more), anthers with ± reflexed subulate connective. Ovary conical, densely pubescent, tapering to a short glabrous 3-branched style. Capsule leathery, enclosed by accrescent calyx and bearing above the oblanceolate calyx lobes, 3 of which are larger than the others, rarely all subequal.

1. S. robusta Gaertner f. Hindi: *Sal;* Nep: *Sakhua* (34). Fig. 29 f–j.

Large tree. Leaves ovate-oblong, 10–20 × 6–12cm, acuminate to a blunt point, base cordate, glabrous, lateral veins 12–14 pairs; petioles 2–2.5cm; stipules lanceolate, c 8mm, densely covered with silvery peltate scales. Panicles branching in unilateral racemes, densely pubescent. Calyx (including lobes) 5mm, densely yellowish pubescent. Petals ovate-lanceolate, 12–14 × 5mm, pubescent outside, glabrous and orange-coloured within. Stamens 2–3mm, filaments broadened at base, connective elongated c 0.5mm beyond anther-cells. Ovary conical, c 4mm, style c 1mm. Capsule ovoid, c 2 × 1.2cm; longer calyx lobes c 6.5 × 1.5cm, shorter ones c 5 × 0.7cm, pubescent, longitudinal veins 8–10.

Bhutan: S – Samchi district (native), Phuntsholing district (cultivated) and Sarbhang district (native), **C** – Punakha district (cultivated); **Sikkim.** Subtropical terai forests, 200–1370m. April.

Often cultivated; 'the most valuable tree in the district' (34); timber very durable, used in bridge-building and for railway-sleepers; resin used for incense and bark for tanning (34). Resin and leaves used medicinally (13).

2. VATICA L.

Panicles axillary and terminal, finely pubescent. Calyx lobes ovate, ± valvate. Petals elliptic, pubescent outside. Stamens 15, anthers oblong, connective elongated, acuminate, erect. Capsule leathery, 1–2-seeded, calyx lobes ovate, subequal.

1. V. lanceifolia Blume. Fig. 29k.

Medium-sized tree. Leaves elliptic-lanceolate, 12–18 × 5–7cm, acuminate, base cuneate, glabrous, lateral veins 10–12 pairs; petioles 1–2cm, thickened above. Calyx c 4mm, lobes ovate, acute. Petals cream-coloured, 2–2.5 × 0.3–0.4cm. Stamens 2–3mm. Ovary ovoid, c 2mm, style ± as long, thickened above, 3-branched. Capsule ovoid, 2–2.5 × 1.5–2cm, finely tomentose, calyx lobes c 1cm long and broad, ± cordate at base.

Bhutan: locality unknown (80). April–May.

Doubtfully recorded from Bhutan, based on a Griffith collection.

Family 56. THEACEAE

by A.J.C. Grierson

Trees or shrubs. Leaves alternate, simple, pinnately veined, mostly evergreen, exstipulate. Flowers solitary or clustered, unisexual or bisexual, actinomorphic, often bracteate. Sepals 5–6, free or shortly connate, imbricate. Petals 5–8, free or connate at base, imbricate. Stamens 5 or 15 to numerous, free or connate at base, sometimes adnate to petals, anthers basifixed or medifixed, sometimes chambered, opening by lengthwise slits. Ovary 3–5-celled, superior. Fruit indehiscent and berry-like or capsular and dehiscent into 3–5 valves from a central axis.

1. Flowers unisexual; fruit subglobose, c 5mm diameter ....**1. Eurya**
\+ Flowers bisexual; fruit ovoid, oblong or subglobose, 1.5–2.5 × 1–1.5cm..2

2. Anthers basifixed; fruit ovoid, c 2 × 1.5cm, berry-like, indehiscent **2. Ternstroemia**
\+ Anthers medifixed; fruit capsular, subglobose or oblong-ellipsoid, splitting into 1–5 valves from a central axis 3

3. Capsule oblong-ellipsoid, splitting into 5 narrowly elliptic valves; seeds with an apical wing ....**3. Gordonia**
\+ Capsule subglobose, splitting into 1–5 broadly elliptic valves; seeds unwinged or ± surrounded by a narrow wing...4

4. Capsule woody, 5-valved; seeds compressed, bearing a semicircular wing ....**4. Schima**
\+ Capsule ± leathery, 1–3-valved; seeds subglobose, unwinged ....**5. Camellia**

1. EURYA Thunberg

Dioecious shrubs or small trees. Flowers 2–4 in axillary clusters, usually from axils of fallen leaves, sometimes solitary; pedicels short (3–5mm), usually bearing

2 small bracteoles. Sepals 5. Petals 5, slightly connate at base. Stamens 5–15, adnate to base of petals, anthers basifixed, sometimes chambered. Ovary 3-celled, styles 3, free or united. Fruit subglobose, somewhat fleshy and berry-like, indehiscent.

1. Male flowers with 5 stamens; female flowers with styles free to base **1. E. cavinervis**
\+ Male flowers with about 15 stamens; female flowers with styles united at base 2

2. Veins of leaf indistinct on upper surface, faintly prominent beneath; male flowers with unchambered anthers; female flowers with style base 0.5–1mm and style branches 1.5–2mm .... **2. E. acuminata**
\+ Veins of leaf± prominent above and beneath; male flowers with chambered anthers; female flowers with style base 2–2.5mm and style branches 1.5–3mm **3. E. cerasifolia**

1. E. cavinervis Vesque; *E. japonica* sensu F.B.I. non Thunberg. Nep: *Jhingni* (34).

Shrub 1–3m. Buds and young shoots usually glabrous, sometimes sparsely appressed-pubescent at first, often with 2 decurrent ridges from each node. Leaves elliptic or oblanceolate, 4–12 × 1.5–4cm, acute or more usually acuminate, base rounded or cuneate, glabrous, margins serrate in upper half or two-thirds, veins ± deeply impressed above; petioles 3–5mm. Sepals broadly ovate, 2.5–3mm long and broad, ciliate, otherwise glabrous. Petals white or cream-coloured, obovate, 4–4.5 × 3–4mm. Male flowers: stamens 5, anther cells not chambered. Female flowers: ovary subglobose, c 1.5mm diameter, styles 0.5–1mm, free. Fruit broadly ellipsoid, c 5mm long and broad.

Bhutan: C –Tongsa district (near Kinga Rapden) and Tashigang district (near Tobrang), **N** – Upper Mo Chu and Upper Kuru Chu districts (Sergong La); **Sikkim**. *Abies/Rhododendron* forests, 3050–3350m. April–May.

A somewhat variable species vegetatively: twigs can be completely glabrous or, as in typical material, sparsely appressed-pubescent and they are usually distinctly ribbed. The Bhutan specimens, which may represent a distinct variety, have more pubescent and less strongly-ribbed branches. Timber valuable (34).

2. E. acuminata DC. Nep: *Sanu Jhingni* (34), *Jingane*. Fig. 29 a,b.

Shrub or tree to 12m, young branches pubescent with ± spreading hairs. Leaves elliptic-lanceolate, 6–12 × 1.25–3.5cm, acuminate or sometimes acute, base cuneate, margins finely serrate, glabrous or sparsely pubescent beneath, veins indistinct on upper surface, faintly prominent beneath; petioles 2–3mm. Sepals broadly elliptic, c 2 × 2mm, glabrous. Petals obovate, c 4–4.5 × 3mm, yellow. Male flowers: stamens c 15, anther cells not chambered. Female flowers: ovary subglobose, c 1.5mm diameter, glabrous, style base 0.5–1mm, branches 1.5–2mm. Fruit c 5mm diameter.

Bhutan: S – Chukka, Gaylegphug and Deothang districts, **C** –Thimphu,

Punakha, Tongsa and Tashigang districts, **N** – Upper Mo Chu district; **Sikkim.** Warm broad-leaved and Subtropical forests, 850–2300m. August–November.

3. E. cerasifolia (D. Don) Kobuski; *E. symplocina* Blume. Sha: *Khuberbu Shing;* Nep: *Bara Jhingni* (34).

Tree 2–7m. Buds and young branches appressed-pubescent. Leaves elliptic-lanceolate, 6–12 × 2–4cm, acuminate, base rounded or cuneate, margins entire or serrate in the upper half, brown when dry, usually blotched with green, lateral veins orange-coloured, ± prominent above and beneath. Sepals broadly elliptic, c 2.5 × 2mm, obtuse, appressed-pubescent externally. Petals yellow, oblong-elliptic, 4 × 2–2.5mm. Male flowers: stamens c 15, anther cells divided into 4–6 rounded chambers. Female flowers: ovary pubescent or glabrous, style base 2–2.25mm, branches 1.5–3mm. Fruit blue-black, broadly ellipsoid or subglobose, 5 × 4–5mm, glabrous or with a few hairs.

Bhutan: S – Chukka, Gaylegphug and Deothang districts, **C** – Punakha, Tongsa and Tashigang districts; **Sikkim.** Warm evergreen broad-leaved forests, 1700–2000m. October–November.

The record of *E. trichocarpa* Korthals from Bhutan (80) is based on a misidentification of this species. Timber valuable for fence posts and fuel (34).

2. TERNSTROEMIA L. f.

Evergreen trees. Leaves coriaceous. Flowers bisexual, solitary, axillary, pedicels usually recurved. Sepals 5. Petals 5, connate at base. Stamens numerous; anthers basifixed. Ovary 2–3-celled; style minutely 2–3-lobed. Fruit berry-like, indehiscent.

1. T. gymnanthera (Wight & Arnott) Sprague

Tree 4–12m. Leaves elliptic, oblanceolate or obovate, 4.5–11.5 × 1.5–5cm, acute or obtuse, base cuneate, glabrous, margins entire; petioles 0.5–1.5cm. Flowers fragrant; pedicels 7–15mm, slightly 2-ridged, bearing 2 minute ovate bracteoles near apex. Sepals broadly elliptic, 4–6 × 3–4mm, leathery, rounded at apex. Petals yellowish, oblong, 6–8 × 5–6mm, fleshy. Fruit ovoid, c 2 × 1.5cm, scarcely juicy.

Sikkim: locality unknown. 2100–2550m. June.

3. GORDONIA Ellis

Evergreen trees. Flowers bisexual, solitary, axillary or several at branch ends, 2–4-bracteolate. Sepals usually 5, grading in size from bracteoles to petals. Petals free or slightly connate at base. Stamens in 5 bundles or all shortly connate and adnate to base of petals. Ovary 3–5-celled; style stout, shortly lobed. Capsule oblong-ellipsoid, woody; seeds ellipsoid, 4–8 per cell, winged above.

1. G. excelsa Blume; *G. dipterosperma* Kurz. Nep. *Hinguwa* (34).

Leaves coriaceous, elliptic, 10–14 × 2.5–4.5cm, acuminate, base attenuate to petiole c 5mm, margins obscurely crenate-serrate, glabrous. Flowers ± sessile, 4–4.5cm across, bracts and sepals suborbicular, 5–12 × 7–10mm, greyish appressed-pubescent. Petals oblong, c 2 × 1.2cm, greyish pubescent outside, yellow or pink within. Capsule c 2.5 × 1.5cm.

Bhutan: S–Deothang district (Deothang), **C**–Tongsa district (Jirgang Chu); **Sikkim:** Mungpu and Sureil. Subtropical and Warm broad-leaved forests, 300–1370m. March–May.

4. SCHIMA Blume

Evergreen trees. Flowers bisexual, solitary or few in axils of uppermost leaves, sometimes forming a short raceme, pedicels bearing 2 deciduous bracteoles. Sepals 5, subequal. Petals 5, connate at base, the outer smallest, almost hooded, enveloping flower-bud. Stamens numerous, adnate to base of petals. Ovary 5-celled, style simple, broadened and lobed at apex. Capsule woody, subglobose, 5-valved; seeds 3 in each cell, compressed, kidney-shaped, bearing a semicircular wing.

1. S. wallichii (DC.) Korthals. Dz: *Puyam;* Sha: *Zala Shing;* Nep: *Chilaune* (34), *Gogra, Aule Chilaune* (34). Fig. 29 c–e.

Tree up to 30m. Leaves ovate-elliptic, 9–16 × 4–6cm, acute or acuminate, base rounded or cuneate, margins entire, glabrous above, usually pubescent beneath; petioles 1.5–2cm. Flowers fragrant, 3.5–5cm across, pedicels 2–3cm, bracteoles oblong, c 5 × 2mm. Sepals suborbicular, c 4mm, ciliate on margins, otherwise glabrous. Petals white, obovate, up to 3 × 1.5cm, minutely pubescent externally at base, otherwise glabrous. Stamens yellow. Capsule subglobose, 2–2.5cm diameter, woody, dehiscing in upper two-thirds into 5 valves. Seeds c 7 × 4mm, surrounded by wing 2–3mm.

Bhutan: S–Samchi to Gaylegphug district, **C**–Punakha district; **Sikkim.** Subtropical and Warm broad-leaved forests, 300–2000m. May–July.

Branches cut and burned as manure; timber durable if well seasoned, used for house-building and ploughshares (34). Contact with bark causes intense itching (16).

2. S. khasiana Dyer

Similar to *S. wallichii* but leaf margins serrate; bracts oblong-elliptic up to 2 × 1cm; flowers up to 6.25cm across; sepals and lower parts of petals densely white-pubescent.

Bhutan: C – Punakha, Tongsa, Mongar and Tashigang districts. Broad-leaved forests, 1900–3000m. July.

5. CAMELLIA L.

Evergreen trees or shrubs. Flowers 1–2, subterminal or axillary, shortly pedicellate, bisexual; bracteoles distinct from sepals or grading into and not distinct from them, deciduous. Sepals 5–6. Petals 7–8, free or united at base. Stamens numerous, free or united at base. Ovary 3-celled, styles 3, free or united at base. Capsule subglobose, leathery, 1–3-celled; seeds subglobose, unwinged.

1. C. kissii Wall.; *C. drupifera* Loureiro. Nep: *Hinguwa* (34).

Shrub or tree to 12m. Leaves elliptic-ovate, 5.5–9 × 1.7–3.5cm, acuminate, base cuneate or rounded, margins finely serrate near apex, pubescent beneath at first; petioles 3–7mm. Pedicels 2–6mm, erect, densely scarred. Bracteoles and sepals ± indistinguishable, suborbicular, 1.25–9mm, caducous. Petals broadly ovate, c 12 × 10mm. Ovary ovoid, c 4mm, densely pubescent, style 5–6mm, divided to middle. Capsule subglobose, 1.5–2.5 × 1.2–2.2cm.

Bhutan: S – Phuntsholing district (Kamji), Chukka district (Gedu) and Gaylegphug district (Dakpai), **C** – Tongsa district (Shamgong); **Sikkim.** Evergreen broad-leaved forests, 1600–2100m. October–February.

Timber hard and strong, used for axe and hoe handles and walking-sticks (34).

2. C. sinensis (L.) O. Kuntze; *C. thea* Link. Sha: *Jashing;* Nep: *Chha* (34); Eng: *Tea Plant.*

Similar to *C. kissii* but pedicels 1–1.5cm, usually recurved, not densely scarred, bearing at first two caducous, subopposite, ovate bracteoles (c 2mm) near middle; sepals 5–6, suborbicular, 3–5mm diameter, persistent; petals 1–2 × 0.8–2.3cm.

Bhutan: S – cultivated at Phuntsholing and Gedu; **Sikkim.** June–January.

There are two varieties in cultivation: the typical var. **sinensis** forming a bush or small tree 1–6m, with leaves 5–9 × 2–3cm, obtuse or with a short rounded point, and var. **assamica** (Masters) Kitamura forming a tall tree c 17m, and having large leaves 8–14(–22) × 3.5–5.5(–7.5)cm, ± acuminate at apex. Cultivated in the Terai for its valuable foliage which is used to prepare tea.

Family 57. GUTTIFERAE

by D.G. Long

Evergreen trees with resinous sap. Leaves opposite, decussate, simple, pinnately veined, coriaceous, glabrous, entire, usually exstipulate. Flowers bisexual or functionally unisexual, actinomorphic, solitary or in axillary fascicles or cymes, or terminal panicles. Sepals 4–5, free, imbricate. Petals 4–5, free, imbricate. Stamens numerous around ovary, filaments connate at extreme base or forming 4–5 bundles. Ovary superior, 1–12-celled, ovules 1–4 per cell, basal or axile, stigma sessile or borne on slender style, often peltate or 4–8-lobed. Male flowers with or without rudimentary ovary; female flowers with free or united staminodes. Fruit 1–8-seeded, fleshy and indehiscent or becoming woody and dehiscent; stigma persistent or not; calyx often persistent, sometimes accrescent around base of fruit.

1. Flowers often in 3–8-flowered axillary fascicles or cymes, rarely in panicles or racemes, functionally unisexual; female flowers with staminodes, males with or without rudimentary ovary; stigma sessile .... **1. Garcinia**
+ Flowers 1–2, axillary or terminal, or numerous in terminal panicles, bisexual; stigma on elongate style 2
2. Leaf veins close, parallel, straight and conspicuous; flowers c 1.5cm across; stigma peltate; fruit 1-seeded .... **2. Calophyllum**
+ Leaf veins widely spaced and arching upwards, or close and parallel but almost invisible; flowers 2.5–6cm across; stigma peltate or minutely 4-fid; fruit 1–4-seeded .... **3. Mesua**

1. GARCINIA L.

Trees with yellow resinous sap. Leaves rarely stipulate; lateral veins close and parallel or distantly spaced. Flowers functionally unisexual, usually in axillary fascicles or cymes, rarely solitary, or in terminal panicles or racemes, strongly scented, opening at night. Sepals concave, 4–5, imbricate. Petals 4–5, imbricate. Male flowers with numerous stamens forming a globose mass or annular ring around a central rudimentary ovary, or united into 4–5 bundles. Female flowers usually apparently bisexual but with 5–many free or connate staminodes; ovary 2–12-celled, ovules 1 per cell, stigma sessile, peltate or deeply lobed. Fruit a coriaceous 1–8-seeded berry.

Many species of the genus yield a gum and a yellow dye (camboge)from the resinous sap (126).

1. Flowers 5-merous; branchlets strongly angular; leaves oblong **5. G. xanthochymus**
+ Flowers 4-merous; branchlets terete or weakly striate; leaves ovate, lanceolate or elliptic 2
2. Lateral veins prominent beneath, widely-spaced 3
+ Lateral veins slender, closely parallel and not prominent 4
3. Leaves ovate-elliptic, coriaceous, with dark thread-like lines on lower surface .... **1. G. stipulata**
+ Leaves usually obovate, membranous, without dark thread-like lines on lower surface .... **2. G. sopsopia**

4. Leaves ovate, often rounded at base, thickly coriaceous, 9–20 × 4–9cm; branches supra-axillary .... **3. G. anomala**
\+ Leaves elliptic-lanceolate, attenuate at base, thinly coriaceous, 8–13 × 3–4cm; branches axillary .... **4. G. cowa**

1. G. stipulata Anderson. Nep: *Dur Lampate* (34).

Tree to 20m. Leaves coriaceous, ovate-elliptic, 12–23 × 4–9cm, shortly acuminate, base cuneate or almost rounded, veins 10–15 pairs, prominent beneath, lower leaf-surface with dark thread-like lines; petioles 1–1.5cm; stipules subulate, 5–7mm, deciduous. Male flowers in 4–6-flowered axillary cymes, pedicels 1–1.5cm; sepals suborbicular, concave, 7–9mm; petals narrowly ovate, c 1.5 × 0.7cm, yellow; stamens numerous in a ring around rudimentary ovary with conspicuous stigma. Female flowers 1–2 per axil, ovary 2-celled with peltate, flat-topped, unlobed stigma. Fruit oblong-ellipsoid, 2.5–3cm long, 2-seeded.

Bhutan: S – Deothang district (Raidong), **C** –Tongsa district (Mangde Chu Valley) and Mongar district (Shongar); **Sikkim:** Darjeeling and Kalimpong. Subtropical forests, 600–1400m. August–September.

Fruit edible (34).

2. G. sopsopia (Hamilton) Mabberley; *G. paniculata* Roxb. Nep: *Kusum.*

Similar to *G. stipulata* but leaves membranous, broadly elliptic or often obovate, 15–25 × 7–14cm, lower surface without dark thread-like lines; exstipulate; male flowers numerous in terminal panicles; female flowers in terminal racemes; fruit globose, 3–3.5cm diameter.

Bhutan: S – Sarbhang district (Phipsoo) and Gaylegphug district (Lodrai Khola near Gaylegphug). Subtropical forests, 300–310m. November–February.

Sometimes cultivated for its edible fruit.

3. G. anomala Planchon & Triana

Small tree; branchlets yellowish, conspicuously supra-axillary. Leaves ovate or elliptic, 9–20 × 4–9cm, bluntly pointed, base rounded, veins numerous, fine and closely parallel; petioles c 1cm. Cymes axillary, 3-flowered; peduncle 0.5–1cm, bearing 2 terminal leafy bracts 1–2cm long; pedicels up to 0.5cm with 2 scale-like bracteoles at base. Male flowers with suborbicular sepals 5–6mm, and oblong

FIG. 30. **Hypericaceae** and **Guttiferae. Hypericaceae**. a–d, *Hypericum uralum*: a, portion of shoot with leaves and flowers; b, petal; c, bundle of stamens; d, ovary. e–g, *Hypericum himalaicum*: e, leaf with gland-dots; f, flower; g, maturing fruit. **Guttiferae**. h–j, *Calophyllum polyanthum*: h, portion of shoot with leaf and inflorescence; i, flower; j, fruit. k–o, *Garcinia xanthochymus*: k, portion of angular stem with leaves showing sheath-like appendage at base of petiole; l, inflorescence; m, female flower, one sepal and two petals removed to show staminodes; n, male flower, two petals removed to show bundles of stamens and alternating fleshy glands; o, fruit. Scale: k × 2/5; h, j, o × ½; l × ⅔; a × 4/5; f × 1; b, c, d, e, i, m, n × 2; g × 2½. Drawn by M. Bates.

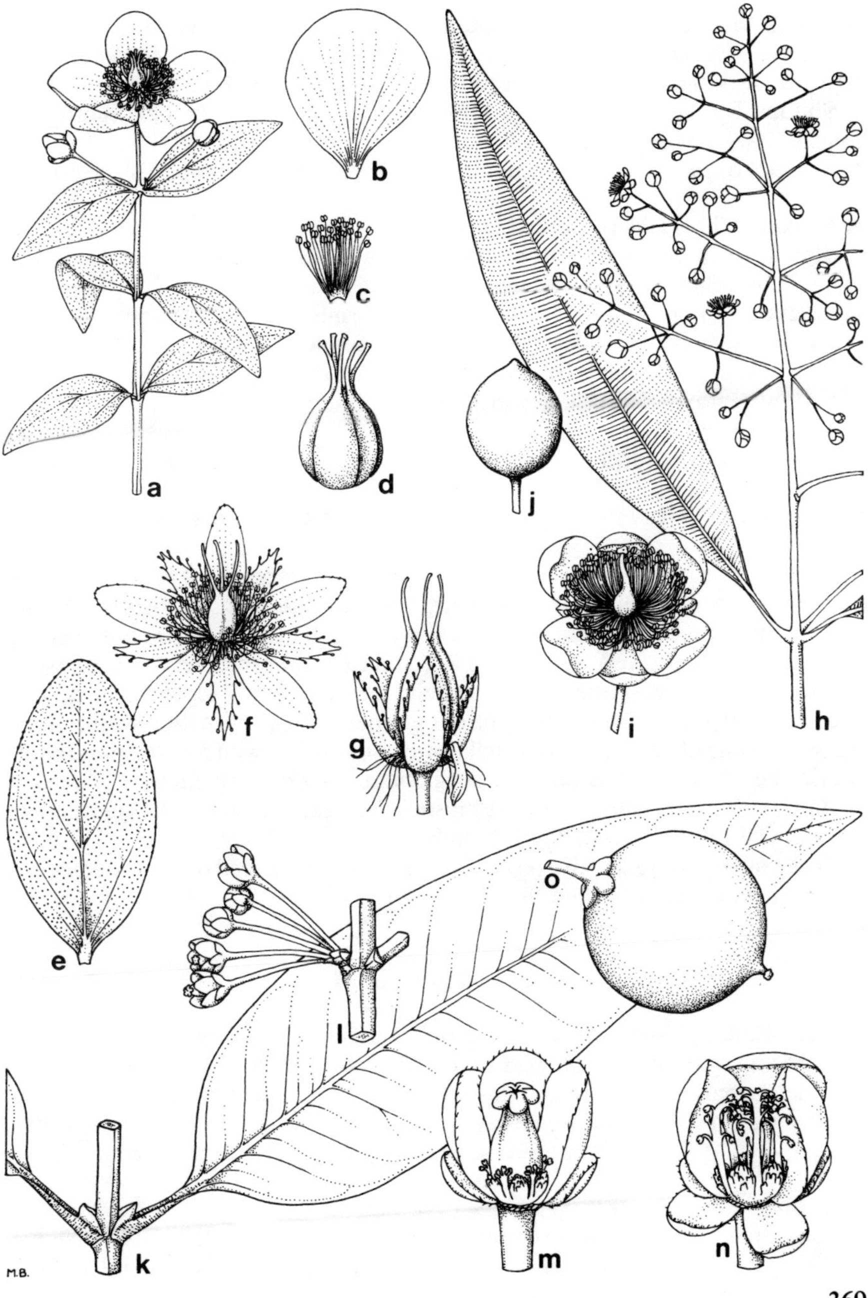
a
b
c
d
e
f
g
h
i
j
k
l
m
n
o
M.B.

yellowish petals 6–10 × 3–5mm; stamens as in *G. stipulata.* Female flowers with numerous staminodes around ovary; ovary 2-celled, stigma flat-topped with reflexed weakly lobed margin. Fruit ovoid, 3–4cm long, 1–2-seeded.

Sikkim: Darjeeling district and Jaldaka Valley on Bhutan border. Subtropical forests, 500–1200m. June–July.

Records of *G. cornea* L. from Sikkim (34) propably refer to this species.

4. G. cowa Choisy. Nep: *Kaphal* (34).

Similar to *G. anomala* but branchlets axillary; leaves more thinly coriaceous, elliptic-lanceolate, 8–13 × 3–4cm, acuminate, base attenuate; male flowers in fascicles of 5–7 on pedicels 3–5mm, stamens forming a dense cluster, rudimentary ovary absent; female flowers with 4 clusters of staminodes around 4–8-celled ovary, stigma sessile, 4–8-lobed; fruit globose, c 4cm, vertically grooved, 4–8-seeded.

Sikkim: Birick, Darjeeling and Gangtok. Subtropical forests, 200–1000m. February–May.

Resin used to produce a varnish (126).

5. G. xanthochymus Hook.f; *G. tinctoria* Dunn. Nep: *Chunyel, Thekra.* Fig. 30 k–o.

Tree with drooping angular branches. Leaves coriaceous, oblong, 15–35 × 5–9cm, shortly acuminate or apiculate, base cuneate, veins numerous, slightly prominent, margins thickened; petioles thick, 1–2cm. Male flowers in axillary fascicles of 4–8, pedicels 1–2cm; sepals 5, concave, suborbicular, 4–6mm; petals 5, white, obovate, 8–11mm long; filaments connate into 5 stout stalks c 5mm, bearing short free filaments and anthers at apex, stalks alternating with 5 fleshy glands. Female or apparently bisexual flowers similar but with 5-celled ovoid ovary bearing deeply 5-lobed stigma. Fruit subglobose, 5–6cm, dark yellow, 1–4-seeded.

Bhutan: S – Sarbhang district (Phipsoo); **Sikkim:** Kurseong, Darjeeling, etc. Subtropical forests, 300–1000m. March–May.

Fruit edible (34); yellow dye produced from bark (126); seed oil used medicinally in treatment of dysentry and diarrhoea (13).

2. CALOPHYLLUM L.

Trees. Leaves glossy, with very numerous, straight, crowded, parallel veins almost at right angles to midrib. Flowers in terminal sometimes leafy panicles, usually bisexual. Sepals 4, sometimes petal-like. Petals 4, imbricate. Stamens numerous, filaments connate only at extreme base. Ovary 1-celled with slender style and peltate stigma, ovule solitary. Fruit a drupe, sepals and stigma not persistent.

1. C. polyanthum Choisy. Nep: *Rate* (34). Fig. 30 h–j.

Large tree to 20(–45)m with compressed, angular branchlets. Leaves elliptic, 8–15 × 2–6cm, acuminate, base cuneate, margins undulate, lateral veins at

70 – 85° to midrib. Panicles broad, up to 15cm across. Pedicels 0.5 – 1cm. Sepals concave, 2 outer small, orbicular, c 3mm, inner 2 obovate, petaloid, c 5mm. Petals white, obovate, 6 – 7mm long. Stamens 3 – 4mm. Ovary globose with style 2 – 3mm. Fruit broadly ellipsoid or ovoid, pointed, 2.5 – 3.5cm long.

Bhutan: S – Samchi district (Tamangdhanra Forest) and Gaylegphug district (Lodrai Khola); **Sikkim:** terai and Darjeeling foothills. Subtropical forests, 300 – 1200m. April – June.

Wood very strong and durable (34, 48), used in boat-building (126).

3. MESUA L.

Trees. Leaf veins close, straight and parallel or more distant and arched. Flowers 1 – 2, terminal or axillary, or numerous in terminal panicles, bisexual. Sepals 4 imbricate. Petals 4, imbricate. Stamens numerous, filaments connate at extreme base. Ovary 1 – 2-celled; ovules 2 – 4 per cell; style elongate, stigma peltate or minutely 4-fid. Fruit 1 – 4-seeded, dehiscent, woody and nut-like or fleshy, indehiscent and drupe-like, surrounded by thickened accrescent calyx; stigma not peristent.

1. M. ferrea L. Nep: *Nagesuri* (34), *Nageswar;* Eng: *Iron Wood* (13).

Tree to 25m. Leaves elliptic-lanceolate, 10 – 15 × 2.5 – 4cm, acuminate, base rounded or acute, green and glossy above, whitish waxy beneath, pinkish when young; lateral veins very inconspicuous, numerous, fine and parallel, at c 80° to midrib; petioles 0.5 – 1cm. Flowers large, fragrant, 4 – 6cm diameter, 1 – 2 terminal or in axils of upper leaves. Pedicels c 5mm. Sepals suborbicular, concave, thickened, 1.5 – 2.5cm, puberulous. Petals white, obovate, 2 – 3 × 1.5 – 2cm. Anthers large, conspicuous, yellow. Style curved, with peltate stigma. Fruit ovoid, dehiscent, woody and nut-like, c 2.5 × 1.5cm, with pointed apex, 1 – 4-seeded.

Sikkim: Darjeeling district, Jaldaka Valley, etc. Subtropical forests, 300 – 1000m. February – April.

The Himalayan plants belong to var. **salicina** (Planchon & Triana) Vesque (var. *ferrea* auct. non L.); sometimes cultivated. Valued for its durable timber and ornamental, fragrant flowers (16,48); oil from the flowers used as perfume, and bark, leaves, flowers and seeds used medicinally (13).

2. M. floribunda (Wall.) Kostermans; *Kayea floribunda* Wall.

Distinguished from *M. ferrea* by its larger leaves, 12 – 18 × 3 – 6cm, green on both surfaces, lateral veins more conspicuous, distant and arching upwards; petioles 1 – 2cm; smaller flowers c 2.5cm diameter, numerous in terminal panicles 7 – 10cm long; sepals c 5mm; petals ovate-oblong, c 10 × 5mm, white or pinkish; anthers minute, stigma minutely 4-fid; fruit subglobose, fleshy, indehiscent and drupe-like, 1 – 2-seeded, c 3.5cm diameter.

Sikkim: Darjeeling. Subtropical forests, 200 – 900m. March – July.

Family 58. HYPERICACEAE

by N.K.B. Robson & D.G. Long

Perennial herbs or low shrubs, glabrous. Leaves opposite, simple, pinnately veined, entire, often pellucid-dotted or -streaked and with blackish gland dots, exstipulate. Flowers bisexual, actinomorphic, axillary or terminal, solitary or in cymes, sometimes corymbose. Sepals 4–5, free, imbricate, green, often glandular. Petals 4–5, free, yellow. Stamens in 5 bundles opposite petals, the bundles sometimes uniting in pairs or into a ring, each bundle with numerous stamens united at base. Ovary superior, 1–5-celled, ovules 2–many per cell, axile or parietal; styles 3–5, free. Fruit a 3–5-valved capsule with numerous seeds, rarely becoming fleshy and indehiscent.

1. HYPERICUM L.

Description as for Hypericaceae.

1. Shrubs (dwarf in *H. reptans*); styles and staminal bundles 5; petals and stamens deciduous 2
\+ Herbs; styles 3–4, stamens united into 3–4 bundles or a ring; petals and stamens persistent 10
2. Prostrate subshrub; stems creeping and rooting at nodes ....**1. H. reptans**
\+ Erect or spreading shrubs or subshrubs; stems not creeping and rooting ...3
3. Leaves cordate at base, reticulate; cymes many-flowered, sepals triangular, 3–4mm ....**2. H. griffithii**
\+ Leaf bases rounded or cuneate; leaves not reticulate; cymes usually 1–7-flowered; sepals elliptic, lanceolate, obovate or orbicular, 5–15mm ...4
4. Sepals spreading in bud and fruit, elliptic, lanceolate or ovate, broadest at or below middle, usually acute or acuminate 5
\+ Sepals erect in bud and fruit, suborbicular or obovate, broadest above middle, usually rounded at apex 7
5. Leaves elliptic to elliptic-oblong, 1–1.4cm........................**3. H. sherriffii**
\+ Leaves lanceolate to narrowly ovate, 2.5–7cm 6
6. Petioles 2–4mm; flowers 4–7cm across; sepals usually elliptic or ovate, 8–18 × 3–7mm ....**4. H. choisianum**
\+ Petioles 0.5–2mm; flowers 2.5–3.5cm across; sepals lanceolate to oblanceolate, 7–8 × 2mm ....**5. H. gracilipes**

7. Flowers usually borne from uppermost node, 4 – 6cm across; sepals 7 – 10mm; superficial gland dots few or none on lower leaf surface **6. H. hookerianum**
+ Flowers usually borne from 2 or more nodes, 2.5 – 4cm across; sepals 4 – 7mm; superficial gland dots abundant on lower leaf surface........................8

8. Flowers 3 – 4cm across; petals 1.5 – 2cm; stamens 9 – 12mm..**8. H. williamsii**
+ Flowers 2.5 – 3cm across; petals 1 – 1.3cm; stamens 5 – 6mm 9

9. Branchlets 4-winged or ribbed; leaves ovate-lanceolate; cymes 3 – 5-flowered **7. H. uralum**
+ Branchlets terete; leaves ovate-elliptic; cymes 1 – 2-flowered..**9. H. tenuicaule**

10. Stems quadrangular; anthers, leaves and sepals without black or dark reddish glands; capsule smooth...11
+ Stems terete or sometimes with 2 slender lines; anthers and usually leaves and sepals with black or dark reddish glands; capsule ribbed....................12

11. Suberect or diffuse herb; leaves broadly elliptic-oblong or ovate, suberect or spreading; flowers 7 – 10mm across; petals 3 – 4mm, shorter than sepals **10. H. japonicum**
+ Slender erect herb; leaves linear-oblong, closely appressed to stem; flowers 10 – 12mm across; petals 5 – 6mm, longer than sepals .....**11. H. gramineum**

12. Leaves lanceolate, rarely ovate-lanceolate, acute or subacute, with numerous conspicuous large pellucid dots; styles long, 1.5 – 2.5 × length of ovary...**12. H. elodeoides**
+ Leaves ovate or broadly elliptic, obtuse; pellucid dots small and inconspicuous (except in *H. ludlowii* and *H. petiolulatum*); styles short, 0.7 – 1.3 × length of ovary 13

13. Flowers 12 – 20mm across; sepals elliptic-lanceolate, glandular-ciliate, rarely entire, with few or no sessile marginal gland dots; erect herbs 14
+ Flowers 8 – 11mm across; sepals linear, entire or ovate and glandular-ciliate but then usually with numerous black sessile glands; prostrate, decumbent or erect herbs...15

14. Flowers solitary or 2 – 3 in cymes, subsessile and sheathed by eciliate leaf-like bracts ....**13. H. monanthemum**
+ Cymes 3 – 12-flowered; flowers on pedicels 2 – 4mm, subtended but not sheathed by glandular-ciliate bracts ....**14. H. himalaicum**

15. Upper leaves often glandular-ciliate near base; flowers numerous in dense corymbs; sepals ovate with glandular-ciliate margins and numerous black sessile gland dots..**15. H. wightianum**
+ Upper leaves entire; flowers solitary or in 2 – 3-flowered cymes; sepals linear-lanceolate, entire or occasionally glandular-ciliate, black or dark red sessile gland dots absent or few 16

16. Prostrate or decumbent much-branched herb; leaves borne on petioles 1–4mm; flowers solitary or 2–3 in cymes ....**16. H. petiolulatum**
+ Slender erect little-branched herb; leaves subsessile; flowers solitary **17. H. ludlowii**

1. H. reptans Dyer

Prostrate subshrub with long creeping stems, branches 15–30cm, suberect or decumbent when young, later rooting. Leaves broadly elliptic, 8–14 × 3–5mm, obtuse, base acute, without blackish glands but with linear pellucid blotches; sessile. Flowers solitary, terminal, 2.5–3cm across. Sepals obovate, leaf-like, 7–9 × 3.5–5mm, eglandular. Petals asymmetrically obovate, 1.2–1.5 × c 1.4cm, yellow often tinged red. Stamens c 5mm. Ovary subglobose, 4–5mm, styles c 4mm, recurved at apex. Fruit globose, c 8mm diameter, somewhat fleshy and indehiscent.

Sikkim: Lachen, Lachung, etc. Mountain rocks and screes, 2700–4000m. July–September.

2. H. griffithii Dyer

Shrub 1–3m. Leaves ovate, 6–9 × 3–4.5cm, obtuse with minute mucro, base cordate, thinly coriaceous, delicately reticulate, pellucid-dotted but without superficial gland dots; sessile. Flowers in terminal many-flowered corymbose cymes 6–12cm across. Sepals triangular, acute, 3–4 × 2mm. Petals spathulate, 1.5–2 × 0.7–0.8cm. Stamens 8–10mm. Ovary ovoid, c 6 × 3mm; styles longer than ovary, 9–10mm. Capsules woody, 1–1.2 × 0.7–1cm.

Bhutan: S – Samchi district (Torsa Valley), **C** – Mongar district (Kuru Chu near Mongar) and Tashigang district (Ghunkara, Bamri Chu and Gamri Chu Valleys.) Scrub in dry valleys, sometimes under Chir pine, 1060–2000m. March–April.

A local E Himalayan endemic, known outside Bhutan only from Arunachal Pradesh. Records for Chumbi are based on a specimen from the Torsa Valley, Bhutan.

3. H. sherriffii Robson

Spreading shrub 0.6–1.2m. Leaves elliptic to elliptic-oblong, 1.0–1.4 × 0.3–0.5cm, acute or subacute-apiculate, base cuneate, glaucous beneath, with pellucid short streaks and dots, sessile or on very short petiole. Flowers solitary, diameter unknown. Sepals spreading (in fruit), linear-elliptic, 5–8 × 1.5–2mm, acute. Petals, stamens and ovary unknown. Capsule ellipsoid-ovoid, 0.8–1.1cm long.

Bhutan: S – Deothang district (Chungkar). On cliffs and steep rocky slopes, 2100m.

Endemic to Bhutan.

4. H. choisianum Robson; *H. hookerianum* Wight & Arnott var. *leschenaultii* sensu F.B.I. non *H. leschenaultii* Choisy, *Norysca hookeriana* (Wight & Arnott) Wight var. *leschenaultii* Kimura (69, 71).

Bushy shrub 1–2m. Leaves lanceolate or narrowly ovate, 2.5–7 × 1–2(–3)cm, tapering to subacute or bluntly apiculate apex, base usually rounded, sometimes broadly cuneate, pale green beneath but not glaucous, pellucid lines and dots present but superficial glands absent or very few; petioles 2–4mm. Flowers solitary or in threes, 4–7cm diameter. Sepals usually spreading, elliptic or ovate, 8–18 × 3–7mm, broadest at middle, tapering to sharp point, rarely subacute. Petals asymmetrically obovate, 1.5–2.5(–3) × 1.2–2(–2.5)cm. Stamens 7–10mm. Ovary ovoid, 7–8mm, bearing styles 0.5–0.7 × as long. Capsules ovoid, c 1.5cm.

Bhutan: C–Thimphu district (Paro, Dochu La and Sinchu La), Bumthang district (N of Byakar and Badar La) and Mongar district (near Sengor and Donga La), **N**–Upper Mo Chu district (N of Gasa) and Upper Kulong Chu district (Lao); **Sikkim.** Cool broad-leaved and Hemlock forests, 2600–3300m (–4000m in Sikkim). June–July.

Records from Sikkim of *H. lysimachioides* Dyer, a native of the W Himalaya (34) probably refer to this species.

5. H. gracilipes Fischer

Similar to *H. choisianum* but leaves without pellucid blotches, with numerous superficial gland dots; petioles 0.5–2mm; cymes usually 3–7-flowered; flowers 2.5–3.5cm across; sepals lanceolate to oblanceolate, 7–8 × c 2mm; petals 1.3–1.5cm; stamens 6–7mm; ovary c 5mm bearing styles ± as long; capsule c 1.2cm.

Sikkim: Ghoom. 2100–2300m. June–July.

6. H. hookerianum Wight & Arnott; *Norysca hookeriana* (Wight & Arnott) Wight. Nep: *Mehandiphul* (34).

Similar to *H. choisianum* but leaves often with a few superficial gland dots beneath; flowers sometimes solitary but often in 3–9-flowered cymes, usually larger, 4–6cm across; sepals erect, obovate or suborbicular, 7–10 × 5–8mm, usually rounded; petals 2.3–3 × 2–2.5cm.

Bhutan: S–Deothang district (N of Tshilingor), **C**–Thimphu, Punakha, Tongsa, Bumthang and Tashigang districts, **N**–Upper Bumthang Chu and Upper Kulong Chu districts; **Sikkim**. Moist forests, especially Cool broad-leaved and Fir, 1800–3300m. June–August.

Some specimens with smaller flowers and mucronulate sepals may be hybrids with *H. uralum.*

7. H. uralum D. Don; *H. patulum* sensu F.B.I. non Murray, *Norysca urala* (D. Don) K. Koch. Nep: *Urilo* (34). Fig. 30 a–d.

Bushy shrub, 1–2m, with numerous spreading branchlets; branchlets compressed and 4-winged or ribbed. Leaves ovate-lanceolate, 1.5–3(–4) × 0.5–1.2(–2)cm, acute or bluntly pointed, base cuneate, sessile, glaucous-white beneath with dense blackish superficial gland dots, pellucid blotches usually inconspicuous. Cymes mostly 3–5-flowered. Flowers 2.5–3cm across. Sepals obovate, 4–7 × 3–4mm, obtuse. Petals asymmetrically obovate, 1–1.3 × 0.8–1cm. Stamens c 5mm. Ovary subglobose, 3–5mm, bearing styles 0.7–0.9 × as long. Capsules subglobose, 7–11mm.

Bhutan: **S** – Deothang district, **C** –Thimphu to Tashigang districts, **N** – Upper Mo Chu district; **Sikkim.** Open hillsides and margins of Blue pine, Cool broad-leaved and Hemlock forests, 2100–3000m. July–August.

Seeds used medicinally (13).

8. H. williamsii Robson

Similar to large-leaved forms of *H. uralum* but without numerous spreading branchlets; branchlets terete; leaves ovate, 2–5 × 1–2.5cm, tapering to obtuse, minutely apiculate apex, base broadly cuneate or rounded; petioles 1–2mm; flowers larger, 3–4cm across; sepals 6–9 × 3–5.5mm; petals 1.5–2 × 1.3–1.5cm; stamens longer, 9–12mm; ovary ovoid, 6–7mm, bearing styles ± as long; capsules ovoid, c 1.3 × 1cm.

Sikkim: Gangtok and Rumtok. 1700–1800m. June–July.

9. H. tenuicaule Dyer; *Norysca urala* (D. Don) K. Koch. var. *angustifolia* Kimura

Similar to *H. uralum* and *H. williamsii* but branchlets terete; leaves ovate-elliptic, 2.5–4cm, subacute but not mucronate; flowers usually 1–2, 2.5–3cm across; sepals 4–5mm; petals 1–1.3cm; stamens 5–6mm; styles as long as ovary.

Sikkim: Kulhait and Lachen valley. 2100–2400m. August.

10. H. japonicum Murray; *H. laxum* (Blume) Koidzumi, *Sarothra laxa* (Blume) Kimura

Suberect or diffuse, sometimes prostrate herb, 7–25cm, stems quadrangular, arching and rooting near base. Leaves suberect or spreading, broadly elliptic-oblong or ovate, 5–10 × 2–6mm; obtuse, base rounded, pellucid-dotted, 3-veined, sessile. Flowers 7–10mm across, in terminal dichotomous often broad cymes. Sepals elliptic-obovate, 3.5–6mm, acute or subobtuse, with pellucid dots and lines. Petals 3–4mm usually distinctly shorter than sepals. Stamens c 3mm, narrowly fused into a ring at base. Ovary ovoid, 1.5–2mm, bearing 3 much shorter straight styles c 0.5mm. Capsules ovoid, c 3mm.

Bhutan: **S** – Samchi and Gaylegphug districts, **C** – Ha to Tashigang districts, common; **Sikkim.** Damp roadsides and grassy slopes, 600–2300m. February–August.

A variable species; some plants from Punakha and Tashigang districts approach *H. gramineum* in their more erect habit and larger flowers, with petals 4.5–5mm and subequal or slightly longer sepals.

11. H. gramineum G. Forster; *H. lalandii* sensu F.B.I. non Choisy

Similar to *H. japonicum* but a more slender, erect, little-branched herb; leaves closely appressed to stem, linear oblong, 2–4mm broad, acute or obtuse; cymes more compact; flowers larger, 10–12mm across; sepals more narrowly elliptic, acute; petals 5–6mm, usually longer than sepals; ovary ovoid-ellipsoid c 2mm, bearing longer styles c 1mm; capsules cylindric-ovoid, c 5mm.

Bhutan: **S** – Chukka district (Chukka), **C** – Thimphu, Punakha, Tongsa, Bumthang and Tashigang districts, **N** – Upper Mo Chu district (Gasa and Laya). Marshy meadows, 1800–3600m. June–August.

12. H. elodeoides Choisy; *H. napaulense* Choisy sensu F.B.I. p.p.

Erect usually unbranched herb 15–35cm. Leaves lanceolate, rarely ovate-lanceolate, 1.5–4 × 0.4–1.2cm, acute or subacute, base cordate and often glandular-ciliate, sessile, veins prominent beneath, with numerous conspicuous larger pellucid dots and smaller blackish marginal glands. Flowers 3–15 in broad terminal corymbose cymes; bracts lanceolate, glandular-ciliate. Flowers 1.8–2cm across, on pedicels 3–12mm. Sepals narrowly elliptic-lanceolate, 6–7 × 1–1.5mm, with conspicuous glandular lines, margins glandular-ciliate and with a few marginal dots. Petals oblanceolate, 9–10mm, with a few black glands. Ovary narrowly ovoid, c 2.5mm, bearing elongate styles 1.5–2 × as long. Capsules ovoid, 5–8mm, longitudinally ribbed.

Bhutan: **C** – Thimphu, Tongsa, Bumthang, Mongar and Tashigang districts, **N** – Upper Mo Chu district (Kohina); **Sikkim.** Clearings and margins of Hemlock and Fir forests, 2100–3300m. June–August.

13. H. monanthemum Dyer; *H. filicaule* (Dyer) Robson, *Ascyrum filicaule* Dyer

Slender, often unbranched, erect herb 9–25cm. Leaves elliptic, 1–1.5 × 0.5–1cm, obtuse, sessile, minutely pellucid-dotted and with scattered blackish marginal glands. Flowers solitary or in 2–3-flowered cymes, sheathed by eciliate leaf-like bracts. Flowers often 4-merous, 1.6–2cm across, subsessile. Sepals elliptic, 4–6 × 2–3mm, acute, margins glandular-ciliate but without sessile gland dots. Petals 8–10mm. Ovary ovoid or subglobose, styles ± equalling ovary.

Bhutan: **N** – Upper Pho Chu (Leji) and Upper Kulong Chu (Me La) districts; **Sikkim:** Lachung; **Chumbi.** Exposed hillsides, 3600–4100m. June–August.

14. H. himalaicum Robson. Fig. 30 e–g.

Similar to *H. monanthemum* but often with short slender branches; leaves larger, 1.2–2.6 × 0.6–1.4cm, pellucid dots usually few and inconspicuous; flowers in 3–12-flowered cymes; pedicels 2–4mm; bracts glandular-ciliate, not leaf-like or sheathing; sepals elliptic-lanceolate, 1–1.5mm broad; petals 6–9mm.

Bhutan: **C** – Thimphu district (Sinchu La); **Sikkim:** Jongri, Tonglo, etc. Clearings in Hemlock and Fir forests, 3000–3900m. July–August.

A specimen from Thimphu district (Chenkaphug) is intermediate between *H. himalaicum* and *H. elodeoides* and may be a hybrid.

15. H. wightianum Wight & Arnott; *H. humifusum* sensu F.B.I. non L., *H. napaulense* sensu F.B.I. p.p. non Choisy

Rigid erect herb 15–45cm, little branched except in upper part. Leaves elliptic to obovate, 2–3 × 1–1.5cm, obtuse, base rounded or cordate and glandular-ciliate, minutely pellucid-dotted, sessile. Flowers numerous in dense terminal corymbs; pedicels short, 1–3mm. Bracts ovate, acute, glandular-ciliate. Flowers 9–11mm across. Sepals ovate, 5–6 × 2.5–3mm, acute, glandular-ciliate and with numerous black sessile gland dots. Petals 4–5mm. Ovary ovoid or globose, styles shorter than or equalling ovary. Capsule ovoid or subglobose.

Bhutan: S – Chukka district (Chukka and Marichong), **C** –Thimphu district (Thimphu Chu Valley) and Mongar district (Sengor); **Sikkim:** Darjeeling and Kalimpong. Grassy hillsides and river beds, 1000–3000m. June–July.

16. H. petiolulatum Dyer

Similar to *H. wightianum* but a slender, prostrate or decumbent herb; leaves ovate, 0.5–1.5 × 0.5–1cm, obtuse, base rounded or cuneate, without marginal cilia; distinctly petiolate with petiole 1–4mm; flowers 1–3, terminal and axillary, on slender pedicels 2–13mm; sepals linear, 3–4mm, entire, with a few black sessile glands.

Bhutan: C –Thimphu, Punakha, Tongsa, Mongar and Tashigang districts; **Sikkim.** Damp clearings and ditches in Cool broad-leaved and Hemlock forests, 2050–3000m. July–August.

17. H. ludlowii Robson

Similar to *H. wightianum* and *H. petiolulatum* but a slender, erect, little-branched herb; leaves elliptic, 5–10 × 3–5mm, subsessile; flowers solitary, on pedicels 2–4mm; sepals linear-lanceolate, entire or weakly glandular-ciliate; capsules ellipsoid.

Bhutan: C – Bumthang district (Dhur). Marshy mountain meadows, 3050m. July.

Family 59. DROSERACEAE

by D.G. Long

Perennial insectivorous herbs. Leaves alternate on stems or in basal rosettes, densely covered with stalked insect-trapping glands, stipulate or exstipulate. Flowers in subterminal or terminal racemes on leafy stems or leafless scapes, bisexual, actinomorphic. Sepals (4–)5, connate at base, persistent. Petals 5, free, persistent. Stamens 5. Ovary superior, 1-locular; ovules parietal, numerous; styles 3–5. Fruit a 3–5-valved capsule, seeds numerous.

1. DROSERA L.

Description as for Droseraceae

1. Leaves borne on elongate stem, exstipulate .... **1. D. peltata**
\+ Leaves in a basal rosette, stipulate .... **2. D. burmannii**

1. D. peltata Thunberg. Fig. 32 j–m.

Stems erect, 7–19cm, simple or with a few short branches, arising from underground tuber. Basal leaves usually absent; stem leaves numerous, alternate, lamina peltate, semicircular with attenuate tails at basal angles, c 4mm across, margins with spreading gland-tipped cilia 1–3mm, upper surface with short-stalked glands; petioles 4–11mm. Racemes subterminal, 1.5–2.5cm, 4–6-flowered; pedicels slender 2–9mm. Sepals ovate, c 2 × 1.3mm, glandular-ciliate. Petals obovate, white, 4–6 × 2–3mm. Styles 3, each deeply divided into filiform segments. Capsules subglobose, c 2mm diameter, enclosed by persistent sepals and petals.

Bhutan: S – Chukka district, **C** –Thimphu, Punakha, Tongsa, Bumthang and Tashigang districts, **N** – Upper Mangde Chu district; **Sikkim.** Marshes and damp meadows, 1800–3600m. June–September.

The E Himalayan plants belong to subsp. **peltata** (var. *lunata* (DC.) Clarke).

2. D. burmannii Vahl

Differs from *D. peltata* in having a persistent basal rosette of leaves, each obovate, 7–10 × 4–6mm, with deeply divided stipules, subsessile; racemes terminal on leafless scapes 5–17cm; sepals not ciliate; styles 5, undivided.

Sikkim: Jalpaiguri terai. Marshy ground, 200m. January–February.

Family 60. FUMARIACEAE

by D.G. Long

Annual or perennial usually glabrous herbs, sometimes climbing, roots often tuberous. Leaves alternate, sometimes the upper opposite, mostly 1–4-pinnatisect or-ternatisect, exstipulate. Flowers in racemes or spikes, sometimes umbellate in *Corydalis,* cymose in *Hypecoum,* corymbose in *Dicentra,* zygomorphic, sometimes disymmetric (see below), bisexual. Sepals 2, small. Petals 4 in 2 dissimilar pairs, outer pair larger, one or both spurred or swollen at base (except *Hypecoum*); inner pair narrower, often coherent. Petal spurs enclosing a nectariferous gland. Stamens usually 6, connate into two bundles opposite outer petals, each bundle with a central 2-celled anther and 2 lateral 1-celled anthers, or stamens 4, free in *Hypecoum.* Ovary superior, 1-celled with 2 parietal placentae, each bearing 1–many ovules; style 1.

Fruit a 2-valved few to many seeded capsule or an indehiscent 1-seeded nutlet, or fragmenting into 1-seeded segments in *Hypecoum.*

In Fumariaceae the flowers are zygomorphic, either with one longitudinal plane of symmetry, or disymmetric, with both transverse and longitudinal planes of symmetry.

1. Petals not spurred or gibbous; stamens 4, free; fruit fragmenting into 1-seeded segments .... **1. Hypecoum**
\+ One or both outer petals gibbous at base or spurred; stamens 6, united into 2 bundles of 3; fruit dehiscent, not fragmenting, or a 1-seeded nutlet 2
2. Extensive climbing herbs; leaves often tendrillar; flowers disymmetric; outer petals both gibbous at base..**2. Dicentra**
\+ Erect, bushy or scrambling herbs; leaves never tendrillar; flowers zygomorphic with only longitudinal plane of symmetry; one of outer pair of petals spurred at base 3
3. Bushy or scrambling, weedy, annual herbs; fruit a 1-seeded nutlet **3. Fumaria**
\+ Erect or diffuse, perennial, mostly alpine herbs, rarely weedy; fruit a few to many-seeded capsule ....**4. Corydalis**

1. HYPECOUM L.

Slender annuals. Leaves in a rosette at base, pinnatisect; stem leaves few. Flowers few in elongate cymes; disymmetric. Outer petals not gibbous or spurred, obovate, entire, inner pair 3-lobed. Stamens 4, free, opposite petals, anthers 2-celled. Ovary linear bearing 2 erect stigmas on short style. Fruit linear, transversely septate, fragmenting into 1-seeded segments.

1. H. leptocarpum Hook. f. & Thomson. Fig. 31 a–d.

Basal leaves bipinnatisect, 7–17 × 1–2cm; pinnae 4–8 pairs, sessile, distant, ovate, deeply divided into elliptic, acute lobes; petiole base winged. Flowering stems slender, 15–35cm, branched; stem leaves few, 1–2cm, with linear lobes. Pedicels

FIG. 31. **Fumariaceae** and **Nymphaeaceae. Fumariaceae** a–d, *Hypecoum leptocarpum*: a, basal leaf; b, inflorescence; c, dissected flower; d, portion of capsule. e–j, *Dicentra scandens*: e, flower (nectariferous glands stippled); f, outer petal; g, coherent inner petals; h, staminal bundle; i, ovary; j, capsule. k–n, *Corydalis leptocarpa*: k, upper portion of stem with leaf and inflorescence; l, flower (basal nectariferous gland, ovary and stigma stippled); m, ovary; n, mature capsule. o & p, *Fumaria indica*: o, flower (nectariferous gland stippled); p, fruit. **Nymphaeaceae**. q–s, *Nelumbo nucifera*: q, flower; r, peltate leaf, lamina; s, mature receptacle with nut-like fruits set in pits. Scale: r × 1/6; q × ¼; s × ½; k × ⅔; a, h, i, j × 1; b × 1¼; e × 1⅓; f, g × 1½; l × 1⅔; n × 1¾; c, d × 2; m × 3; o × 5; p × 7. Drawn by M. Bates.

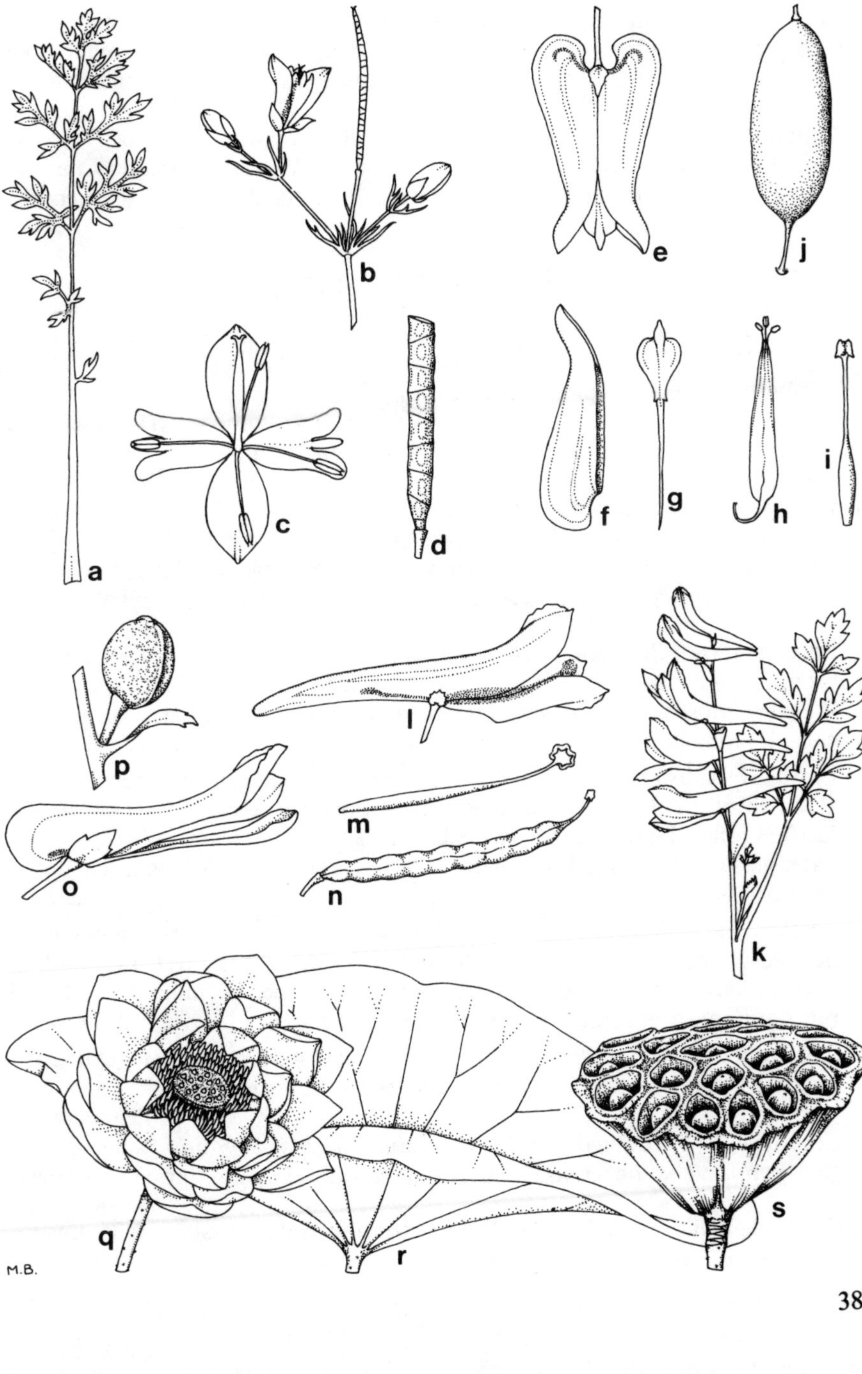
a
b
c
d
e
f
g
h
i
j
k
l
m
n
o
p
q
r
s
M.B.

slender, bracteoles filiform. Sepals c 2.5mm. Petals pinkish or white, often yellow inside; outer pair 5–7mm; inner pair 5–6mm, 3-lobed in upper third, outer lobes obtuse, median narrower and hooded. Fruit 10–30 × 1mm, fragmenting into 8–18 segments.

Bhutan: C – Ha district (Ha) and Bumthang district (Ura), **N** – Upper Mo Chu district (Lingshi) and Upper Bumthang Chu district (Waitang); **Sikkim; Chumbi.** Field margins, stony hillsides and river banks, 2750–4000m. April–September.

2. DICENTRA Bernhardi

Climbing herbs with perennial rootstock. Leaves repeatedly ternate with terminal leaflet usually replaced by a branched tendril; leaflets entire, 3–7-veined from base, petiolulate. Flowers in pendulous leaf-opposed racemes, often corymbose, disymmetric. Outer petals coherent and keeled, forming a compressed tube, gibbous at base, tips hooded and divergent; inner petals exposed at mouth, with long slender claw and connivent, hooded, winged limb enclosing stigma and anthers. Stamens 6 in 2 bundles, filaments adherent to outer petals, each bundle expanded at base into a nectariferous gland enclosed by swollen base of outer petal. Ovary elongate, style slender, stigma flattened; ovules many. Capsule ellipsoid or cylindric, dehiscing by 2 valves leaving placentae attached to persistent style. Seeds papillate.

1. Racemes mostly 7–14-flowered; bracts and sepals entire at base; capsules short, broad, 1.5–3.5 × 0.5–1cm 2
\+ Racemes mostly 2–6-flowered; bracts and sepals toothed, at least at base; capsules long and narrow, 3.5–5 × 0.2–0.5cm 4

2 Leaflets broadly ovate, 2.5–5 × 2–3cm, 3–6-veined from base; nectariferous gland short, thick, 4–5 × c 1.7mm, not hooked at apex .. **3. D. paucinervia**
\+ Leaflets ovate-elliptic or ovate-lanceolate, 1–3.5 × 0.5–2cm, 5–7-veined from base; nectariferous gland narrower, 3.5–5 × 0.6–1mm, deflexed, hooked at apex or not 3

3. Leaflets ovate-elliptic; nectariferous gland 4–5mm long, deflexed and strongly hooked at apex; capsules ovoid-ellipsoid .... **1. D. scandens**
\+ Leaflets narrowly ovate-lanceolate; nectariferous gland c 3.5mm long, deflexed but not hooked at apex; capsules narrowly ellipsoid .... **2. D. ventii**

4. Corymbs 2–6-flowered; bracts 4–8mm; sepals 4–6mm; outer petals 13–15mm, c 2mm broad at insertion; capsules 2–3mm broad **4. D. lichiangensis**
\+ Corymbs 2–3-flowered; bracts 9–12mm; sepals 7–9mm; outer petals 16–19mm, c 3–5mm broad at insertion; capsules 3–5mm broad **5. D. roylei**

1. D. scandens (D. Don) Walpers; *D. thalictrifolia* (Wall.) Hook. f. & Thomson, *Dactylicapnos scandens* (D. Don) Hutchinson. Fig. 31 e–j.

Climber to 3m. Leaves 2–3-ternate; leaflets ovate-elliptic, 1–3.5 × 0.5–2cm, obtuse or acute, base cuneate, 5–7-veined from base. Racemes 7–10-flowered, on peduncles 2–7cm. Bracts lanceolate, 4–10mm, entire. Sepals triangular, 2–3mm, entire. Petals yellow; outer pair 1.8–2cm. Nectariferous glands 4–5 × 0.8–1mm, strongly hooked at apex. Capsules ovoid-ellipsoid, 14–20 × 6–8mm.

Bhutan: **S**–Deothang district, **C**–Punakha, Tongsa, Mongar and Sakden districts; **Sikkim.** Climbing on shrubs in Cool broad-leaved forests, 2000–2750m. June–September.

2. D. ventii Khanh

Closely allied to *D. scandens* but leaflets more narrowly ovate-lanceolate, 2.2–3 × 1–1.4cm; flowers with short nectariferous glands, c 3.5 × 0.6–1mm, deflexed but not hooked; capsules more narrowly ellipsoid, 17–22 × 4–5mm.

Sikkim: Tonglo, Bijan, etc. Climbing on shrubs, 1500–2700m. July–August.

3. D. paucinervia Stern

Similar to *D. scandens* and *D. ventii* but leaflets larger, broadly ovate, 2.5–5 × 2–3cm, 3–6-veined from base; flowers with short, thick nectariferous glands, 4–5 × c 1.7mm, not hooked, but often producing a waxy secretion; capsules 12–20 × 4–8mm.

Sikkim: Kurseong, Darjeeling, etc. Warm broad-leaved forests, 900–1500m. May–August.

4. D. lichiangensis Fedde

Slender climber. Leaves biternate; leaflets narrowly ovate or elliptic, 1–2 × 0.5–1cm, obtuse or acute, base cuneate, 7–10-nerved. Flowers 2–6 in corymbose cymes on peduncles 2–4cm. Bracts linear-lanceolate, 4–8 × c 1mm, ciliate-dentate. Sepals similar to bracts but shorter, 4–6mm. Outer petals yellow, 13–15mm, each c 2mm broad at insertion, with large auricle at base. Nectariferous glands 2.5–3mm, slender, curving upwards. Capsules cylindric, 3–6 × 0.2–0.3cm.

Bhutan: **C**–Thimphu district (Tashi Cho Dzong). On shrubs in dry valley, 2400m. April–August.

5. D. roylei Hook. f. & Thomson

Closely allied to *D. lichiangensis* but corymbs 2–3-flowered; bracts 9–12 × c 2mm; sepals 7–9 × c 2mm; outer petals 16–19mm, broader (3–5mm) at insertion; capsules 3.5–4.5 × 0.3–0.5cm.

Bhutan: **S**–Chukka district (near Chukka); **Sikkim:** Tista Valley. 1400–1500m. June–September.

3. FUMARIA L.

Erect or scrambling annual herbs. Leaves 2–4-pinnatisect. Flowers in leaf-opposed racemes, zygomorphic. Petals white or purplish; outer pair dissimilar, upper with swollen spur at base and hooded apex, lower shorter, not spurred; inner pair connivent, clawed. Stamens 6 in two bundles, upper bundle produced into a basal nectariferous gland enclosed by petal spur. Ovary subglobose; ovules 1 or 2; style filiform. Fruit a 1-seeded nutlet.

1. Leaf lobes linear, 1–1.5mm broad; peduncles 0.5–1.5cm; petals purple, darker at tips, upper outer petal 6–7mm; nutlets rugose, on erect pedicels **1. F. indica**
+ Leaf lobes elliptic or oblong, 2–4mm broad; peduncles 2–5cm; petals white, dark red at tips, upper outer petal 10–12mm; nutlets smooth, on recurved pedicels .... **2. F. capreolata**

1. F. indica (Haussknecht) Pugsley. Fig. 31 o,p.

Erect or scrambling herb 20–50cm. Leaves up to 12cm, lobes linear, 3–10 × 1–1.5mm, acute. Racemes dense, 15–20-flowered, 2.5–4cm, on peduncles 0.5–1.5cm. Petals purple, darker at tips; upper outer petal 6–7mm, including swollen spur 1.5mm. Nutlets subglobose, 2mm, rugose, on erect pedicels.

Bhutan: C–Tongsa district (Shamgong) and Tashigang district (Shungphu, Gamri Chu). Weed of cultivated ground, 1980m. March–May.

2. F. capreolata L.

Similar to *F. indica* but a more extensive scrambler to 1m; leaf-lobes elliptic or oblong, 2–4mm broad; racemes on longer peduncles 2–5cm; petals white, dark red at tips, upper outer petal 10–12mm; nutlets 2.5mm, smooth, borne on recurved pedicels.

Sikkim: Tosa. Weed of cultivation, 4400m. September.

Probably introduced from Europe.

4. CORYDALIS Ventenat

Perennial erect or decumbent glabrous herbs. Roots tuberous, fibrous or woody. Leaves pinnatisect or ternatisect, alternate or stem leaves opposite. Flowers in terminal racemes, sometimes condensed and subumbellate, zygomorphic (with longitudinal plane of symmetry). Petals yellow, blue, purple or white; outer pair dissimilar, upper spurred; inner pair clawed, often cohering. Stamens 6 in 2 bundles; upper bundle with a basal nectariferous gland projecting into petal spur. Style filiform; stigmas 2. Capsule dehiscing to leave seed-bearing placentae attached to style.

1. Flowers blue, purple or rarely white 2
+ Flowers yellow 13

2. Stems 15–100cm; stem leaves large, 6–15cm, sessile (basal pinnae arising close to stem)....3
+ Stems mostly 5–20cm, (sometimes 30–50cm in *C. leptocarpa, C. dorjii* and *C. trifoliata*); stem leaves small, 1–5cm, long-petiolate to subsessile (stem-leaves 10–13cm but long-petiolate in *C. dorjii*)....4

3. Robust herb, 50–100cm; stem leaves numerous, pinnatisect; petal spur broad, 6–8mm long....**1. C. flaccida**
+ Slender herb, 15–30cm; stem leaves few, ternatisect; petal spur slender, c 12mm long....**2. C. kingii**

4. Stems branched, bearing long-petiolate leaves throughout....5
+ Stems simple, erect, naked or with a few long-petiolate basal or sub-basal leaves and sometimes with a sessile or shortly-petiolate leaf above middle....8

5. Stem leaves alternate; flowers in racemes; spur of upper petal equalling or longer than limb....6
+ Stem leaves opposite; flowers subumbellate; spur of upper petal much shorter than limb....7

6. Decumbent herb, 15–40cm; roots fibrous; leaves biternate with crenately lobed segments; bracts 6–10mm, lower ones toothed; upper petal distinctly crested....**3. C. leptocarpa**
+ Erect herb, 30–50cm; roots tuberous; leaves triternate with entire segments; bracts 3–6mm, entire; upper petal narrowly crested at apex....**4. C. dorjii**

7. Leaflets divided into obovate, obtuse teeth; lower bracts 2–3-fid; upper petal strongly constricted behind hooded tip, 16–19(–22)mm (up to 25mm in var. *physodes*), including 3.5–6mm spur; lip of lower petal deflexed **5. C. gerdae**
+ Leaflets finely divided into linear or oblanceolate, acute segments; lower bracts entire; upper petal weakly constricted behind hooded tip, 13–15mm, including short 1.5–3mm spur; lip of lower petal not deflexed....**6. C. latiflora**

8. Upper part of stem leafless but with leaf-like bracts at base of umbel....9
+ Upper part of stem bearing a single shortly-petiolate or sessile leaf a short distance below bracts....10

9. Plants 6–15cm; basal leaves 1.5–5cm with oblong to oblanceolate segments; umbels 4–6-flowered; bracts entire; upper petal 15–20mm including long 10–13mm spur....**7. C. alpestris**
+ Plants 5–7cm; basal leaves 1–1.5cm with obovate segments; umbels 2–4-flowered; lowest bract deeply 3-fid; upper petal c 10mm including short c 3mm spur....**8. C. oxalidifolia**

10. Stem leaf sessile (sometimes absent); lip of lower petal broad and almost flat 11
+ Stem leaf petiolate; lip of lower petal boat-shaped....12

11. Umbels 2–4-flowered; upper petal 13–22mm, not or very narrowly crested, tip of spur deflexed; lip of lower petal usually broader than long, suborbicular, obtuse, 6–11mm broad, distinctly (4–8mm) exserted beyond upper petal **9. C. ecristata**

\+ Umbels 4–6-flowered; upper petal 11–14mm, distinctly crested, tip of spur weakly deflexed; lip of lower petal longer than broad, rhombic, acute, 4–6mm broad, shortly (1–3mm) exserted beyond upper petal **10. C. cashmeriana** subsp. **brevicornu**

12. Roots tuberous; stems 15–30cm; basal leaves absent, stem leaf trifoliate, with elliptic entire leaflets; bracts simple, elliptic; petal spur broad **11. C. trifoliata**

\+ Roots fibrous, stems 8–12cm; basal leaves usually present, stem leaf bipinnatisect with toothed segments; bracts leaf-like; petal spur slender **12. C. oligantha**

13. Densely tufted herb, 3–5cm; bracts exceeding flowers; leaves ternate **13. C. mucronifera**

\+ Erect or diffuse herbs, mostly 10–120cm (shorter and caespitose in *C. chasmophila* and *C. dubia* but then leaves pinnate); bracts never exceeding flowers; leaves ternate or pinnate 14

14. Stems simple (sometimes with axillary branches in *C. polygalina*), without persistent petiole bases; stem leaves 1–2 (or more in *C. polygalina*) 15

\+ Stems branched, often surrounded at base by persistent petiole bases; stem leaves numerous 17

15. Racemes 3–5-flowered; stem leaves 1-pinnate, with suborbicular, entire or bluntly 3-fid pinnae .... **14. C. lathyroides**

\+ Racemes 8-to many-flowered; stem leaves simple and entire (rarely none), or pinnatisect, or ternatisect with linear parallel-veined segments 16

16. Stem leaf solitary, linear, entire; upper petal 9–10(–12)mm ... **15. C. juncea**

\+ Stem leaves 2(–5), pinnatisect or ternatisect into linear-lanceolate segments; upper petal c 15mm .... **16. C. polygalina**

17. Leaves ternately dissected (divided into 3 subequal ternately or pinnately subdivided leaflets) (in *C. crispa*, *C. bowes-lyonii* & *C. filicina* with larger terminal leaflet and appearing somewhat pinnate); basal leaves broadly ovate to orbicular in outline; petal spur often upwardly curved, sometimes straight or deflexed 18

\+ Leaves pinnately subdivided into 5 or more equal or unequal pinnately subdivided leaflets; basal leaves narrowly ovate or oblong in outline; petal spur straight or deflexed 30

18. Erect, robust, little-branched herbs, mostly 40–120cm; lower leaves 6–10cm broad; racemes often more than 20-flowered (4–8-flowered in *C. delicatula*) ... 19
+ Diffuse or suberect, weak, much-branched herbs, mostly 15–30cm (–40cm in *C. iochanensis*); lower leaves 2–5cm broad; racemes 4–20-flowered ... 23

19. Rootstock a simple rhizome ... 20
+ Roots numerous, densely fascicled ... 22

20. Rachis of leaflets unwinged, ultimate segments obovate, 3–5mm broad, obtuse or truncate with minute mucro; outer petals broadly crested ... **23. C. bowes-lyonii**
+ Rachis of leaflets conspicuously winged with decurrent bases of pinnae, ultimate segments oblong, 1–3mm broad, bluntly or sharply pointed; outer petals very narrowly crested ... 21

21. Ultimate leaf segments oblong, bluntly pointed; bracts lanceolate, 3–5mm, entire, equalling pedicels; upper petal 11–13mm; capsules obovoid ... **17. C. chaerophylla**
+ Ultimate leaf segments acuminate, sharply pointed; bracts obovate, 8–10mm, deeply 3–5-fid, much longer than pedicels; upper petal 14–17mm; capsules cylindric ... **18. C. geraniifolia**

22. Ultimate leaf segments obovate, 2–7mm broad, obtuse or truncate; racemes 4–8-flowered; upper petal 13–14mm, crest not projecting forward beyond lip ... **20. C. delicatula**
+ Ultimate leaf segments linear or elliptic, 2–5mm broad, acute or subacute; racemes many-flowered; upper petal 22–25mm, with crest projecting forwards beyond lip ... **19. C. laelia**

23. Petal spur strongly deflexed ... **21. C. iochanensis**
+ Petal-spur strongly upwardly curved to almost straight ... 24

24. Upper petal c 18mm, not crested, spur straight ... **24. C. filicina**
+ Upper petal 10–16mm, crested, spur upwardly curved ... 25

25. Stems erect or decumbent, not slender and wiry; leaves 2–5cm, terminal leaflet often larger than laterals; racemes 6–20-flowered ... 26
+ Stems diffuse, very slender and wiry; leaves 1–3cm, terminal leaflet similar to laterals; racemes 3–6-flowered (8–16-flowered in *C. cavei*) ... 27

26. Leaf segments obovate, mucronate; racemes 10–20-flowered; bracts equalling pedicels; crests of outer petals not projecting forwards beyond lips ... **22. C. crispa**
+ Leaf segments linear-oblanceolate, acute; racemes 6–12-flowered; bracts longer than pedicels; crests of outer petals projecting forwards beyond lips ... **25. C. drepanantha**

27. Upper petal 10–12mm, with slender spur; nectariferous gland one eighth to half the length of spur 28
\+ Upper petal 13–16mm, with broad spur; nectariferous gland half to two thirds the length of spur 29

28. Lip of lower petal not gibbous near insertion; nectariferous gland one third to half the length of spur; capsules linear ....**26. C. casimiriana**
\+ Lip of lower petal gibbous near insertion; nectariferous gland one sixth to one eighth the length of spur; capsules obovoid ....**27. C. longipes**

29. Lip of lower petal with short broad spur 1.2–1.4mm .....**28. C. changuensis**
\+ Lip of lower petal not spurred or gibbous ....**29. C. cavei**

30. Petal spur short, 1.5–5mm long, not or scarcely longer than broad, deflexed 31
\+ Petal spur long, 4–12.5mm, 2–5 × as long as broad, straight or deflexed 34

31. Leaves coarsely dissected into obovate, obtuse or mucronate segments; flowers slender, 10–12 × c 2mm; pods undulate ....**30. C. ophiocarpa**
\+ Leaves finely dissected into linear or elliptic, acute segments; flowers large, 13–21 × 3–8mm; pods flat 32

32. Robust herbs 15–40cm, stems stout; flowers 13–15mm, numerous in dense racemes ....**31. C. meifolia**
\+ Weak herbs 4–22cm, stems slender; flowers 16–22mm, 4–9 in short subumbellate racemes 33

33. Stems 8–22cm; leaves 4–6 × 1.5–2.2cm, ultimate segments linear-elliptic, 1.3–2.2 × 0.5–0.8mm; sepals ovate, large, c 4 × 3mm, acute; upper petal 16–20mm ....**32. C. calliantha**
\+ Stems 4–14cm; leaves 2.5–3 × 1–1.5cm, ultimate segments obovate or oblanceolate, 2–4 × 1mm; sepals rounded, broader than long, c 1.5 × 2mm; upper petal 20–22mm ....**33. C. aurantiaca**

34. Stem bases thickly clothed with persistent petiole bases 35
\+ Stem bases with few or no persistent petiole bases 36

35. Stems 15–35cm; basal leaves 6–11cm, ultimate segments elliptic, acute; upper petal 15–19mm, including broad 6–8mm long spur ......**34. C. sikkimensis**
\+ Stems 5–7cm; basal leaves 2.5–4cm, ultimate segments obovate, obtuse; upper petal 20–22mm, including slender 10–12mm long spur...**35. C. chasmophila**

36. Slender herbs 10–20(–30)cm; leaves finely dissected, ultimate segments 0.5–1.5mm broad; upper petal 11–18mm 37
\+ Robust herbs 30–45cm; leaves more coarsely dissected, ultimate segments 2–6mm broad; upper petal 19–23mm 38

37. Leaves thin, not crisped, ultimate segments oblanceolate; pedicels 2–8mm; upper petal 11–12mm .... **36. C. stracheyi**
+ Leaves rather thick and crisped, ultimate segments linear; pedicels 10–13mm; upper petal 15–18mm .... **37. C. dubia**

38. Leaf segments 2–3mm broad; pedicels 14–20mm; outer petals very narrowly crested; spur straight .... **38. C. franchetiana**
+ Leaf segments 3–5mm broad; pedicels 3–8mm; outer petals broadly crested; spur deflexed at apex .... **39. C. denticulato-bracteata**

1. C. flaccida Hook. f. & Thomson. Med: *Nangtsee Trey.*

Erect branched herb 0.5–1m, roots elongate, fibrous. Basal leaves few, up to 10cm long and broad, on petioles 10–15cm, 3–4-pinnatisect, segments ovate, terminal largest, bluntly toothed, obtuse or apiculate. Stem leaves numerous, sessile. Racemes 3–8cm, 10–20-flowered. Bracts 5–15mm, lower pinnatifid, upper linear, toothed. Pedicels equalling bracts. Sepals broadly ovate, 2–2.5mm, cordate, acute, dentate. Petals purplish, outer pair broadly crested, upper 14–16mm, including slightly upwardly curved spur 6–8 × 2mm. Capsules linear, 20–30 × 2mm.

Bhutan: C–Thimphu, Punakha, Mongar and Tashigang districts, **N**–Upper Mangde Chu, Upper Bumthang Chu and Upper Kulong Chu districts; **Sikkim.** Clearings in Fir/Rhododendron forests, 3350–4200m. June–July.

Similar to *C. ophiocarpa* which differs in its yellow flowers and petiolate stem leaves.

2. C. kingii Prain

Similar to *C. flaccida* but a smaller herb 15–30cm with a slender rhizome; stem leaves few, 1–3, ternately divided, c 7 × 5cm, ultimate segments 3-toothed, acute; sepals acuminate, subentire; upper petal c 15mm, including long spur c 12mm.

Chumbi: Phari. Rocky slopes on mountains.

3. C. leptocarpa Hook. f. & Thomson. Fig. 31 k–n.

Diffuse herb 15–40cm, with several stems arising from a cluster of slender fibrous roots. Basal leaves biternate, 5–8 × 3–5cm, petioles 5–9cm, terminal and lateral segments ovate-elliptic, up to 2×1cm, crenately lobed, pale beneath. Racemes 2–6cm, 5–8-flowered. Bracts obovate, 6–10mm, lower toothed, upper narrower and entire. Petals purple or mauve, sometimes whitish, outer pair crested, upper 26–30mm, including spur 13–15 × 3–3.5mm with tip deflexed. Nectariferous gland 6–7mm. Capsules linear, 20–30 × 1.5–2mm.

Bhutan: S–Chukka and Deothang districts, **C**–Ha, Thimphu, Punakha, Tongsa, Mongar and Tashigang districts, **N**–Upper Mo Chu and Upper Kuru Chu districts; **Sikkim.** Damp streamsides and shady ravines in Warm and Cool broad-leaved forests, 1000–2500m. March–June.

4. C. dorjii Long

Similar to *C. leptocarpa* but an erect herb 30–50cm, arising from a cluster of spindle-shaped tuberous roots 1.5–2cm; leaves triternatisect, broadly triangular, 7–11cm long and broad, ultimate segments elliptic-obovate, 1.2–2.2 × 0.5–1.3cm, obtuse, entire; petioles of basal leaves 10–14cm, of stem leaves 2–6cm; flowers blue or mauve, 2–4 in short terminal racemes; outer petals with low crest near apex, upper petal 25–28mm, including slender deflexed spur 16–18mm; lip of lower petal gibbous near insertion.

Bhutan: N – Upper Kuru Chu (Dunkhar) and Upper Kulong Chu (Lao) districts. Cool broad-leaved forests, 1980–2560m. April–May.

Similar in foliage to *C. delicatula* which has yellow flowers.

5. C. gerdae Fedde. Med: *Tongri Zilpa.*

Decumbent herb 7–20cm, with long scaly rhizome. Stems surrounded by persistent petiole bases. Basal leaves bipinnatisect, 2–4 × 1.5–2cm, on slender petioles 4–8cm, leaflets broadly ovate or orbicular, rather deeply divided into obovate, obtuse teeth. Racemes umbellate, with 3–6 horizontal flowers exserted above bracts. Bracts 6–12mm, lower ones 2–3-fid. Pediceis 10–15mm. Petals purple or mauve and white, tipped dark green or blackish, outer pair broadly crested, constricted behind hooded tip, upper 16–19(–22)mm, including short broad deflexed spur 3.5–6 × 2.5–3mm, lower petal with deflexed lip.

Bhutan: N – Upper Mo Chu district (Yale La near Lingshi and slopes below Chomo Lhari); **Chumbi:** Chomo Lhari and above Phari. Mountain screes, 4400–4900m. July–Sptember.

Endemic to W Bhutan and Chumbi. Closely related to *C. aurantiaca* and *C. calliantha* which differ in their yellow, suberect flowers. The above description refers to the typical var. **gerdae;** var. **physodes** Long from Ha district (between Kang La and Ha) differs in its larger size, longer pedicels 2.5–3.5 cm, more deeply incised bracts and larger flowers with upper petal c 25mm, including very broad spur 4.5–5.5 × 4–4.5mm.

6. C. latiflora Hook. f. & Thomson

Similar to *C. gerdae* but smaller plant, 7–11cm; leaves more finely bipinnatisect, leaflets more deeply divided into linear or oblanceolate, acute segments; flowers scarcely exserted above bracts; bracts linear, entire, 6–12mm; pedicels slightly shorter than bracts; flowers smaller, less inflated; upper petal 13–15mm including very short spur as broad as or broader than long, 1.5–3 × 2–3.5mm, upper and lower petals winged but only weakly constricted; lower petal almost straight, lip not deflexed; capsules ellipsoid, c 5.5 × 2.5mm.

Sikkim: Yak La, Cho La and Thanka La; **Chumbi:** E of Phari. Rocky slopes and streamsides, 4400–4900m. July–October.

7. C. alpestris Meyer

Erect herb 6–15cm, with swollen tuberous roots, stems simple. Leaves borne near base of stem below ground, ternate, suborbicular, 1.5–5cm, on petioles 5–15cm; leaflets orbicular or obovate, deeply divided into 3–5 oblong or oblanceolate minutely mucronate segments. Stem naked in upper two-thirds bearing a dense 4–6-flowered terminal umbel. Bracts ovate or obovate, 5–7mm, acute, entire. Pedicels short, 3–5mm. Petals blue, upper 15–20mm, including long spur 10–13 × 3–4mm, slightly deflexed at tip, upper lip hooded, erect. Capsules ellipsoid.

Bhutan: C –Thimphu district (Gyep Chu, Paro Chu), **N** – Upper Mo Chu district (Lingshi, Shinje La and Yale La) and Upper Bumthang Chu district (Kantanang). Alpine screes and grassy slopes, 4500–4700m. May–June.

8. C. oxalidifolia Ludlow

Similar to *C. alpestris* but smaller, 5–7cm, with leafless stems and few suborbicular ternate basal leaves 1–1.5cm, on petioles 5–7cm; leaflets divided into broader obovate, obtuse or mucronate lobes, umbel 2–4-flowered, partly hidden by bracts; lowest bract suborbicular, c 10mm, deeply 3-lobed, others obovate, entire; pedicels 2–6mm; upper petal c 10mm, including short spur c 3 × 1.5mm.

Bhutan: N – Upper Kulong Chu district (Me La). Grassy alpine hillsides, 4250–4570m. June–August.

Endemic to Bhutan.

9. C. ecristata (Prain) Long; *C. cashmeriana* Royale var. *ecristata* Prain, *C. cachemiriana* sensu F.B.I. p.p. non Royle

Slender herb with 1–3 simple stems 10–15cm, arising from a cluster of narrow tuberous roots. Basal leaves few, soon withering, biternate, 2–3cm diameter, on slender petioles 3–6cm. Leaflets obovate, entire or bluntly toothed. Stem leaf solitary, bract-like, 0.5–1.5cm, sessile, deeply pinnatifid into linear segments. Flowers 2–4 in terminal umbel, subtended by deeply 3–5-toothed bracts. Pedicels slender, 1–2cm. Petals blue, upper 13–22mm, including spur 6–16 × 2.5–3mm, deflexed at tip; upper lip hooded and subacute, not or very narrowly crested; lower lip long-exserted (4–8mm) beyond upper, spathulate, 8–13 × 6–11mm, limb flat, suborbicular, obtuse or shallowly 3-lobed. Capsules ellipsoid, c 7 × 2mm.

Two varieties occur in Sikkim and Bhutan:

var. **ecristata:** Upper petal 13–18mm, including short spur 6–10mm, weakly deflexed at tip.

Bhutan: C – Punakha district (Black Mountain), Tongsa district (Rinchen Chu) and Sakden district (Orka La), **N** – Upper Pho Chu (Chesha La) and Upper Kuru Chu (Narim Thang) districts; **Sikkim:** common in W Sikkim; **Chumbi.** Alpine grassy slopes, screes, cliff-ledges and in Juniper/Rhododendron scrub, 3650–4880m. June–September.

var. **longicalcarata** Long: Upper petal 18–22mm, including long spur 11–16mm, strongly deflexed at tip.

Bhutan: C – Ha district (Chelai La) and Thimphu district (Barshong); **Sikkim:** confined to E Sikkim; **Chumbi.** In alpine screes, turf slopes and in Juniper/Rhododendron scrub, 3200–4500m. May–July.

10. C. cashmeriana Royle; *C. cachemiriana* sensu F.B.I. p.p.

Similar to *C. ecristata* but stems often more numerous, stem leaf sometimes absent; umbels denser, 4–6-flowered; bracts often 4–6-fid; upper petal 11–14mm including short spur 4–7mm; upper lip with crest c 0.5mm broad; lower lip rhombic, acute,smaller, 9–11 × 4–6mm, shortly exserted (1–3mm) beyond upper, keel distinctly crested.

Bhutan: C –Thimphu district (Pajoding and between Tremo La and Kang La), **N** – Upper Bumthang Chu district (Pangotang); **Sikkim:** confined to N Sikkim; **Chumbi.** Alpine screes and turf slopes, 4000–4900m. June–September.

The above description and records apply to subsp. **brevicornu** (Prain) Long (*Corydalis cashmeriana* Royle var. *brevicornu* Prain, *C. jigmei* Fischer & Kaul). A more local and higher altitude plant than *C. ecristata,* to which many of the old records of *C. cashmeriana* belong.

11. C. trifoliata Franchet; *C. trifoliolata* auct.

Slender herb similar to *C. ecristata* and *C. cashmeriana* but often taller, 15–30cm; basal leaves absent; stem leaf solitary, trifoliate, on petiole 0.5–1.5cm; leaflets elliptic, 1–1.5cm, sessile, acute, entire; umbels 2–4-flowered; bracts simple, elliptic, acute, the lowest 1–2 × 0.5–1cm; petals blue, rarely white, not crested, upper 12–15mm including straight spur 5–7mm, lower lip boat-shaped, shortly exserted beyond upper.

Bhutan: C – Punakha, Bumthang and Mongar districts, **N** – Upper Mangde Chu and Upper Kulong Chu districts; **Sikkim.** Damp ground in Fir forests and in Rhododendron scrub, 3650–4250m. June–July.

Very distinctive in its entire bracts and trifoliate leaves with entire leaflets.

12. C. oligantha Ludlow

Similar to *C. ecristata* but stems 8–12cm, basal leaves usually several, ovate, bipinnatisect, 1–2cm, on petioles 1.5–3cm, deeply divided into oblong or obovate lobes; stem leaf similar but borne on petiole 5–10mm, subtending small axillary bulbil; flowers 2–3, subtended by sessile leaf-like bracts; petals pale blue or white, not crested, upper 14–15mm including slender spur 8–10mm; lower lip boat-shaped, acute.

Bhutan: N – Upper Kuru Chu (Sergong La) and Upper Kulong Chu (Lao) districts. Clearings in Fir/Rhododendron forest, 2900–3050m. May.

Endemic to NE Bhutan and adjacent Arunachal Pradesh.

13. C. mucronifera Maximowicz

Short, tufted herb 3–5cm with slender tap-root. Stems branching near base. Basal leaves ternatisect, c 1cm, on petioles 1–2cm; segments obovate, bluntly 3-fid; stem leaves smaller. Flowers in simple bracteate racemes. Bracts large, obovate, c 12 × 8mm, laciniate, exceeding flowers. Upper petal 6–8mm, including spur c 3mm; outer petals keeled or narrowly crested. Capsules oblong, 2-seeded.

Chumbi: Phari district.

14. C. lathyroides Prain

Slender herb 9–20 cm; rootstock and basal leaves unknown. Stem leaves 1–2(–3), distant on upper part of stem, 3–6 × 1–2cm, simply pinnate on petioles 1–3cm; leaflets 3–4 pairs, broadly ovate or suborbicular, 7–16 × 5–14mm, obtuse and mucronate or bluntly 3-fid, sometimes purplish beneath. Racemes 3–5-flowered, on slender axillary peduncles 1.5–7cm. Bracts linear or 3-fid, 3–6mm. Pedicels 6–11mm. Petals not crested, upper c 12mm including spur c 7mm; lower lip spathulate, exserted beyond upper.

Bhutan: C –Thimphu district (Barshong) and Tongsa district (Thita Tso, Rinchen Chu); **Sikkim**: Zemu valley. Mountain cliff ledges, 3350–4100m. July–August.

15. C. juncea Wall.

Slender simple-stemmed herb 8–30cm, borne on a dense cluster of elongate tubers. Basal leaf usually solitary, biternate or trifoliate, on long petiole 8–20cm; leaflets variable, sometimes simple, elliptic, more often deeply palmatisect, up to 3.5cm across, with linear to obovate segments, petiolules 1–3cm. Stem leaf usually solitary, linear, 0.5–1.5cm, entire. Racemes 5–20-flowered. Bracts linear, entire, up to 2cm. Pedicels equalling or exceeding bracts. Upper petal 9–10(–12)mm, including short, broad spur c 3.5 × 2mm, keel broadly crested. Lower lip deflexed, crested.

Bhutan: C –Thimphu district (Barshong and Pajoding) and Tongsa district (Padima Tso, Rinchen Chu), **N** –Upper Bumthang Chu district (Lubsing La); **Sikkim.** Alpine turf and scree slopes, 3600–4600m. June–September.

16. C. polygalina Hook. f. & Thomson; *C. graminea* Prain

Similar to *C. juncea* but sometimes with short axillary shoots; basal leaves pinnatisect or ternatisect into linear-lanceolate segments 1.5–2cm; stem leaves 2(–5), similar but shortly petiolate or sessile; lower bracts usually pinnatisect; upper (rarely all) bracts simple, entire, 6–12mm; pedicels much exceeding bracts; upper petal c 15mm including broad spur half its length, keel with broad crest narrowly decurrent along spur.

Bhutan: C –Thimphu district, **N** –Upper Mo Chu, Upper Pho Chu, Upper Mangde Chu and Upper Kuru Chu districts; **Sikkim.** Alpine screes and gravel, 3650–4880m. June–September.

A variable species, especially in foliage. *C. graminea* represents a variant with few leaves and entire bracts.

17. C. chaerophylla DC.

Robust herb 0.6–1.2m, with stout rootstock and persistent stem and petiole bases. Basal leaves broadly ovate, 9–15 × 7–14cm, on long petioles 15–25cm; primary division ternate into 3 petiolulate, bipinnatisect major divisions; pinnae of each major division sessile, decurrent, segments oblong, blunt-tipped, pale beneath. Stem leaves similar but smaller, becoming sessile upwards. Racemes terminal, forming branched panicles 10–20cm; flowers numerous, secund. Bracts lanceolate, 3–5mm, entire, equalling pedicel. Petals yellow, outer pair apiculate, very narrowly crested, upper 11–13mm, including slender usually straight spur 8–9mm. Capsules obovoid, c 8×3mm, spreading.

Bhutan: C – rare, Tashigang district (Donga La); **Sikkim:** widespread. Streamsides in Fir forests, 2750–3650m. July–October.

18. C. geraniifolia Hook. f. & Thomson; *C. chaerophylla* var. *geraniifolia* (Hook. f. & Thomson) Hara

Similar to *C. chaerophylla* but ultimate leaf segments acuminate, sharply pointed; bracts obovate, 8–10mm, deeply 3–5-fid, much longer than pedicels; flowers larger, upper petal 14–17mm, including spur 9–11mm; capsules cylindric, c 10 × 2mm.

Bhutan: C – rare, Tongsa ditrict (Chendebi) and Tashigang district (Donga La); **Sikkim:** local, Tonglo, Chowbhanjan, etc. Moist forests, 2400–3350m. August–October.

19. C. laelia Prain

Robust herb 30–50(–100)cm, stems stout, little-branched, arising from dense cluster of slender fibrous roots, stem base surrounded by old petiole remains. Basal leaves ovate, 10–14 × 8–14cm, biternatisect, on long, sheathing petioles 15–25cm, leaf divisions deeply divided into linear or elliptic, acute or subacute segments or teeth 2–5mm broad. Stem leaves on shorter petioles 1–3cm. Racemes stout, many-flowered, forming branched panicles. Lower bracts 1–2cm, deeply pinnatisect into linear teeth, upper shorter, linear, entire. Petals yellow, outer pair very broadly crested with crests projecting forwards beyond lips, upper 22–25mm including straight spur 10–11mm with somewhat deflexed tip. Capsules obovoid-cylindric, c 12 × 2mm.

Bhutan: C – Ha district (Tare La), Punakha district (Black Mountain and SW of Wangdu Phodrang), and Tongsa district (Chendebi); **Sikkim:** Changu, Nathu La, etc.; **Chumbi.** Hemlock and Fir forests, and on streamsides in Juniper/Rhododendron scrub, 2740–3960m. May–September.

The common plant represents subsp. **laelia** which has finely divided leaves with linear, acute segments 2–3mm broad, and a narrow petal spur 2–2.5mm broad. The collections from Wangdu Phodrang and Chendebi represent subsp. **bhutanica** Long, differing in their coarsely divided leaves with oblanceolate-elliptic, subacute or minutely mucronate segments 3–5mm broad, and petal spur 2.5–3.5mm broad. In foliage these plants are very similar to the Chinese *C. yunnanensis* Franchet, but in that species the petal crests are narrower and do not project forwards.

20. C. delicatula Long

Similar to *C. laelia* in its densely fasciculate roots but differing in its smaller size, 26–45cm, triternatisect leaves dissected into delicate obovate, obtuse or truncate and minutely mucronate segments 3–10 × 2–7mm; flowers 4–8 in short racemes; bracts strongly dimorphic, the lowermost leaf-like, 1–2cm, upper bracts oblanceolate, 2–5mm, acute, entire; outer petals crested but crests not projecting beyond lips, upper petal 13–14mm, spur somewhat curved upwards.

Bhutan: **C** – Punakha district (Ritang and W slope of Pele La) and Mongar district (W side of Donga La), **N** – Upper Kulong Chu district (Lao). Fir and Hemlock forests, on damp shady banks, 2740–3200m. April–June.

Endemic to Bhutan. Similar in foliage to *C. dorjii* which has blue or mauve petals.

21. C. iochanensis Léveillé

Slender erect or decumbent herb 15–30(–40)cm, stems branching only at base. Basal leaves ovate or suborbicular, 3–4.5 × 2–4cm, deeply bipinnatisect, on slender petioles 5–15cm, ultimate segments oblanceolate to obovate, 1.5–5mm broad, obtuse or mucronate. Stem leaves 2–4, similar but on shorter petioles 1.5–5cm. Racemes compact, 2–3cm, few- to many-flowered. Lower bracts 6–12mm, deeply toothed, upper bracts linear, entire. Pedicels 4–7mm. Petals pale yellow, upper 13–15mm including strongly deflexed spur 7–8mm, upper lip keeled but not crested, lower lip equalling upper, boat-shaped, straight, acute, with gibbous swelling near base. Capsules ellipsoid, 13–15 × 3–4.5mm, seeds 9–15 in 2 rows.

Bhutan: **C** –Thimphu district (Dotena Chu and Pemithanka), Tongsa district (Phobsikha) and Bumthang district (Takhung, Kyikyi La and Ura La). Mossy ground in Spruce forests, 2500–3500m. May–June.

22. C. crispa Prain; *C. stracheyioides* Fedde. Med: *Khamlug Bashaka.*

Decumbent much-branched herb, stems 10–23cm, arising from stout woody rootstock. Leaves numerous, ovate, 2–3 × 2–2.5cm, finely biternatisect, but terminal leaflet larger than laterals and leaf appearing somewhat pinnate; ultimate segments obovate, mucronate, 1–2mm broad; petioles 1–2.5cm. Flowers 10–20 in dense rounded racemes 2–4cm. Lower bracts 5–7mm, pinnatisect; upper bracts linear, 2–4mm, entire. Pedicels slender, 4–8mm. Flowers yellow; outer petals with broad rounded crests; upper petal 10–14mm including upwardly curved spur 5–8mm. Nectariferous gland about three-quarters length of spur.

Bhutan: **C** –Thimphu district (Naro, Wasa La), **N** – Upper Mo Chu district (Lingshi and Yale La); **Chumbi:** Phari and below Chomolhari. Damp mountain screes and gravel, 3350–4570m. July–September.

Endemic to W Bhutan and the Chumbi Valley.

23. C. bowes-lyonii Long

Similar to *C. crispa* but a more robust, erect, less-branched plant, stems 45–90cm; leaves unequally biternatisect, larger, 7–12 × 8–12cm, ultimate segments 3–5mm broad; racemes 4–8cm, many-flowered; flowers larger, upper petal (11–)14–16mm, with slightly broader spur.

Bhutan: **C** –Thimphu district (W side of Thimphu Valley), **N** – Upper Bumthang Chu district (Chamka) and Upper Mangde Chu district (Ju La). Marshy alpine meadows and clearings in Juniper/Rhododendron scrub, 3800–4140m. July–September.

24. C. filicina Prain

Similar to *C. crispa* but a small, slender plant 10–15cm; flowers larger, upper petal c 18mm, not crested, spur straight.

Sikkim: Singalila range.

25. C. drepanantha Long

Erect herb, stems 12–29cm, branched at base, borne on slender rhizome. Lower stem leaves broadly ovate, ternatisect, 3.5–5 × 4–5.5cm, ultimate segments and teeth linear-oblanceolate, 0.5–1.5mm broad, acute; petioles 5–9cm. Upper leaves smaller, shortly petiolate. Racemes 3.5–5cm, 6–12-flowered. Lower bracts leaf-like, 1.1–2.1cm, upper linear, 4–6mm, longer than pedicels. Sepals ovate, c 0.5mm, dentate. Outer petals broadly crested, with apex of crest irregularly dentate and projecting forwards beyond lips; upper petal 10–13mm, including upwardly curved spur 6–9mm. Nectariferous gland about half length of spur. Capsules obovate-elliptic, c 9 × 3mm, obtuse.

Bhutan: C –Tashigang district (Dengchung, Khoma Chu); **Arunachal Pradesh:** Nyam Jang Chu. Damp streamsides in Cool broad-leaved forests, 2130–2285m. April.

Endemic to E Bhutan and adjacent Arunachal Pradesh.

26. C. casimiriana Duthie & Prain; *C. sibirica* sensu F.B.I. p.p. non (L.f.) Persoon, *C. prainiana* Kanodia & Mukerjee *nom. superfl.*

Diffuse, much branched herb 15–30cm, rootstock slender. Leaves numerous, mostly borne on stems, broadly ovate in outline, 1–3 × 1–3cm, deeply and equally biternatisect, on slender petioles 3–6cm; ultimate segments obovate or oblanceolate, 1.5–3mm broad. Racemes lax, few-flowered, 1.5–3cm. Lower bracts deeply toothed, upper entire. Sepals suborbicular, 0.6–1mm, dentate. Petals yellow, upper 10–12mm, including upwardly curved slender spur 5.5–7mm long and 0.8–1.7mm broad at base; upper lip with low, entire crest not decurrent on spur; lower lip equalling upper. Nectariferous gland 2–3mm, one-third to one-half the length of spur. Capsules linear, 6–10mm, with a single row of seeds.

Bhutan: C – Ha, Thimphu and Bumthang districts, **N** – Upper Mangde Chu, Upper Bumthang Chu and Upper Kuru Chu districts; **Sikkim.** Streamsides in Juniper/Rhododendron scrub and alpine meadows 2750–4700m. July–September.

27. C. longipes DC.; *C. sibirica* sensu F.B.I. p.p. non (L.f.) Persoon, *C. longipes* var. *phallutiana* Fedde, var. *burkillii* Fedde, var. *smithii* Fedde

Similar to *C. casimiriana* but bracts all deeply toothed; upper petal with broad often crenately toothed crest decurrent along spur; spur slightly longer and broader, more tapering; lower lip with rounded swelling near base; nectariferous gland much shorter, 0.7–1.4mm, one-sixth to one-eighth the length of spur; capsules obovoid, 4–6mm with 2 rows of seeds.

Bhutan: C – Bumthang district (Tashiling); **Sikkim:** widespread. Streamsides, screes and disturbed soil, sometimes a weed of cultivation, 2800–4200m. June–Setptember.

28. C. changuensis Long

Closely allied to *C. longipes* but flowers larger, upper petal 13–15mm, less strongly curved upwards, spur longer and broader, 7.7–10 × 2–3.4mm, with crest extending almost to tip of spur; lower lip with short spur 1.2–1.4mm near insertion; nectariferous gland much longer, 4.8–5.7mm, two-thirds to three-quarters the length of spur.

Sikkim: Changu. Gravelly slopes, 3600–3900m. September.

Endemic to Sikkim.

29. C. cavei Long

Similar to *C. casimiriana* and its allies in habit and foliage, but racemes longer, 4–7cm, 8–16-flowered; sepals larger, suborbicular, 1.8–2mm; outer petals broadly crested; upper petal 13.5–15mm, including broad spur 7–8.5 × 2.5–3mm; nectariferous gland 4–6mm, one-half to two-thirds the length of spur.

Sikkim: Phallut. 2740–3350m. September.

30. C. ophiocarpa Hook. f. & Thomson

Erect herb 0.5–1m. Basal leaves numerous when young, ovate-triangular, 10–20 × 7–12cm, bipinnatisect, on flattened petioles 6–15cm; ultimate segments obovate, obtuse or mucronate. Stem leaves numerous, smaller, on petioles 1–2cm. Racemes 5–15cm, many-flowered. Bracts lanceolate, 2–5mm, finely-pointed, entire. Sepals suborbicular, c 1mm. Upper petal 10–12mm, including short rounded spur c 2.5mm, lip winged but not crested. Lower lip equalling upper. Capsules linear, 20–35 × 2mm, strongly undulate.

Bhutan: C – Bumthang district (near Tang), **N** – Upper Bumthang Chu district (Bada La); **Sikkim.** Blue pine forests, 3000–3300m. June.

Similar to *C. flaccida* which differs in its purple flowers and sessile stem leaves.

31. C. meifolia Wall.

Robust herb 20–40cm, with stout fibrous roots, stout stems and persistent leaf bases. Basal leaves large, up to 20 × 12cm, deeply bipinnatisect, on flattened petioles up to 15cm; ultimate segments linear, very narrow, up to 10 × 0.5mm. Stem leaves becoming shorter and subsessile. Flowers numerous in dense rounded racemes. Lower bracts leaf-like, 2–3cm, with flattened linear base and deeply-divided linear segments. Pedicels 10–15mm. Sepals oblong c 1.5 × 1mm. Petals yellow sometimes tipped red or purple, upper 12–13mm including short deflexed spur c 1.5mm, crest rounded, broad; lower petal broadly crested, slightly longer than upper. Capsules narrowly obovoid, c 8 × 3mm.

Sikkim: Yak La, etc. Damp streamsides and gravel, 4200–4500m. July–August.
The above description refers to var. **meifolia.**

var. **sikkimensis** Prain differs from var. *meifolia* in its shorter stems, up to 20cm, and leaves mostly 5–10cm long, much less finely divided with short, narrowly elliptic segments up to 5 × 1mm; bracts shorter, 1.5–2cm, less finely divided; sepals larger, 2–3 × 1–1.5mm; capsules shorter and broader, c 6 × 4mm.

Bhutan: N–Upper Mo Chu district (Laya); **Sikkim:** Gocha La, Tankra La, Chulong; **Chumbi.** Moist alpine scree slopes, 3960–5190m. June–September.

32. C. calliantha Long

Similar to *C. meifolia* var. *sikkimensis* but stems shorter, 8–22cm, much more slender; leaves smaller, 4–6 × 1.5–2.2cm, on narrower petioles 1–2.5cm, main petiolules longer, 3–9mm, ultimate segments linear-elliptic, shorter, 1.3–2.2 × 0.5–0.8mm; flowers 4–9 in short condensed racemes; bracts 1.1–1.7cm; pedicels 12–25mm; sepals much larger, broadly ovate, c 4 × 3mm, acute, margins serrate; petals larger, upper 16–20mm including broad deflexed spur 3–5mm.

Bhutan: C–Thimphu district (Barshong), Punakha district (Maru and Ritang), Tongsa district (Thita Tso, Rinchen Chu), **N**–Upper Mo Chu district (Lingshi) and Upper Mangde Chu district (Namda La). Alpine cliff ledges, damp screes and streamsides, 3800–4730m. June–October.

Endemic to Bhutan. Related to *C. gerdae* which differs in its purple or mauve flowers.

33. C. aurantiaca Ludlow

Similar to *C. meifolia* var. *sikkimensis* and *C. calliantha* but a smaller, more slender plant 4–14cm; leaves smaller, 2.5–3 × 1–1.5cm, on long petioles 2–6cm; ultimate segments obovate or oblanceolate, 2–4 × 1mm, subacute or obtuse; flowers 4–6 in subumbellate congested racemes; bracts large, 1–2.5cm, deeply divided, pedicels long, 1.5–3cm; sepals rounded, broader than long, c 1.5 × 2mm, dentate; upper petal 20–22mm, with narrower, deflexed spur 3–4mm.

Bhutan: N–Upper Mangde Chu (Saga La) and Upper Bumthang Chu districts (Dhur Chu, Dole La, Marlung and Waitang). Alpine cliff ledges and gravelly slopes, 4500–4700m. June–July.

Endemic to Bhutan.

34. C. sikkimensis (Prain) Fedde; *C. duthiei* Maximowicz var. *sikkimensis* Prain

Herb 15–35cm, with numerous stems thickly clothed with persistent petiole bases arising from stout woody rootstock. Basal leaves 6–11 × 1.5–4cm, on long petioles 7–15cm; ultimate segments elliptic, 3–5 × 1mm, sharply pointed. Upper stem leaves smaller, subsessile. Flowers 8–15 in short dense racemes 3–4cm. Lower bracts large, broadly ovate, 15–20 × 12–14mm, deeply pinnatisect; upper bracts becoming smaller and less divided. Pedicels 10–15mm. Upper petal 15–19mm including broad straight spur 6–8mm, crest large and decurrent along spur; lower petal broadly winged and crested. Capsules oblong, c 1cm; seeds smooth, glossy.

Bhutan: N – Upper Kuru Chu district (Singhi) and Upper Kulong Chu district (Me La); **Sikkim.** Rocky streamsides and cliff ledges, 4100–4600m. May–September.

35. C. chasmophila Ludlow

Similar to *C. sikkimensis* but a smaller plant, 5–7cm, stems densely clothed with persistent petiole bases; basal leaves small, 2.5–4 × 1–1.5cm, on slender petioles 3–7cm; segments few, obovate, obtuse, 3–4 × 1–2mm; flowers 3–12 in short racemes; bracts large, 12–15mm, with broad base and deeply laciniate apex; pedicels 10–40mm, slender; upper petal 20–22mm including straight slender spur 10–12mm; crest not decurrent on spur.

Bhutan: C – Punakha district (Black Mountain). Alpine cliff ledges, 4400m. June.

36. C. stracheyi Prain; *C. ramosa* Hook. f. & Thomson p.p., *nom. illeg.* non Prain

Weak much-branched herb 10–18cm. Basal leaves few. Lower stem leaves oblong, 3–7 × 2–2.5cm, on petioles 3–5cm; ultimate segments oblanceolate, 2–5 × 0.5–1mm, acute. Flowers few to numerous in terminal racemes 2–5cm. Lower bracts 7–10 × 5–7mm, deeply pinnatisect, upper bracts shorter and with fewer teeth. Pedicels short, 2–8mm. Upper petal 11–12mm including spur 4–5mm; upper and lower petals broadly crested. Capsules narrowly obovoid, 7–10 × 3mm, seeds glossy but minutely papillate.

Bhutan: N – Upper Pho Chu district (head of W branch of Pho Chu), Upper Kuru Chu district (Narim Thang and Gong La) and Upper Kulong Chu district (Me La); **Sikkim.** Damp mountain screes and streamsides, 4000–4600m. June–August.

This description applies to the typical var. **stracheyi.**

var. **ecristata** Prain differs in its larger size, up to 30cm or more, larger leaves up to 15 × 5cm with broader segments, and flowers with very low or no crests on upper and lower petals.

Bhutan: N – Upper Kulong Chu district (Shingbe, Me La); **Sikkim**: Changu, Guicha La, etc. Damp mountain screes and streamsides, 3650–4570m. July–Setptember.

37. C. dubia Prain. Med: *Tongri Zilpa.*

Similar to *C. stracheyi* var. *stracheyi* but leaves of a thicker texture, more finely dissected with numerous linear segments; lower bracts larger, 10–12mm; pedicels longer, 10–13mm; upper petal 15–18mm including spur 7–9mm, with low crest.

Bhutan: N – Upper Mo Chu district (Ya La) and Upper Kulong Chu district (Me La); **Sikkim; Chumbi.** Alpine scree slopes, 3900–4800m. July–September.

38. C. franchetiana Prain

Robust herb 25–32cm; rootstock unknown. Basal leaves bipinnatisect, 6–7cm long, dissected into oblong-lanceolate segments 2–3mm broad, acuminate; petioles

6–12cm, winged. Flowers 7–15 in racemes 5–8cm, yellow with dark purple tips. Lower bracts c 2mm, pinnatifid; upper bracts elliptic, c 1.5cm, finely pointed. Pedicels 1.4–2cm, becoming reflexed in fruit. Outer petals very narrowly crested, upper 20–23mm including straight spur 12–13mm. Lower lip deflexed. Nectariferous gland half the length of spur. Capsules (immature) pendulous, elliptic, c 13 × 3mm.

Bhutan: C–Thimphu district (between Tremo La and Kang La, Paro Chu); **Chumbi:** Sham Chen and Meto Sebo. 4260–4570m. July–August.

A poorly-known species endemic to W Bhutan and Chumbi.

39. C. denticulato-bracteata Fedde

Similar to *C. franchetiana* but leaves more coarsely dissected into segments 3–5mm broad; racemes denser, many-flowered; bracts much exceeding shorter pedicels 3–8mm; outer petals broadly crested; upper petal 19–21mm including spur 8–10mm, strongly deflexed at tip.

Sikkim: Lungma Chu. 4880m. August.

Known only from the type collection; a specimen from W Bhutan (between Barshong and Lingshi) is very similar and may be conspecific, but has more finely dissected foliage.

Family 61. PAPAVERACEAE

by D.G. Long

Annual, biennial or perennial herbs with milky juice (latex); indumentum of simple, barbellate or stellate hairs. Leaves mostly basal in a rosette, simple, pinnatifid or pinnatisect, rarely palmately lobed, exstipulate, stem leaves usually few, alternate. Flowers on leafless scapes or in leafy racemes or panicles, actinomorphic, hermaphrodite, nodding in bud, showy. Sepals 2–3, caducous. Petals 4–8, free, biseriate, imbricate, often crumpled at first. Stamens numerous, free. Ovary superior, 1-celled or apparently 2–10-celled by intrusive placentae; ovules numerous, placentation parietal; stigmas usually connate, capitate on distinct style, rarely free or sessile on ovary or lobed disc. Fruit a capsule dehiscing by 2–10 pores or valves; seeds numerous.

1. Plants prickly 2
+ Plants not prickly 3

2. Leaves pinnatifid, dentate, sepals 3 , horned; petals yellow .....**2. Argemone**
+ Leaves entire or shallowly crenately-lobed; sepals 2, not horned; petals blue **3. Meconopsis** (*M. horridula, M. sinuata*)

3. Leaves palmately lobed; stigmas sessile on ovary; capsule valves 4–7, dehiscing to base ....**4. Cathcartia**

\+ Leaves simple or pinnately lobed; stigmas on distinct style (sessile but on conspicuous disc in *Papaver*); capsule valves 3–10 dehiscing at apex or in upper third (dehiscing to base but 2-valved in *Dicranostigma*) 4

4\. Ovary and capsule flat-topped with conspicuous lobed disc, stigmas sessile or not 5

\+ Ovary and capsule without disc, tapering into distinct style 6

5\. Stigmas sessile on disc ....**1. Papaver**

\+ Stigmas on distinct style ....**3. Meconopsis** (*M. discigera*)

6\. Sepals not horned; stigmas connate, capitate, 3 –10-lobed; capsule 3–10-valved, dehiscing only at apex or in upper third ....**3. Meconopsis**

\+ Sepals horned at apex; stigmas 2, erect, free; capsule 2-valved, dehiscing to base **5. Dicranostigma**

1. PAPAVER L.

Annual or perennial herbs with white juice. Leaves usually pinnatifid or pinnatisect, sometimes in a rosette at base. Flowers solitary, terminal and axillary or on long leafless scapes. Sepals 2, falling early, not horned. Petals 4 in 2 whorls, suborbicular. Ovary 1-celled with 4–20 placentae; stigmas 4–20, sessile, radiating on a lobed disc. Capsule dehiscing by subapical pores under lobes of persistent disc.

1\. Plants glaucous, glabrous; stem leaves unlobed or shallowly pinnatifid, sessile **1. P. somniferum**

\+ Plants green, bristly-hairy (at least sepals); leaves deeply pinnatisect, on long or short petioles 2

2\. Flowers borne on leafy stems ....**2. P. rhoeas**

\+ Flowers borne on leafless scapes ....**3. P. nudicaule**

1. P. somniferum L. Eng: *Opium Poppy.* Fig. 32 h.

Annual herb with tap root and erect little-branched stem 0.5–1m, glabrous. Basal leaves ovate-oblong, shallowly pinnatifid, 15–25 × 8–15cm, acute, base rounded or cordate, lobes triangular, coarsely dentate, glaucous green; petioles short, 1–2cm; stem leaves smaller, more shallowly lobed or unlobed, sessile. Flowers terminal and axillary on leafless pedicels 10–20cm. Petals 5–9cm, white or pale purple, sometimes with a dark basal blotch. Ovary globose with 5–12-lobed apical disc bearing sessile stigmas. Capsule globose, 2–4cm, flat-topped with enlarged disc.

Bhutan: C–Thimphu district (Tashi Chho Dzong) and Tongsa district (Chendebi); **Sikkim:** Darjeeling. Cultivated and possibly naturalised, 2285–3050m. March–July.

Formerly widely cultivated to produce the drug opium, formed by drying of latex bled from unripe capsules. Other forms are cultivated to produce poppy-seed, used as food, and to produce an edible oil (13, 125).

2. P. rhoeas L. Eng: *Field Poppy*

Annual branched herb 30–50cm. Basal leaves ovate, deeply pinnatisect, 10–15 × 5–9cm, segments linear, acute, dentate, bristly hairy or glabrous, on petiole 2–3cm; stem leaves smaller, on short petioles. Flowers terminal on pedicels 9–15cm, bristly hairy. Sepals bristly hairy. Petals 3–4cm, red with black basal blotch. Ovary obconic with 8–12 stigmatic lobes, each with a rounded apical lobe. Capsule subglobose, 1–2cm.

Bhutan: S – Gaylegphug district (Gaylegphug). Cultivated in garden as an ornamental, 300m. March–June.

3. P. nudicaule L.

Similar to *P. rhoeas* but bristly hairy throughout; basal leaves 3–6 × 1–2.5cm with elliptic-obtuse lobes and longer petioles 5–12cm; each rosette with 1–5 leafless scapes; petals 2–3.5cm, red, purple or yellow; capsule obovoid, c 1.5 × 1cm.

Chumbi.

A very variable species; the Chumbi plants, cultivated as ornamentals, were described as var. *grandiflorum* Prain but probably come within the range of subsp. *rubro-aurantiacum* Fedde.

2. ARGEMONE L.

Prickly herbs with yellowish juice. Leaves glaucous, pinnatifid and dentate. Flowers solitary, terminal and axillary. Sepals 3, caducous, bearing a horn below apex. Petals 6 in 2 whorls, yellow, obovate. Ovary 1-locular with 3–5 placentae. Stigmas 3–5, sessile on ovary. Capsules ellipsoid, 4–6-ribbed, prickly, splitting along sutures in upper third.

1. A. mexicana L. Eng: *Mexican Poppy, Prickly Poppy.* Fig. 32 a,b.

Annual branched herb, 0.3–1m. Leaves elliptic-obovate, 7–30 × 4–9cm, pinnatifid with 3–7 pairs of oblong, acute lobes, base cordate, margins coarsely spinous-dentate, glaucous-green with pale markings along veins, prickly on veins beneath, glabrous above; sessile and clasping stem. Sepals elliptic, c 1.5 × 1cm, prickly and with a subterminal spine-tipped horn 5–10mm. Petals 2–3 × 1.5–2.5cm. Capsules 2.5–3.5 × 1–1.5cm.

FIG. 32. **Papaveraceae** and **Droseraceae. Papaveraceae**. a & b, *Argemone mexicana*: a, portion of shoot with leaves, flower and bud; b, capsule. c & d, *Meconopsis paniculata*: c, portion of shoot with leaves, flower and buds; d, capsule. e, *Meconopsis discigera*: capsule. f & g, *Cathcartia villosa*: f, portion of shoot with leaf and flower; g, capsule. h, *Papaver somniferum*: capsule. i, *Dicranostigma lactucoides*: capsule. **Droseraceae**. j–m, *Drosera peltata*: j, whole plant in flower; k, leaf, lower surface; l, flower; m, ovary with branched styles. Scale: a, c, f, g, h, j × ⅔; i × 4/5; b × 1; e × 1¼; d × 1½; l × 2; k × 4; m × 4½. Drawn by M. Bates.

a
b
c
d
e
f
g
h
i
j
k
l
m
M.B.

Sikkim: Tista valley. Weed of waste ground in subtropical zone, 300m. January–December.

Native of tropical America, widely naturalised. Seed produces an oil used for burning in lamps and also to treat skin diseases (125).

3. MECONOPSIS Viguier

Med: *Ajak* refers to some species of *Meconopsis* used medicinally.

Biennial or perennial, monocarpic or polycarpic herbs (see below) with yellow juice and stout rootstock, bearing erect scapes or stems, indumentum often of barbellate bristles or hairs. Leaves mostly basal in a rosette, simple or pinnatifid, sometimes pinnatisect, elliptic, oblong or lanceolate, rarely ovate; stem leaves several or none. Flowers solitary on leafless scapes, or on simple or branched leafy stems, or in lateral cymes on stout stems. Sepals 2, ovate or elliptic, falling early. Petals showy, mostly 4, sometimes up to 10 (often more numerous in terminal flowers), obovate or suborbicular. Ovary ellipsoid or cylindric, often bristly, rarely with apical disc; style conspicuous, elongate, stigmas united, rarely free. Capsule cylindric to subglobose, 4–10-valved, dehiscing by small subapical pores or longer slits in upper third to leave persistent placentae attached to style.

Several *Meconopsis* species have a 'monocarpic' life-form in which they grow for several years prior to flowering, then flower, set seed and die; others have persistent perennial rosettes or rhizomes and are described as 'polycarpic'; withered remains of old flowering shoots can usually be found on the latter. All the species are highly prized as ornamental garden plants, although some are difficult to cultivate, eg. *M. bella* and *M. sherriffii;* most widely grown are *M. grandis, M. napaulensis, M. paniculata, M. simplicifolia* and *M. superba.*

1. Ovary and capsule flat-topped, bearing a lobed epigynous disc **13. M. discigera**
\+ Ovary and capsule tapering at apex into style, without epigynous disc......2

2. Plants prickly with slender rigid prickles..3
\+ Plants never prickly (sometimes softly bristly)....................................4

3. Flowering stem leafless except at base, often surrounded by leafless scapes; leaves entire or shallowly crenate; petals 2.5–3.5cm ....**1. M. horridula**
\+ Flowering stem leafy throughout, leafless scapes absent; leaves strongly crenately lobed; petals 1.5–2cm..**2. M. sinuata**

4. Flowering stems 1–2.5m, each with 6–many flowers; monocarpic; plants retaining a dense leaf rosette in winter 5
\+ Flowering stems mostly 0.15–0.75m (up to 1.5m in *M. grandis*), each with only 1–5 flowers; polycarpic or monocarpic; basal rosette not persisting in winter 7

5. Leaves shallowly serrate; flowers white ....**5. M. superba**
+ Leaves deeply toothed or lobed; flowers yellow, red, purple or blue 6

6. Flowers yellow, plants softly yellowish villous ....**3. M. paniculata**
+ Flowers red, purple or blue; plants stiffly brownish villous **4. M. napaulensis**

7. Plants robust, 30–70cm (up to 150cm in *M. grandis*); leaves and stems usually densely bristly or sericeous; leaves entire or serrate (rarely shallowly lobed in *M. simplicifolia*); flowers (4–)6–10cm across 8
+ Plants slender, 5–30(–45)cm; leaves and stems subglabrous or with few bristles; leaves entire, crenate, pinnatifid or pinnatisect; flowers 2–5(–7)cm across 10

8. Flowering stems leafless; basal rosette with few withered leaf remains; petals (2–)3–4cm ....**6. M. simplicifolia**
+ Flowering stems bearing several leaves (rarely leafless in *M. sherriffii*); basal rosette with conspicuous withered leaf remains; petals 4–7cm 9

9. Basal leaves elliptic, brownish bristly; flowers 1–4 on stem up to 1.5m; petals 4–7cm, blue or purplish..**7. M. grandis**
+ Basal leaves oblanceolate, brownish sericeous; flowers solitary on stem 15–30cm; petals 4–5cm, pinkish-red ....**8. M. sherriffii**

10. Plants polycarpic; stem base including very numerous withered petiole bases 1.5–2.5cm broad; flowers solitary on leafless scapes; petals 2–3cm **9. M. bella**
+ Plants monocarpic; stem base naked or surrounded by few persistent petiole bases, total breadth 0.4–0.8cm; stems leafy, 1–5-flowered (except in *M. concinna* which has 1-flowered leafless scapes); petals 1.5–2cm 11

11. Leaves ovate, often deeply pinnatisect, lower ones completely withering before flowering; flowering stems leafy; petals pale blue, c 1.5cm ......**12. M. lyrata**
+ Leaves oblong or elliptic, entire, crenate, or shallowly and bluntly lobed, basal ones persisting at flowering time, or only petiole bases persisting; flowering stems leafless (except towards base in *M. primulina*); petals blue or dark purplish 12

12. Stems with persistent leaf bases; leaves entire or shallowly crenate; flowering stems leafy in lower part; petals blue..........................**10. M. primulina**
+ Stems without persistent leaf bases; leaves pinnately lobed; flowering stems leafless; petals dark purplish...**11. M. concinna**

1. M. horridula Hook. f. & Thomson; *M. racemosa* Maximowicz. Med: *Tsher Gnoin.*

Monocarpic herb with long slender tap root, stems 15–30(–80)cm, with dense, pale, spreading prickles throughout. Leaves mostly basal, rosetted, oblanceolate, 5–12 × 1–3cm, obtuse or subacute, base long attenuate, margins entire or shallowly crenately lobed near apex; petioles 2–6cm; stem leaves absent or a few near stem base. Flowers 3–10 on pedicels 2–10cm borne in a raceme on stout erect stem, often surrounded by 3–15 single-flowered leafless scapes 5–20cm arising from basal rosette or sometimes only single-flowered scapes present. Petals 4–8, usually blue, 2.5–3.5cm. Ovary ellipsoid, densely prickly, bearing elongate style. Capsules broadly ellipsoid, c 1.5 × 1cm, prickly, 5–6-valved, dehiscing only near apex.

Bhutan: C –Thimphu to Sakden districts, **N** – Upper Mo Chu to Upper Kulong Chu districts, widespread; **Sikkim.** Alpine cliffs and screes, 3800–4700m. June–August.

The plant is used medicinally.

2. M. sinuata Prain

Similar to racemose forms of *M. horridula* but more slender, with a single elongate leafy flowering stem and few basal leaves; stem leaves numerous, oblanceolate, 4–12 × 1–2cm, obtuse, margins strongly crenately lobed throughout, upper sessile, lower petiolate; flowers 4–8, axillary or not, on pedicels 2–5cm; petals smaller, 1.5–2cm; capsules longer and narrower, 3.5–5 × 0.7–0.9cm, 3–4-valved.

Bhutan: C –Tongsa district (Black Mountain) and Bumthang district (Penge La), **N** – Upper Mangde Chu (Ju La), Upper Bumthang Chu (Kantanang) and Upper Kulong Chu (Me La) districts; **Sikkim.** Rocky alpine hillsides, amongst dwarf Rhododendron and in Fir forests, 3800–4400m. June–August.

3. M. paniculata Prain; *M. napaulensis* sensu F.B.I. non DC. Med: *Upel Serpo, Utpal, Upal.* Fig. 32c,d.

Large monocarpic herb, 1–2.5m, with stout rootstock, softly yellowish villous and minutely stellate-hairy throughout. Rosette leaves elliptic, 30–50 × 8–20cm, deeply pinnatifid, sometimes pinnatisect near base; lobes ovate or oblong, acute or obtuse, coarsely crenate or serrate; petioles 15–25cm; stem leaves smaller, becoming sessile on upper part of stem. Flowering stem solitary, with short branches; flowers numerous, pendulous, borne singly in upper part and in 2–6-flowered lateral cymes in lower part; pedicels 2–6cm. Petals 4, yellow, 3–5cm. Ovary subglobose, densely hairy, bearing style 7–10mm with capitate 6–10-lobed stigma. Capsules ellipsoid, 2.5–3 × 1.2–1.5cm, 6–10-valved, dehiscing only near apex, borne on elongated pedicels 5–15cm.

Bhutan: C – Ha, Thimphu, Tongsa, Bumthang and Sakden districts, **N** – Upper Mo Chu, Upper Mangde Chu, Upper Bumthang Chu and Upper Kulong Chu districts; **Sikkim.** Grassy and rocky alpine hillsides, amongst Juniper/ Rhododendron scrub and in Fir forests, 3350–4260m. June-August.

4. M. napaulensis DC.; *M. wallichii* Hooker, non *M. napaulensis* sensu F.B.I.

Similar to *M. paniculata* but more stiffly brownish villous throughout; basal leaves often more strongly pinnatisect, segments oblong with large rounded teeth; petals red, purple or blue; capsules oblong-ellipsoid, 1.5–3 × 1cm.

Bhutan: C – rare, Thimphu district (Chelai La, Paro) (117); **Sikkim** and **Chumbi:** common. Fir forests and by mountain streams, 2700–3600m. July–September.

Rootstock reported to be highly poisonous (81).

5. M. superba Prain

Similar to *M. paniculata* and *M. napaulensis* but basal leaves simple, unlobed, obovate, shallowly serrate, densely whitish sericeous; stem leaves oblanceolate, sessile; flowering stem a simple raceme, bearing only 6–20(–30) flowers on long stout pedicels; petals white, larger, 5–6cm; capsules larger, 4–5 × 1.5–2cm, borne on very stout elongated pedicels 15–25cm.

Bhutan: C – Ha district (Ha La and Kyu La). Amongst shrubs on scree slopes, 4100-4250m. June–August.

Endemic to W Bhutan. The original locality, 'Ho-Ko-Chu' has sometimes been given as Chumbi, but is probably from the above district of Bhutan.

6. M. simplicifolia (D. Don) Walpers. Med: *Upel Gnoinpo.*

Polycarpic or sometimes monocarpic herb, 30–70cm, brownish bristly throughout. Basal rosette with few withered leaf remains, leaves all basal, oblanceolate, 4–10 × 1–2cm, obtuse or subacute, base long-attenuate, margins entire or with a few teeth or shallow rounded lobes; petioles 4–15cm. Flowers solitary, nodding, on 1–5 scapes 30–70cm. Petals 5–8, purple or blue (2–)3–4cm. Filaments coloured as petals. Ovary ellipsoid, glabrous or bristly; style slender with stout capitate stigma. Capsules oblong-ellipsoid, 5–7 × 1–1.5cm, slightly constricted above base, thinly bristly, with 4–9 valves dehiscing in upper third.

Bhutan: C – Ha, Thimphu, Punakha and Tongsa districts, **N** – Upper Mo Chu to Upper Kulong Chu districts, common; **Sikkim.** Juniper/Rhododendron scrub and on rocky alpine hillsides and screes, 3350–4600m. May–July.

7. M. grandis Prain

Similar to *M. simplicifolia* but a taller plant up to 1.5m; basal rosettes swollen with withered leaf remains; basal leaves elliptic, 15–25 × 5–7cm, acute, base cuneate or shortly attenuate, margins with several shallow, broad teeth; petioles 10–25cm; stem leaves becoming sessile and whorled (bract-like) at top of stem, sometimes cordate; flowers 1–4 per stem, on pedicels 10–15cm; petals usually 4, larger, 4–7cm, usually blue or tinged purple; filaments white; capsule 4–6-valved.

Bhutan: C – Sakden district (Orka La and Nyuksang La), **N** – Upper Mo Chu district (Yale La) and Upper Kulong Chu district (Me La); **Sikkim:** Jongri. Rocky hillsides, 3650–4200m. June–August.

Cultivated in Sikkim around shepherds' huts for extraction of oil from seeds.

8. M sherriffii Taylor

Similar to *M. simplicifolia* and *M. grandis* but withered leaf bases abundant and conspicuous; basal leaves oblanceolate, attenuate at base, usually brownish sericeous; flowering stems 15–30cm, bearing several sessile stem leaves (bracts) usually in a whorl in lower half; flowers always solitary; petals 4–5cm, pinkish-red; filaments white; ovary broadly ellipsoid, densely sericeous with short style and large capitate stigma; capsules ellipsoid, c 3 × 1.5cm, hairy.

Bhutan: N – Upper Pho Chu (Chojo Dzong and Gafoo La) and Upper Mangde Chu (Passu Sefu) districts. On alpine cliffs and screes and in Juniper /Rhododendron scrub, 4200–4600m. June–July.

Endemic to Bhutan and adjacent SE Tibet. Some specimens from Chojo Dzong have leafless scapes.

9. M. bella Prain

Dwarf herb with perennial tap root and rosette with numerous withered petiole bases and up to 18 stems 5–15cm. Basal leaves numerous, very variable, simple, elliptic, lanceolate or oblong, 1.5–4 × 0.5–1cm, obtuse, base attenuate, sometimes deeply pinnatisect or bipinnatisect, 1–3 × 1–1.5cm, ultimate segments bluntly 3-fid, glabrous or with a few hairs; petioles 2–10cm, bases persistent. Flowers solitary on leafless scapes 4–15cm. Petals 4, blue, 2–3cm. Ovary ellipsoid, glabrous or bristly, style short, stout. Capsules narrowly obovoid, c 1.5 × 0.7cm, 4–7-valved, dehiscing only near apex, borne on elongated thickened recurved scapes 10–15cm.

Bhutan: C – Ha, Thimphu, Punakha, Tongsa and Bumthang districts, **N** – Upper Mo Chu to Upper Kulong Chu districts, widespread; **Sikkim.** Alpine cliff ledges and steep grassy slopes, 3700–4600m. June–August.

10. M. primulina Prain

Slender monocarpic herb, 10–30cm, with slender tap root, sparsely hairy throughout. Basal leaves mostly withering, few persisting at flowering time, simple, oblong or elliptic, 2–5 × 0.5–1cm, obtuse, base attenuate, margins entire or shallowly crenate; petioles 2–3cm, bases persistent around stem base; stem leaves similar, few, borne only on lower part of stem. Flowering stems erect, usually with 1–3 long branches arising from near base, each branch with a terminal flower. Petals 4–8, blue, 1.5–2cm. Stamens numerous. Ovary oblong with a few bristles, style elongate, stigma 4-lobed. Capsules cylindric, 3–4 × 0.4cm, 4-valved, dehiscing in upper third.

Bhutan: C –Thimphu district (Tremo La), **N** – Upper Mo Chu district (Lingshi, Lingshi La, Shinje La, Phile La, Yale La, Kangla Karchu La), locally common; **Chumbi:** Phari. Alpine slopes, amongst dwarf *Rhododendron* and in screes and moraines, 3190–4600m. May–July.

Endemic to W Bhutan and the neighbouring Chumbi valley.

11. M. concinna Prain; *M. lancifolia* (Franchet) Prain var. *concinna* (Prain) Taylor

Similar to *M. primulina* but stem without old persistent petiole bases; leaves all basal, persistent at flowering time, elliptic, pinnately lobed with broad rounded lobes, base cuneate; petioles 2–7cm; flowers solitary on slender scapes 5–10cm; petals 4, dark purplish.

Bhutan: **C** – Sakden district (Orka La). Alpine grassy slopes, 4110m. July. A disjunct species known elsewhere only from W China.

12. M. lyrata (Cummins & Prain) Prain; *M. polygonoides* (Prain) Prain, *Cathcartia lyrata* Cummins & Prain, *C. polygonoides* Prain

Similar to *M. primulina* but a weaker herb with swollen tap root; basal leaves soon withering, usually pinnatifid or pinnatisect, ovate, rarely simple, oblong and entire, 2 – 5 × 1 – 3cm, subacute, base truncate, cordate or cuneate; lateral lobes 2 – 4, elliptic, much smaller than terminal lobe; petioles longer, 3 – 8cm, bases not persistent; stem leaves 2 – 5, distant, uppermost sessile; flowers up to 5, terminal and in axils of upper leaves, on slender curved pedicels 3 – 8cm; petals pale blue, usually 4, c 1.5cm; stamens fewer, 12 – 36; ovary glabrous with short or long style.

Sikkim: Phallut, Jongri, etc; **Chumbi.** Rocky alpine slopes and cliffs, 3000 – 4500m. June – September.

13. M. discigera Prain. Fig. 32 e.

Monocarpic herb, 30 – 70cm, with slender tap root. Aerial parts yellowish-bristly throughout. Basal leaves numerous, in a dense rosette; lamina oblanceolate, 4 – 10 × 1 – 2.5cm, obtuse, base long-attenuate, margins bluntly toothed or shallowly pinnatifid in upper part; petioles 3 – 12cm, winged, densely spreading bristly, bases persisting in a dense bristly mass; stem leaves few, sessile, lanceolate. Flowering stem solitary, ribbed, bearing 10 – 15 flowers on slender pedicels 1 – 4cm. Petals 4, blue or purple, 4 – 5cm. Ovary oblong, yellowish bristly, truncate at apex with a broad glabrous 6 – 10-lobed disc, style elongated bearing 6 – 10 linear, erect lobes. Capsules oblong, 3 – 4 × 1.5cm, thick and woody with enlarged disc and persistent style, dehiscing by 6 – 10 small valves under lobes of disc, borne on enlarged pedicels 3 – 10cm.

Bhutan: **C** –Thimphu district (Barshong and Kuma Thang, Upper Paro Chu), **N** – Upper Mo Chu district (Gyengo Thang, E of Chomo Lhari); **Sikkim:** Gucha La. Alpine cliffs, screes and rocky ravines, 3600 – 4100m. June – August.

4. CATHCARTIA Hooker

Similar to *Meconopsis* but always perennial and polycarpic; leaves broadly ovate, palmately lobed; flowers on erect branched stems; petals always 4; ovary cylindric, glabrous, bearing sessile stigma; capsules cylindric, 4 – 7-valved, dehiscing almost to base.

1. C villosa Hooker; *Meconopsis villosa* (Hooker) G. Taylor. Fig. 32. f,g.

Perennial herb with stout rootstock and persistent withered leaf bases, flowering stems erect, usually unbranched, 0.6 – 1.5m; brownish villous throughout. Basal leaves few, broadly ovate or suborbicular, 6 – 12 × 8 – 15cm, palmately 3-or 5-lobed, base cordate, 3 – 5-veined, lobes coarsely and bluntly toothed; petioles 15 – 25cm; stem leaves smaller, upper ones sessile. Flowers solitary, terminal and axillary, 1 – 5 per stem, borne on slender pedicels 3–12cm. Petals always 4, suborbicular, yellow, 2.5–3cm. Ovary cylindric, glabrous, 1.5–2cm bearing sessile stigma with 4–7 radiating lobes. Capsules cylindric, 4–8 × 0.5–0.7cm, 4–7-valved, dehiscing almost to base leaving persistent placentae attached to stigma.

Bhutan: C–Thimphu district (Dochu La), Punakha district (W of Pele La), Tongsa district (Rinchen Chu and Longte Chu) and Sakden district (Sakden), **N**–Upper Kulong Chu district (Lao); **Sikkim.** Rocky streamsides in Hemlock and Fir forests, 2700–4000m. May–July.

Endemic to E Himalaya, valued horticulturally as an ornamental woodland-garden plant.

5. DICRANOSTIGMA Hook. f. & Thomson

Perennial herbs with yellow juice and stout taproot bearing several little-branched flowering stems. Basal leaves numerous, in a dense rosette, deeply pinnatifid. Stem leaves few, alternate, sessile. Flowers few, on slender pedicels, terminal and in axils of stem leaves. Sepals 2, horned at apex, falling early. Petals 4 in 2 whorls, orange. Ovary narrowly ellipsoid, 1-celled, with 2 parietal placentae; style stout, erect with 2 ascending stigmas. Capsule elongate-cylindric, dehiscing to base by 2 valves; placentas persistent, attached at apex to style.

1. D. lactucoides Hook. f. & Thomson; *Stylophorum lactucoides* (Hook. f. & Thomson) Baillon, *Chelidonium dicranostigma* Prain. Fig. 32 i.

Decumbent rosette herb with flowering stems 8–35cm, subglabrous or thinly hairy throughout. Basal leaves elliptic-oblanceolate, 6–15 × 1.5–7cm, later in season becoming larger, up to 25cm, deeply pinnatifid with 4–7 pairs of oblong or squarish lobes becoming larger towards leaf apex, margins irregularly dentate, glaucous green with green or purplish blotches, tapering at base into winged ciliate petiole 1–6cm; stem leaves much smaller, with 1–3 pairs of lobes, sessile. Flowers 3–5cm across, borne on slender curved pedicels 1.5–12cm. Sepals broadly elliptic, 1–1.5cm, with short blunt horn. Petals obovate, 1.5–2.5 × 1–2cm. Ovary puberulous, c 1cm including stigmas. Capsules 3–8cm, often slightly curved, borne on elongated pedicels up to 20cm.

Chumbi: Phari. Mountain screes, 4000m. June.

Family 62. CAPPARACEAE

by A.J.C. Grierson

Trees, shrubs, sometimes scandent, or herbs; indumentum absent or of simple or stellate hairs. Leaves alternate, simple, pinnately veined, entire or digitately compound, stipulate or exstipulate. Inflorescence racemose or paniculate, terminal, or axillary or supra-axillary. Flowers bisexual, rarely unisexual, actinomorphic or somewhat zygomorphic. Sepals 4–6, free. Petals 4 or absent. Stamens 6–numerous, anthers splitting longitudinally. Ovary superior, frequently borne on an elongated gynophore, 1-celled. Fruit capsular or berry-like, several to many-(rarely 1-)seeded.

1. Leaves simple 2
\+ Leaves digitately compound, 3–7-foliate 3
2. Sepals 6; petals absent; stamens adnate to gynophore near its base and leaving a visible scar at fruiting time above the scar of perianth attachment **1. Stixis**
\+ Sepals 4; petals 4; stamens not united to gynophore which bears only scars of perianth attachment at fruiting time .... **3. Capparis**
3. Trees; ovary ellipsoid or subglobose; fruit berry-like, seeds immersed in pulp **2. Crateva**
\+ Herbs; ovary ± cylindric; fruit a capsule .... **4. Cleome**

1. STIXIS Loureiro

Unarmed woody climber. Leaves simple. Flowers in clustered axillary racemes or panicles. Sepals 6. Petals absent. Stamens numerous borne near base of short gynophore. Ovary 1-celled, styles 3, minute. Fruit 1-seeded.

1. S. suaveolens (Roxb.) Pierre; *Roydsia suaveolens* Roxb. Nep: *Kasouli Lahara* (34)

Leaves coriaceous, oblong or obovate, 10–25 × 5–9cm, abruptly acuminate or apiculate, base cuneate, glabrous, lateral veins 5–8 pairs, prominent; petioles 1–4cm. Inflorescence 10–20cm, flowers fragrant. Sepals ovate-elliptic c 5 × 2.5mm, olive-green, pubescent, becoming reflexed. Stamens 8–12mm, borne on gynophore 2–3mm above its base. Ovary ellipsoid c 2mm, borne on gynophore c 5mm. Fruit obovoid, 2.5–3 × 2–2.5cm (according to some up to 4–5.5 × 3–4cm), fruit-wall 4–5mm thick, spongy, surface covered with flat whitish warts.

Bhutan: S – Samchi district (Samchi); **Sikkim.** Subtropical forests, 300–650m. September–April.

2. CRATEVA L.

Deciduous trees. Leaves palmately 3-foliate; stipules minute, caducous. Flowers unisexual, in terminal corymbose racemes, appearing with young leaves. Sepals 4, valvate. Petals 4, subequal, clawed at base. Stamens 12–50, adnate to base of gynophore, anthers infertile in female flowers. Ovary 1-celled, ellipsoid or subglobose, borne on a slender gynophore, caducous in male flowers. Fruit ± globose, wall spongy, covered with flat whitish warts, seeds immersed in pulp.

1. C. religiosa Forster f. agg.; *C. unilocularis* Hamilton, *C. nurvala* Hamilton. Sha: *Lado Shing;* Nep: *Chiple* (34), *Shipley Kane, Chiple Kath.*

Tree 3–15m, young shoots brown, bearing a few long pale lenticels. Leaflets subcoriaceous, ovate, obovate or elliptic, 7–15 × 4–7cm, shortly acuminate, middle leaflet largest, symmetrically cuneate at base, lateral ones oblique, rounded at base, glabrous; petioles 4–13cm, usually glandular-papillate at apex; petiolules 3–7mm. Petals cream-coloured, claw 4–6mm, blade 1.25–2 × 0.5–1.5cm. Filaments purplish, 3–4.5cm. Ovary ellipsoid or subglobose, 2.5–3 × 1.75–2mm, including short rounded style; gynophore 3–4.5cm. Fruit 3–5.5cm diameter, fruit wall 6–7mm thick. Seed broadly ellipsoid, somewhat compressed 14–15 × 10–12 × 7–8mm, smooth.

Bhutan: S – Samchi, Sankosh, Sarbhang, Gaylegphug and Deothang districts, **C** –Tongsa district (Shamgong); **Sikkim.** Subtropical and Warm broad-leaved forests, 300–1000m. March.

The above description is based on East Himalayan material which is intermediate in the overlapping series of characters used by Jacobs to separate the constituent species of this aggregate. These specimens best accord with *G. unilocularis,* although, due to the thick fruit wall, they approach *C. nurvala* but lack the crested seeds of that species.

3. CAPPARIS L.

Trees or shrubs, sometimes scrambling or climbing by hooked spines, indumentum of simple, medifixed or stellate hairs. Leaves simple, cataphylls (modified first leaves) sometimes present at base of shoots. Flowers 2–10 in vertical rows at or above leaf axils, rarely solitary or in lateral or terminal racemes, sometimes subumbellate or corymbose. Sepals 4 in two unequal pairs. Petals 4 in two pairs, the upper pair erect and closely coherent at base, the lower longer and obliquely descending. Stamens 10–numerous, ± as long as gynophore. Ovary ellipsoid, borne on an elongate gynophore, stigma inconspicuous. Fruit ellipsoid or globose, seeds 1–many, embedded in pulp.

FIG. 33. **Cruciferae, Capparaceae** and **Moringaceae. Cruciferae**. a–d, *Brassica campestris*: a, inflorescence with upper leaf; b, flower; c, the same, sepals and petals removed to show stamens and ovary; d, pod, one valve removed to show replum and seeds arranged on false septum. e & f, *Draba elata*: e, pod; f, the same with one valve removed to show replum and seeds arranged on false septum. g, *Parrya nudicaulis*: pod. h, *Torularia humilis*: pod. i, *Dilophia salsa*: pod, one valve displaced to show seeds. j–l, *Lepidium capitatum*: j, pod; k, the same, with one valve removed to show replum and false septum; l, seed. m, *Capsella bursa-pastoris*: pod. n, *Thlaspi arvense*: pod. o, *Raphanus raphanistrum*: pod. **Capparaceae**. p, *Capparis acutifolia*: portion of shoot with leaf and inflorescence. q, *Cleome gynandra:* flower. **Moringaceae**. r, *Moringa oleifera*: flower, from above. Scale: a, g, p × ⅔; d × 1; n, o × 1⅔; b, c, h, r × 2; e, f × 2½; m, q × 3; i × 3½; j, k × 8; l × 10. Drawn by M. Bates.

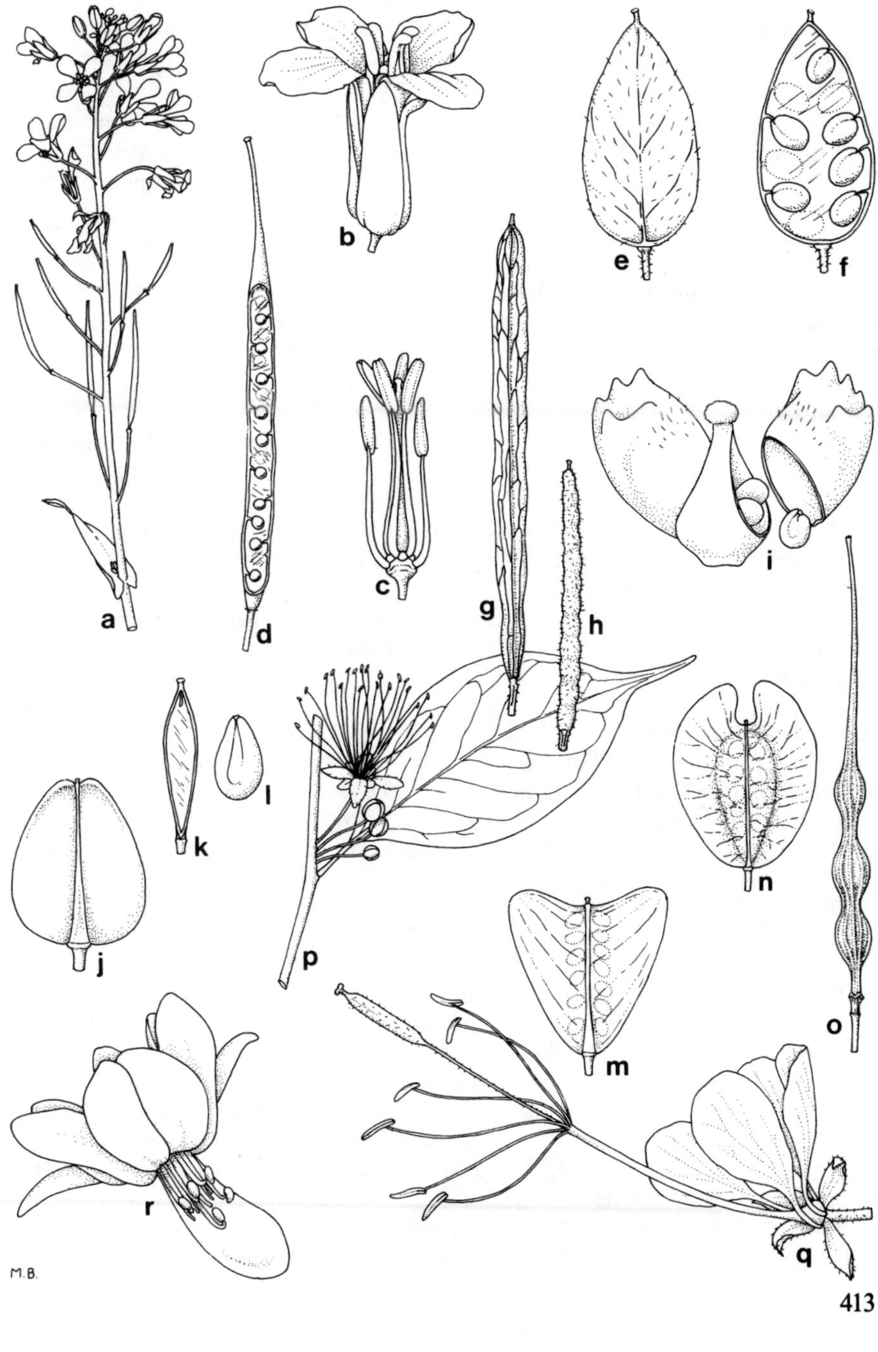
a
b
c
d
e
f
g
h
i
j
k
l
m
n
o
p
q
r
M.B.

1. Flowers 2–10 in vertical rows at or above leaf axils, rarely solitary, sometimes borne on leafless parts of shoots .... **Species 1–3**
+ Flowers not in rows but in mostly terminal inflorescences 2

2. Inflorescence a slender raceme .... **6. C. assamica**
+ Inflorescences of subumbellate corymbs sometimes forming panicles **Species 4 & 5**

1. C acutifolia Sweet; *C. viminea* sensu F.B.I. non Oliver. Nep: *Chila Pati.* Fig. 33 p.

Climbing shrub, branches unarmed or with short (2–5mm) straight stipular spines, shoots minutely brownish stellate at first, without cataphylls at base. Leaves ovate-elliptic, 5.5–12 × 2–4cm, abruptly acuminate or acute, base rounded or cuneate, minutely stellate pubescent at first; petioles 5–7mm. Flowers 2–4 in vertical rows at or above leaf-axils, sometimes solitary, pedicels 1–2cm. Sepals broadly elliptic, 5–8 × 3–4mm, obtuse, concave, the outer pair ± covering the inner ones in bud. Petals white, obovate, 9–12 × 3–5mm. Stamens numerous, filaments 2.5–3cm. Ovary ovoid, c 2 × 1mm, glabrous. Fruit subglobose, 1–1.5cm, blackish.

Bhutan: S–Gaylegphug and Deothang districts, **C**–Tongsa, Mongar and Tashigang districts; **Sikkim.** Subtropical and Warm broad-leaved forests, 300–1400m. March–May.

There are two subspecies in both Sikkim and Bhutan: subsp. **bodinieri** (Léveillé) Jacobs with spiny branches and subsp. **viminea** Jacobs in which branches are usually unarmed.

2. C. olacifolia Hook. f. & Thomson. Nep: *Naski, Hais* (34).

Similar to *C. acutifolia* but shoots brownish stellate-tomentose at first and with subulate cataphylls c 1–1.5mm at base; leaves ovate, 6–14 × 2–4cm, gradually acuminate; flowers 2–3 in axillary rows, sometimes solitary; sepals ovate, acute, 8–10 × 4–5mm, concave; petals oblanceolate, 17–22 × 5mm, white; gynophore 2.75–3.5cm, ovary tomentose; fruit globose, 0.7–1cm diameter, red.

Bhutan: S–Samchi to Gaylegphug districts; **Sikkim.** Subtropical forests, 200–650m. March–April.

3. C. multiflora Hook. f. & Thomson

Similar to *C. acutifolia* but shoots whitish pubescent at first with medifixed hairs and bearing subulate cataphylls 2–3mm at base; leaves elliptic or obovate, 11–30 × 4–8cm, acuminate, base cuneate; flowers numerous, arranged in rows of 7–10 on leafless parts of shoots 10–20cm long, between old and new leaves, pedicels c 1.5cm; sepals ovate, 3–4 × 1.5–3mm; petals c 6 × 3mm, white; stamens 10–12; gynophore 6–12mm, ovary 1.25 × 1mm, glabrous or minutely pubescent; fruit subglobose, c 1cm diameter.

Bhutan: S–Phuntsholing, Gaylegphug and Deothang districts; **Sikkim.** Subtropical forests, 150–1000m. March–May.

4. C. sikkimensis Kurz; *C. cathcartii* Hemsley

Armed scandent shrub, spines hooked, shoots brownish pubescent with simple hairs, lacking cataphylls at base. Leaves coriaceous, ovate-elliptic, 7–12 × 3.5–7cm, acute, base rounded, margins ± inrolled, lateral veins ± conspicuous, 7–8 pairs; petioles 1–2cm. Flowers 3–8 in terminal or axillary corymbs, pedicels 1–2cm. Sepals suborbicular, 7–12mm. Petals obovate, 12–25 × 5–8mm, white. Stamens numerous. Ovary ellipsoid, 2–4 × 1.5–2mm, glabrous, borne on a reddish gynophore 3.5–5.5cm. Fruit globose or ellipsoid, (2.5–)5–6.5 × (2.5–)5–5.5cm, purple.

Bhutan: C–Tongsa district (Dakpai); **Sikkim.** Warm mixed broad-leaved forests, 1750m. June.

5. C. cantoniensis Loureiro; *C. pumila* Bentham

Similar to *C. sikkimensis* but leaves subcoriaceous, elliptic, 7–12 × 2.5–3.5cm, acuminate, base rounded or cuneate, veins inconspicuous; corymbs axillary to uppermost leaves, forming terminal panicles, pedicels 0.5–2cm; sepals suborbicular, 3–6mm; petals obovate, 5–6.5 × 2–3mm, white; gynophore 4–12mm; fruit subglobose, 1–1.5cm diameter.

Bhutan: S–Gaylegphug district (near Surey); **Sikkim.** Warm broad-leaved forests, 300–1000m. July–September.

6. C. assamica Hook. f. & Thomson

Shrub or small tree 2–3m; shoots unarmed or with ascending c 1mm spines, sparsely appressed-pubescent with medifixed hairs and bearing subulate cataphylls c 1mm at base. Leaves elliptic, 12–15(–21) × 3.5–4.5(–6.5)cm, acuminate, base cuneate, glabrous. Racemes elongate, 10–30cm, terminal or subterminal, pedicels 1.5–2cm. Sepals broadly ovate-elliptic, 3–4mm long and broad, ciliate. Petals ovate, 4×1.5mm, white. Stamens 16–20, c 10mm. Ovary subglobose, c 1.5mm, on gynophore 6–9mm. Fruit globose, c 0.75cm, red.

Bhutan: S–Gaylegphug district (near Gaylegphug). Subtropical forests, 340m. March–April.

4. CLEOME L.

Rank-smelling annual herbs. Leaves digitately 5–7-foliate, sometimes with short spine-like stipules. Flowers in terminal bracteate racemes, actinomorphic, or zygomorphic by displacement of petals. Sepals 4. Petals 4, regularly arranged or ascending. Stamens 6–numerous, sessile or borne on gynophore. Style short; stigma capitate. Capsule ± cylindrical, 2-valved.

1. Stamens 6; capsule borne on elongated gynophore .... **Species 1 & 2**
\+ Stamens 12–25; capsule sessile .... **3. C. viscosa**

1. C. gynandra L.; *Gynandropsis pentaphylla* DC. Fig. 33 q.

Stems erect, 15–45cm, glandular-pubescent. Leaflets 5, obovate, 0.7–5 × 0.3–3cm, lowest pair smallest, middle largest, acute, base attenuate, margins entire, glandular-pubescent beneath; petioles 2–10cm. Bracts leaf-like, trifoliate, 3–25mm; pedicels 1–3cm. Sepals oblanceolate, 3–4 × 1mm. Petals white, 8–10 × 3–4mm, narrowly clawed at base. Stamens 6, borne near apex of gynophore, filaments 1–2cm. Ovary at first c 2mm, sessile among stamens, gynophore later elongating up to 7mm above stamens. Capsule 5–6.5 × 0.4–0.6cm, valves papery.

Bhutan: C – Punakha district (Kancham and Wangdu Phodrang); **Sikkim.** Roadsides and waste ground, 300–1800m. March–August.

2. C. hassleriana Chodat

Similar to *C. gynandra* but stems up to 1m; leaves 5–7-foliate, petioles 4–7cm bearing a pair of short spine-like stipules at base, leaflets oblanceolate, 2.5–12 × 1–3.5cm, often spiny on midribs beneath; bracts simple, ovate, c 10 × 5mm; pedicel 2–5cm; sepals lanceolate, 8–10cm, reflexed; petals purplish, c 2.5 × 1cm; stamens sessile, filaments c 4cm; gynophore at first c 3cm, later 5–7.5cm; capsule 4.5–9 × 0.3cm.

Sikkim: locality unknown.

Native of tropical S America, sometimes cultivated for its showy flowers.

3. C. viscosa L.

Stems up to 60cm, glandular-pubescent. Leaves 5-foliate, exstipulate, petioles 1.5–5cm, leaflets elliptic-oblanceolate, 1–4 × 0.4–1.5cm, acute or acuminate, base cuneate, glandular-pubescent. Bracts leaf-like, 3-foliate. Pedicels 1–2cm. Sepals lanceolate, c 8 × 2mm. Petals oblanceolate, c 9 × 3.5mm, yellowish. Stamens 12–25. Ovary sessile. Capsule 6–9 × 0.3–0.4cm.

Bhutan: C – Tashigang district (Dangme Chu); **Sikkim.** Roadsides in warm open valley, 600–1050m. June–July.

Family 63. CRUCIFERAE

by A.J.C. Grierson

Annual, biennial or perennial herbs, rarely woody, glabrous or bearing simple, branched or stellate hairs. Leaves alternate, exstipulate, simple, sometimes pinnately divided. Flowers solitary, scapose, or more commonly in racemes or corymbs, actinomorphic, bisexual. Sepals 4, free, in two opposite pairs, sometimes, especially the inner pair, saccate at base. Petals usually 4, rarely absent, free, alternating with sepals, often with a narrow basal claw, limb spreading to form a characteristic cross-like corolla. Stamens 6 (rarely fewer by suppression), the outer pair shorter than inner 4. Ovary superior, usually divided into 2 cells by a membranous false septum formed by placental outgrowth, sometimes 1-celled or with 2 or more superposed cells, ovules 1–many on two parietal placentae. Style simple or ± absent; stigma often capitate. Fruit pod-like, usually dehiscent, opening by 2 valves to leave the seeds attached to the placental framework (replum) and false septum, sometimes breaking transversely into 1-seeded segments. Seeds 1–many in 1–2 rows in each cell.

1. Petals yellow or orange (sometimes absent in *Rorippa*) 2
+ Petals white, pink or purple 9
2. Pod ending in a slender beak 1–2cm; beak seedless or with 1–2 seeds **1. Brassica**
+ Pod unbeaked or sometimes with conical style 1–2mm 3
3. Dwarf caespitose perennials; flowers apparently solitary on naked pedicels, later racemose by growth of inflorescence axis ....**29. Chrysobraya**
+ Annuals, biennials or perennials; flowers few to numerous in obvious corymbs or racemes 4
4. Leaves pinnate, pinnatisect or deeply pinnatifid 5
+ Leaves entire or toothed...7
5. Leaves finely bi- or tripinnatisect, ultimate segments 1–2mm broad **32. Descurainia**
+ Leaves pinnatisect, pinnatifid to pinnate, segments or leaflets mostly 5–25mm broad 6
6. Petals distinctly longer than sepals; seeds in 1 row................**15. Barbarea**
+ Petals scarcely longer than sepals (sometimes absent); seeds in 2 rows **18. Rorippa**
7. Pods elliptic or ovate, compressed ....**10. Draba**
+ Pods linear, cylindrical or angular...8
8. Pubescence ± spreading; pods less than 1cm ....**19. Lepidostemon**
+ Pubescence closely appressed; pods 1.5–8cm....................**22. Erysimum**
9. Lower leaves pinnate or ternate or deeply pinnatifid or ternately lobed ...10
+ Lower leaves entire, toothed or shallowly pinnatifid 18
10. Pods ending in a slender conical seedless beak 2.5–3cm ....**2. Raphanus**
+ Pods without beak, sometimes with a short conical style-base 1–2mm ...11
11. Slender herbs with ternately lobed leaves; stems never bulbiliferous; flowers purplish ....**9. Lignariella**
+ Robust or slender herbs, leaves variously lobed or divided; flowers mostly white (purplish in *Loxostemon* and some *Cardamine*; rarely leaves ternately lobed in *Loxostemon* but then stem bulbiliferous at base) 12
12. Pods compressed in opposite plane to septum, valves boat-shaped in section 13
+ Pods ± compressed parallel to septum, valves angular or convex in section 15

13. Pods ovate, c 2.5mm, with 1 seed in each valve..**3. Lepidium** *(L. capitatum)*
+ Pods oblong or obcordate, 6–10mm, with several seeds in each valve 14

14. Lower leaves deeply runcinate-pinnatifid; pods obcordate ....**5. Capsella**
+ Lower leaves pinnate with 5–7 pairs of leaflets; pods oblong.....**6. Hedinia**

15. Slender herbs with bulbiliferous rhizomes; filaments broadened and toothed above ....**12. Loxostemon**
+ Robust or slender herbs without bulbiliferous rhizomes (except in *Cardamine loxostemonoides*); filaments slender or flattened below and unappendaged 16

16. Stems, leaves and pods coarsely glandular-papillose ....**20. Dontostemon**
+ Stems and leaves glabrous, smooth or bearing simple eglandular hairs 17

17. Plants of drier ground, not or seldom rooting at nodes; capsules straight, seeds in 1 row, dehiscence explosive by coiling of valves.............**11. Cardamine**
+ Marsh plants, copiously rooting at lower nodes; pods often curved, seeds in 2 rows, dehiscence not explosive, valves remaining ± flat.....**17. Nasturtium**

18. Plants glabrous or bearing only simple or glandular hairs 19
+ Plants bearing medifixed, branched or stellate hairs these sometimes mixed with simple hairs 31

19. Pods compressed in opposite plane to septum, valves ± boat-shaped......20
+ Pods ± compressed parallel to septum, valves flat, keeled or convex.......23

20. Stem-leaves sessile, auriculate; pods 8–15mm long ....**4. Thlaspi**
+ Stem-leaves attenuate or petiolate at base, not auriculate; pods 2–3.5mm long 21

21. Pods obcordate, valves toothed on the back ....**7. Dilophia**
+ Pods ellipsoid, broadly ovate or orbicular, valves without toothed appendage 22

22. Pods notched and winged at apex ....**3. Lepidium** (*L. virginicum)*
+ Pods acute or rounded at apex, unwinged ....**8. Cochlearia**

23. Leaves all basal; flowers in leafless racemes or solitary on scapes...........24
+ Stems leafy 26

24. Flowers solitary on scapes.......................................**13. Pegaeophyton**
+ Flowers 5–15 or more in racemes or corymbs.................................25

25. Medium-sized herbs, 6–30cm; leaves entire, distantly dentate or shallowly pinnatifid; pods linear, 2–10 × 0.5–1.5cm..........................**21. Parrya**
+ Dwarf herbs, 3–5cm; leaves entire or with a few teeth; pods ovoid, 4–10 × 1.5–2mm ....**28. Braya**

26. Low-growing herbs, stems 3–12cm 27
+ Medium-sized or tall herbs, stems 20–50cm 30

27. Glabrous herb; basal leaves ovate.............**24. Eutrema** (*E. heterophyllum)*
+ Pubescent herbs; leaves oblanceolate or obovate 28

28. Flowers purple; leaves 3–5-toothed or -lobed at apex ....**31. Christolea**
\+ Flowers white; leaves entire, sinuous or with a few teeth......................29

29. Basal leaves 4–12cm ....**26. Arcyosperma**
\+ Basal leaves 1–1.5cm ....**28. Braya**

30. Flowers deep reddish-purple; pods cylindrical, valves rounded with inconspicuous midribs, seeds 10–20..............**11. Cardamine** (*C. violacea*)
\+ Flowers white or purplish; pods squarish in section, valves keeled with well-developed midribs, seeds few (c 2–6)...............................**24. Eutrema**

31. Leaves coarsely toothed in upper half; flowers solitary on scapes **23. Pycnoplinthopsis**
\+ Leaves entire or shallowly-toothed throughout; flowers several in racemes or on leafless scapes...32

32. Flowers on leafless scapes; limb of petals white or mauve, claw purple; pods densely greyish pubescent or white tomentose ....**14. Phaeonychium**
\+ Flowers on leafy peduncles; petals ± uniformly coloured; pods glabrous or sparsely pubescent 33

33. Longer filaments broadened into a tooth at apex on both sides **19. Lepidostemon**
\+ Filaments slender, not toothed 34

34. Pods elliptic or ovate 35
\+ Pods linear 36

35. Flowers white or sometimes purplish in *D. williamsii;* pods compressed, valves flat ....**10. Draba**
\+ Flowers purplish; pods ovoid or subcylindrical ....**28. Braya** (*B. tibetica)*

36. Pods ± constricted between seeds...................................**27. Torularia**
\+ Pods smooth...37

37. Weak, decumbent herbs branched from base, with leafy stems 6–10cm **25. Microsisymbrium**
\+ Robust, erect herbs with leafy stems 20–60cm, (in some *Arabis* species 3–20cm but then stems with only 1–3 leaves) 38

38. Flowers medium-sized; petals 5–13mm...............................**16. Arabis**
\+ Flowers small; petals 3.5–4mm..**30. Arabidopsis**

1. BRASSICA L.

Glabrous or sparsely hirsute annuals or biennials, hairs simple. Basal leaves entire or pinnatifid. Flowers in racemes, yellow. Sepals erect or spreading , lateral pair usually pouched. Pods linear, terete or angular, apex prolonged into a seedless or 1–2-seeded narrow beak; seeds few to numerous in 1 row.

1. Plants glaucous; leaves usually thick; flowers large, 10–25mm . . **1.B.oleracea**
\+ Plants green, scarcely glaucous; leaves mostly thin; flowers small, 10–12mm **Species 2 & 3**

1. B. oleracea L.

Stems 30–100cm. Lower leaves lyrately pinnatipartite, 15–40cm, terminal segment large, broadly ovate or suborbicular, rounded, base cordate, margin broadly dentate or crenate, lateral lobes 2–3(–5) pairs, much smaller than terminal; upper leaves simple, oblong, sessile and semiamplexicaul. Racemes ebracteate, elongating up to 30cm in fruit. Pedicels 1.5–2cm. Sepals erect, oblong, 8–10 × 2–3mm, obtuse. Petals 15–20 × 4–6cm, obovate, clawed. Pods 6–11cm with beak 1–2cm.

The typical variety is a European plant, but two varieties are cultivated in Bhutan: var. **capitata** L., Eng: *Cabbage,* in which the concave leaves are densely crowded together forming an enlarged terminal bud; and var. **botrytis** L. , Eng: *Cauliflower* in which the racemes of abortive flowers are densely crowded to form a fleshy head.

2. B. campestris L.; *B. rapa* L. Fig. 33a–d.

Stems simple or branched, 30–100cm. Lower leaves lyrate-pinnatifid, 10–20cm, with large terminal lobe and 1 or more pairs of lateral lobes; upper leaves oblong, obtuse or acute, base cordate, amplexicaul, margins entire. Flowers on pedicels c 10mm. Sepals oblong-ovate, c 5.5×1.5mm. Petals obovate, 7–10 × 4–5mm. Pods 3–6 × 0.4–0.7cm, with beak 1–2cm.

Bhutan: S–Gaylegphug and Chukka districts, **C**–Punakha, Tongsa and Bumthang districts, **N**–Upper Bumthang Chu district. Weed of gardens and roadsides, 400–3800m. May–June.

The following are cultivated for their oil-bearing seeds: subsp. **dichotoma** (Roxb.) Olsson, Hindi: *Toria,* with narrow (c 4mm) 2-valved pods; and subsp. **trilocularis** (Roxb.) Olsson (*B. trilocularis*(Roxb.) Hook. f. & Thomson), Hindi: *Sarson,* with broader (7–8mm) 4-valved pods. Subsp. **rapifera** (Metzger) Sinskaia (*B. napus* L.), Eng: *Turnip,* with tuberous roots, is cultivated in Bhutan.

3. B. juncea (L.) Czernjaev

Similar to *B. campestris* but upper leaves oblong-lanceolate or linear, narrowed and not amplexicaul at base, margins usually toothed.

Bhutan: S–Chukka district (Raidak Valley), **C**–Tashigang district (near Tashigang); **Sikkim.** Cultivated and often growing as an escape, 1300m. February–March.

This species is also a source of oil seed. The cabbage-like var. **rugosa** (Roxb.) Bailey, Nep: *Sag,* is probably more widely cultivated, the young leaves being eaten as a pot herb.

2. RAPHANUS L.

Annual or biennial herbs, ± hirsute with simple hairs. Leaves lyrately pinnate. Flowers in ebracteate racemes. Sepals erect, inner pair shallowly pouched. Petals oblanceolate or obovate, clawed, white or purplish. Pods 2-chambered with a narrow seedless beak, lower chamber inconspicuous, stalk-like; upper one cylindrical and indehiscent or constricted between seeds and breaking into 1-seeded portions.

1. R. raphanistrum L. Eng: *Wild Radish.* Fig. 33o.

Root thin, fusiform.Stems c 1m, erect. Lower leaves pinnate, 15–20cm with 4–7 pairs of oblong leaflets 2–3.5 × 1–2cm, rounded at apex and base, margins shallowly toothed, terminal leaflet broadly ovate up to 5.5cm; upper leaves with fewer leaflets, uppermost simple, elliptic or oblanceolate. Sepals oblong, 7–8 × 1.5–2mm. Petals white or mauve, 1.3–1.7 × 0.5cm (including claw). Pods 4–6 × 0.5cm (or longer), including beak 2.5–3cm, constricted between each of the 1–5 seeds.

Bhutan: C –Tongsa district (Tongsa); **Sikkim.** Weed around cultivated ground, 2350m. June.

2. R. sativus L. Eng: *Radish.*

Similar to *R. raphanistrum* but root swollen and fleshy; pods not constricted between seeds, 7–10mm thick.

Bhutan: S – Chukka district (near Tabji), **C** –Thimphu district (Dotena). Cultivated, 1300–2750m. February–July.

Apparently var. **rotundus** (Miller) DC. with roots up to 15 × 5cm is preferred for cultivation.

3. LEPIDIUM L.

Erect or prostrate annuals with simple, sometimes glandular hairs. Leaves pinnatifid or toothed. Flowers small in ebracteate racemes. Sepals not pouched at base. Petals white, scarcely longer than sepals. Pods ovate or orbicular, laterally compressed, notched and narrowly winged at apex; style minute, valves boat-shaped, septum narrow; seed solitary in each cell.

1. L. capitatum Hook. f. & Thomson. Fig. 33 j–l.

Prostrate herb. Stems up to 30cm, glandular-papillose. Basal leaves pinnatisect, up to 4 × 0.75cm; lobes 3–5 per side, toothed. Racemes short (c 3cm or less), corymbose. Sepals ovate, c 0.75mm, with scarious margins. Petals shorter, spathulate. Pods ovate, c 2.5 × 2mm, acute and minutely notched at apex.

Bhutan: N – Upper Mo Chu district (Lingshi and Barshong); **Sikkim.** Damp shingle and grassy areas, 3500–3960m. May–July.

2. L. virginicum L.

Similar to *L. capitatum* but erect, stems 40–60cm, finely pubescent; lower leaves oblanceolate, 4–7 × 0.75–1.5cm, acute, base attenuate, margins coarsely serrate; upper leaves linear, ± entire; pods broadly ovate or orbicular, 3–3.5 × 3mm, apex rounded and broadly notched.

Bhutan: C–Thimphu district (Paro) and Tashigang district (Khaling). Weed of roadsides and farmland, 2250–2500m. June.

L. ruderale L. has been doubtfully recorded from Bhutan (the locality, Dongdong, may be Tibetan); it differs from *L. virginicum* in having bipinnatifid lower leaves.

4. THLASPI L.

Annual or perennial herbs. Leaves entire or shallowly toothed, stem leaves sessile, auriculate. Flowers in short dense ebracteate racemes becoming elongated in fruit. Sepals often with membranous margins, inner not pouched at base. Petals obovate or spathulate, white or purplish. Pods oblong-elliptic, obcordate or orbicular, laterally compressed; valves boat-shaped sometimes winged, septum narrow; seeds several in each cell.

1. Annual; pods obcordate or orbicular, winged........................**1. T. arvense**
\+ Perennials; pods oblong or elliptic, unwinged......................**Species 2 & 3**

1. T. arvense L. Med: *Deyga, Drega.* Fig. 33n.

Glabrous annual. Stems 15–45cm, erect. Leaves oblong-lanceolate or oblanceolate, 2.5–5 × 0.5–1.5cm, obtuse or acute, base with acute auricles, margins shallowly serrate. Sepals ovate-elliptic, c 2 × 1mm. Petals white, c 3 × 1.25mm. Pods 1–1.5 × 1–1.25cm, deeply notched at apex; wing 2–5mm broad; style minute.

Bhutan: S–Chukka district, **C**–Thimphu to Bumthang districts, **N**–Upper Mo Chu district; **Sikkim.** Weed of roadsides and cultivation, 2200–4000m. April.

2. T. cochlearioides Hook. f. & Thomson

Erect or decumbent, glabrous perennial; stems up to 15cm. Basal leaves when present ovate or orbicular, 7–10 × 3–6mm, obtuse, rounded at base, entire or shallowly toothed; petioles up to 5 cm; stem leaves oblong-elliptic, 7–15 × 4–8mm, acute, base with short acute auricles. Sepals ovate, 2–2.5 × 1mm. Petals white tinged with purple, c 5 × 2mm. Pods elliptic, 8–10 × 5–6mm, rounded at apex and base, often becoming twisted; style c 2mm.

Bhutan: N–Upper Mo Chu district (Lingshi) and Upper Bumthang Chu district (Dole La); **Sikkim.** Alpine screes, 4265–4570m. June–July.

3. T. andersonii (Hook. f. & Thomson) Schulz; *Iberidella andersonii* Hook. f. & Thomson

Similar to *T. cochlearioides* but sometimes with short leafy stolons; pods narrowly oblong-elliptic, c 10 × 3mm, acute at both ends, not becoming twisted.

Bhutan: C –Thimphu, Punakha, Tongsa and Bumthang districts, **N** – Upper Mo Chu, Upper Bumthang Chu and Upper Kulong Chu districts; **Sikkim.** Open steep hillsides, 3400 – 4570m. May – August.

T. cochleariforme DC. has been doubtfully recorded (as *T. alpestre* L.) from Sikkim. It differs from both the above perennial species by having narrowly obcordate pods notched at the apex.

5. CAPSELLA Medikus

Erect annual or biennial herbs with simple and stellate hairs. Basal leaves in a rosette; stem leaves sessile, auriculate. Flowers in ebracteate racemes. Sepals spreading, not pouched at base. Petals obovate, white. Pods obconical or obcordate, laterally compressed, valves deeply boat-shaped, septum narrow. Seeds numerous in 2 rows.

1. C. bursa-pastoris (L.) Medikus. Eng: *Shepherd's Purse.* Fig. 33m.

Stems 10 – 50cm. Basal leaves runcinate-pinnatifid, 3 – 10 × 0.6 – 3cm, acute, base shortly petiolate; lateral lobes 5 – 6 pairs, ovate, acute, sparsely pubescent; stem leaves ovate-lanceolate, 1.5 – 5 × 0.5 – 1cm, acute, sharply auriculate at base. Sepals ovate, 1.5 × 1mm. Petals c 2.5 × 1.25mm. Pods 6 – 7mm long and broad, septum elliptic, c 6 × 1.25mm.

Bhutan: S – Gaylegphug district, **C** –Thimphu, Tongsa, Mongar and Tashigang districts, **N** – Upper Mo Chu district; **Sikkim.** Weed of roadsides and alpine meadows, 1980 – 3960m. March – July.

6. HEDINIA Ostenfeld

Prostrate or spreading perennials, greyish pubescent with simple hairs. Leaves pinnate. Racemes short, corymbose, elongating in fruit, bracteate. Sepals spreading, not pouched at base. Petals spathulate, clawed, white. Pods oblong, laterally compressed, valves keeled, unwinged; seeds 10 – 15 per cell.

1. H. tibetica (Thomson) Ostenfeld; *Capsella thomsonii* Hook. f.

Stems 5 – 25cm. Leaves elliptic in outline, 2 – 5 × 0.7 – 2cm, not or shortly petiolate; leaflets 5 – 7 pairs, 3 – 10 × 1 – 2mm, oblong, entire or pinnatisect, obtuse or acute. Sepals oblong, c 2 × 0.5mm. Petals c 3 × 1mm. Pods 7 – 10 × 3 – 4mm; septum 1 – 1.5mm broad.

Chumbi: Phari; **Sikkim:** Cholamo. 4265 – 4880m. June – July.

7. **DILOPHIA** Thomson

Biennial or perennial, dwarf fleshy glabrous herbs. Stems branching, tufted or prostrate. Leaves entire or obscurely toothed. Flowers in leafy clusters at branch ends. Sepals spreading, inner not pouched. Petals narrowly spathulate, white or pinkish. Pods slightly compressed at right angles to septum, replum broadened at base, valves boat-shaped with a double line of tooth-like projections on the back; style short, thick; seeds large, 2–4 per cell.

1. D. salsa Thomson. Fig. 33 i.
Stems 2–3cm, bearing clusters of oblanceolate or spathulate leaves 5–7 × 1mm at branch ends. Sepals elliptic, c 1.75 × 1mm, obtuse. Petals c 2.25 × 0.5mm. Pods ± cordate in outline, c 2 × 3 mm.
Bhutan: N–Upper Mo Chu district? (Laycho); **Sikkim:** Lhonakh. Riverside shingle, 4400m. June–August.

8. **COCHLEARIA** L.

Biennial or perennial herbs with slender tap root, ± prostrate, subglabrous or with simple hairs. Stems leafy, branched from base. Leaves simple, petiolate. Flowers in corymbose racemes. Sepals spreading, inner pair not pouched. Petals oblanceolate, scarcely clawed. Pods narrowly ellipsoid, valves slightly swollen, venation ± prominent, septum absent, 3–5-seeded.

1. C. himalaica Hook. f. & Thomson
Leaves ovate, 5–12 × 3–7mm, obtuse, base rounded, margins irregularly crenate; petioles 1–2.5cm. Sepals oblong, c 1.75 × 0.75mm. Petals greenish white, c 3 × 1mm. Pods 5–6 × 2–3mm, style c 1mm.
Bhutan: N–Upper Mo Chu district (Linghshi) and Upper Bumthang Chu district (Dole La); **Sikkim.** In moist gravel, 4570m. June–July.

9. **LIGNARIELLA** Baehni

Prostrate slender annuals, glabrous or with simple hairs. Leaves 3-lobed or 3-partite, upper ones sometimes entire. Flowers solitary, axillary or few in bracteate racemes. Inner sepals ± pouched at base. Petals purplish. Pods oblong-ellipsoid, subterete, ± falcate, septum absent; seeds 2–5 in 1 row.

1. L. hobsonii (Pearson) Baehni
Stems up to 30cm. Leaves broadly ovate or suborbicular, 4–7(–15)mm long and broad with 3(–5) short teeth, or deeply 3-lobed; lobes linear or elliptic, obtuse;

petioles 3–15mm. Peduncles 10–25mm. Sepals oblong-ovate, 2–2.5mm. Petals obovate, 3.5–6.5 × 1.5–5mm. Pods 10–13 × 1.5–2mm, style 1–4mm.

Bhutan: C–Thimphu district (near Barshong and Jato La), Tongsa district (Black Mountain), **N**–Upper Mangde Chu district (Ju La) and Upper Kulong Chu district (Me La); **Sikkim.** Gravel at streamsides, 2740–3650m. May–September.

There are two supspecies: subsp. **hobsonii** is larger in habit, leaves and flowers, styles (2.5–)3–4mm; subsp. **serpens** (W. W. Smith) Hara (*Cochlearia serpens* W. W. Smith) is smaller in all respects and with styles 1–1.5(–1.8)mm. All but one of the Bhutanese specimens seen so far are subsp. *hobsonii* and all those from Sikkim are subsp. *serpens;* the only record of subsp. *serpens* from Bhutan is from Tare La near Ha.

10. DRABA L.

Annual, biennial or perennial herbs bearing a mixture of simple, forked and stellate hairs. Leaves entire or shallowly toothed, basal ones in a rosette. Flowers yellow or white, rarely purplish, in corymbose bracteate or ebracteate racemes, rarely solitary or few. Sepals elliptic or ovate, not or slightly pouched. Petals rounded or emarginate, shortly clawed at base. Pods compressed, elliptic, ovate or linear; seeds few to numerous in 2 rows; style short, stigma rounded.

1. Stems leafy 2
\+ Stems leafless (sometimes with 1–2 leaves in *D. sikkimensis)* 6

2. Flowers yellow...3
\+ Flowers white or purplish 4

3. Racemes compact, well-developed, with many flowers...........**Species 1–3**
\+ Racemes lax, few-(2–4- or rarely up to 12-) flowered ....**Species 4 & 5**

4. Flowers purplish or sometimes white; basal leaves usually withered at flowering time...**10. D. williamsii**
\+ Flowers pure white; basal leaves usually present at flowering time 5

5. Annual...**9. D. ellipsoidea**
\+ Perennials ....**Species 6–8**

6. Flowers yellow ....**Species 11–13**
\+ Flowers white ....**Species 14–16**

1. D. elata Hook. f. & Thomson. Fig. 33 e, f.

Slender perennial, stems 5–45cm. Basal leaves oblanceolate or spathulate, 2.5–4 × 0.3–1cm, acute, base attenuate, margins entire, denticulate or shallowly serrate, pubescent with a mixture of stellate, forked and simple hairs; stem leaves 2–7(–10), similar to basal but amplexicaul. Racemes ebracteate, or with narrowly-ovate bracts c 5–10 × 2–3mm. Sepals c 2.5 × 1.5mm, margins scarious. Petals obovate, yellow, c 5(–8) × 2.5(–5)mm, emarginate. Pods elliptic, acute, 7–15 × 2–3mm, glabrous or pubescent, on pedicels 1–2(–2.5)cm; style 1–2mm.

Bhutan: C –Thimphu to Mongar districts, **N** –Upper Mo Chu, Upper Pho Chu, Upper Bumthang Chu, Upper Kuru Chu and Upper Kulong Chu districts; **Sikkim**. In moist gravel, 3350–4570m. May–August.

A variable species so far as petal, pod and pedicel lengths are concerned.

2. D. polyphylla Schulz

Similar to *D. elata* but more robust; stems 20–60cm in flower, up to 90cm in fruit, bearing up to 14 leaves; pods 7–14 × 4–6mm, ± obtuse, on pedicels 1–2.3cm.

Sikkim. 3600–5000m.

3. D. eriopoda Turczaninow

Similar to *D. elata* but annual; stems 3–30(–50)cm, simple or branched; leaves ovate-elliptic, 7–15 × 3–8mm, acute, base sessile, semiamplexicaul, margin with few shallow teeth; sepals 1.5–2mm; petals 2.5–3mm; pods ovate-elliptic, 5–10 × 2.5–3mm, on pedicels 3–10mm, stigma ±sessile.

Bhutan: N –Upper Mo Chu district (Lingshi); **Sikkim.** 4110m. June–July.

4. D. gracillima Hook. f. & Thomson; *D. wardii* W. W. Smith

Slender annual or short-lived perennial. Stems 6–20cm. Basal leaves oblanceolate, 1–1.5 × 0.3–0.6cm, acute, base attenuate, margins entire or denticulate; stem leaves ± elliptic, 5–8 × 2–3mm, sessile. Pedicels 5–15mm in flower, up to 30mm in fruit. Sepals ovate-elliptic, 2.5 × 1.5mm. Petals yellow, obovate, c 3 × 1.5mm. Pods linear-elliptic, 6–15 × 2–2.5mm; stigma ± sessile.

Bhutan: C –Tongsa and Sakden districts, **N** –Upper Mo Chu, Upper Pho Chu and Upper Kulong Chu districts; **Sikkim.** Moist open situations, 3200–4250m. May–June.

5. D. cholaensis W. W. Smith

Similar to *D. gracillima* but petals c 7 × 3mm; styles 1.5–2mm.

Sikkim: Tanka La. 4300m.

6. D. lasiophylla Royle; *D. incana* sensu F.B.I. p.p. non L.

Stems 5–30cm. Basal leaves oblanceolate, 10–15 × 2–5mm, acute, base attenuate, margins entire, ± densely stellate-pubescent, ciliate near base; stem leaves similar, sessile. Flowers in dense racemes. Sepals oblong-elliptic, c 1.5 × 0.75mm. Petals white, oblanceolate, c 3 × 1mm. Pods 6–8 × 2mm, elliptic, on pedicels 3–5mm; style c 0.5mm.

Bhutan: N –Upper Pho Chu district (W branch of Pho Chu) and Upper Kulong Chu district (Me La); **Sikkim.** Streamsides, 4265–4570m. June–August.

7. D. oariocarpa Schulz

Similar to *D. lasiophylla* but racemes lax; pods ovate, 3 × 2.5mm; style c 1mm.

Bhutan: N – Upper Bumthang Chu district (Dole La); **Sikkim:** Lhonakh. 4570–4650m. June–July.

8. D. sherriffii Grierson

Similar to *D. lasiophylla* but larger; rootstocks straggling; stems slender, 7–10cm in flower, up to 30cm in fruit; basal leaves oblanceolate, 2.5–4 × 0.5–0.8cm, acute, base attenuate, margins entire; stem leaves elliptic, shallowly 2–3-toothed otherwise similar; racemes lax, 2–9-flowered; sepals ovate 2.5–3 × 1.5mm; petals obovate, 6–7 × 3–4mm; pods narrowly elliptic, 15–25 × 1.5mm; style c 0.75 mm; pedicels 3–8mm in flower, up to 2cm in fruit.

Bhutan: C – Punakha district (Tang Chu/Sefu Chu ridge), **N** – Upper Mangde Chu district (Namdating, Shashona and Saga La). Sandy soil and cliff ledges, 4400–4730m. June–July.

9. D. ellipsoidea Hook. f. & Thomson

Annual herb. Stems 5–10mm, branched from base, flexuose. Leaves obovate or oblanceolate, 1–3 × 0.4–1.5cm, acute or obtuse, base attenuate, margins entire or shallowly toothed. Racemes lax, 3–8-flowered. Sepals oblong 1–1.5mm. Petals white, narrowly spathulate, c 1 × 0.5mm. Pods elliptic, c 8 × 4mm, apex rounded; style minute.

Sikkim: Jongri and Guicha La. Weed of cultivation, 4265m. September.

10. D. williamsii Hara

Perennial herb. Stems 2–10cm, simple or branched from base. Basal leaves usually withered at flowering time; stem leaves broadly lanceolate or oblanceolate, 5–25 × 1.5–5mm, acute, base attenuate, margins entire or shallowly serrate. Flowers few or numerous in corymbs or racemes; pedicels 2–7mm in flower, up to 15mm in fruit. Sepals oblong, 2–3mm. Petals obovate, 4–5 × 2.5–3.5mm, white or pinkish. Pods (immature) 5–12 × 1–2mm; style 1mm.

Bhutan: N – Upper Mangde Chu district (Saga La and Shashona) and Upper Bumthang Chu district (Waitang). 3960–4570m. June–July.

Specimens from Bhutan are smaller than those from Nepal and with fewer flowers.

11. D. oreades Schrenk; *D. alpina* sensu F.B.I. non L.

Usually densely caespitose with rather thin rootstocks. Leaves oblanceolate, 10–15 × 2–5mm, acute, base attenuate, margins entire or shallowly serrate, ciliate or sparsely pubescent with simple and forked or stellate hairs. Peduncles 1–10cm, bearing 3–12 flowers in ebracteate corymbose racemes. Sepals oblong, 2–2.5 × 1–1.5mm, obtuse. Petals yellow, oblanceolate or obovate, 3–4 × 0.75–1.5mm.

Pods ovate, 6–8 × 3–5mm, acute, base rounded, ± glabrous; style c 0.75mm; seeds 6–12.

Bhutan: C–Thimphu and Bumthang districts, **N**–Upper Mo Chu to Upper Kuru Chu districts; **Sikkim.** On cliffs and open hillsides, 3960–4730m. July.

A specimen from Upper Bumthang Chu district (Pangotang) may represent a distinct species; it has more thickly woody rootstocks, sepals c 3.5mm, petals 5–6mm, pods densely stellate-pubescent, 10 × 4mm, which are acute at base and apex, style 1.5mm.

12. D. bhutanica Hara

Similar to *D. oreades* but racemes bracteate, bracts oblong-elliptic, 5–10 × 1–4mm; sepals oblong, 3–3.5mm; petals obovate, 7–8 × 3.5–5mm; style 2–2.5mm.

Bhutan: N–Upper Mangde Chu district (Waitang and Saga La) and Upper Kuru Chu district (Narim Thang). Alpine pastures, 4265–4400m. June–July.

13. D. humilima Schulz

Similar to *D. oreades* but capsules oblong-elliptic, 6–8 × 3mm, ± obtuse or rounded at both ends; seeds c 24.

Sikkim: Cheumsanthang and Zemu Valley. 4650–5000m.

14. D. sikkimensis (Hook. f. & Thomson) Pohle; *D. tibetica* Hook. f. & Thomson var. *sikkimensis* Hook. f. & Thomson

Densely caespitose, with leaves from previous years persisting ± unchanged at base. Leaves oblanceolate, 5–12 × 2–4mm, subacute, base attenuate, margins entire, densely greyish pubescent with stipitate stellate hairs. Racemes ± elongate, 4-to many-flowered, (2–)5–15cm, sometimes bearing 1–2 elliptic bracts. Sepals oblong, c 3 × 1.5mm, obtuse. Petals white, obovate, c 6 × 3mm. Pods ovate-elliptic, 6–12 × 3mm, acute at each end, sometimes becoming twisted, glabrous or pubescent; style 0.75–1mm.

Bhutan: N–Upper Pho Chu district (Kangla Karchu La); **Sikkim.** Rocky hill slopes, 4880m. June.

15. D. altaica (Meyer) Bunge

Similar to *D. sikkimensis* but with only midribs and stumps of leaves from previous years persisting at base; sparsely pubescent mainly with simple and forked hairs but with a few stellate hairs; racemes corymbose, bracteate at base, 2–5cm in flower, elongating up to 20cm in fruit; sepals c 1.25mm; petals c 2mm; pods ovate, 2.5–5 × 1.5–2mm, acute, base rounded.

Sikkim: Chhortenema La. 5600m. June.

16. D. hicksii Grierson

Similar to *D. sikkimensis* and *D. altaica* but moderately pubescent with minute stellate hairs and few simple or forked hairs; racemes 1–3-flowered, ebracteate, 0.5–2cm; sepals c 2.5mm; petals 4.5–5 × 1.5–2mm; pods ovate, c 3.5 × 2.5mm.

Bhutan: **N** – Upper Kulong Chu district (Me La). Screes and open grassy hillsides, 4250–4570m. June–August.

11. CARDAMINE L.

Annual or perennial herbs, glabrous or with a few simple hairs. Leaves simple, ternate or pinnate. Flowers in ebracteate racemes. Inner sepals somewhat pouched at base. Petals white, pink or purple, obovate, rounded or emarginate, base narrowed or clawed, rarely petals absent. Filaments not broadened above. Pods narrow, linear, valves rounded, venation indistinct, dehiscing explosively and becoming coiled at maturity, replum broad; style short, stigma rounded or ±bilobed; seeds few to many in 1 row.

1. Leaves simple, sessile .... **1. C. violacea**
\+ Leaves compound, usually petiolate, sometimes simple in *C. circaeoides* ... 2

2. Leaves trifoliate (leaflets sometimes 5 in *C. trifoliata*) 3
\+ Leaves pinnate or deeply pinnatisect 4

3. Robust herb; stems c 45cm; leaflets 5–10cm .... **2. C. scoriarum**
\+ Slender herbs; stems 15–30cm; leaflets 0.7–4cm .... **Species 3 & 4**

4. Robust herb; stems 45–150cm .... **5. C. macrophylla**
\+ Slender herbs; stems up to 45(–60)cm 5

5. Leaves auriculate at base (narrowly so in *C. impatiens*) .... **Species 6 & 7**
\+ Leaves not auriculate at base 6

6. Leaflets ovate, margins crenate .... **8. C. yunnanensis**
\+ Leaflets elliptic, oblanceolate, obovate or suborbicular, entire, sinuate or trilobed 7

7. Petals c 3mm (sometimes up to 6mm in *C. multijuga*) .... **Species 9–11**
\+ Petals 5–10mm .... **Species 12 & 13**

1. C. violacea (D. Don) Hook. f. & Thomson

Perennial, stems 20–45cm. Leaves narrowly ovate or oblanceolate, 3–6 × 0.75–2cm, acute or acuminate, base amplexicaul, auricles rounded, margins shallowly dentate, sparsely pubescent beneath. Racemes 5–12-flowered. Sepals oblong-ovate, c 5 × 2mm. Petals c 10 × 4mm, purple. Pods 4–6 × 0.2cm.

Bhutan: **C** – Ha (Damthang), Thimphu (Dochu La) and Punakha (Phobsikha) districts. Streamsides and open forests, 2750–3300m. May–June.

The above description and records relate to subsp. **bhutanica** Grierson which is a smaller plant with fewer-flowered racemes than the typical Nepalese subspecies; in addition basal leaves are usually present and the stem-leaves are more broadly auriculate in subsp. *bhutanica*.

2. C. scoriarum W. W. Smith; *C. smithiana* Biswas

Stems 0.5–1m, glabrous. Leaflets ovate-elliptic, (3.5–)5–10 × (1.5–)2–4cm, acuminate, base rounded or cuneate, margins irregularly crenate-serrate; petioles 2–4cm. Flowers numerous in corymbose racemes. Sepals oblong-elliptic, c 3 × 2mm, with 3–4 veins converging below apex. Petals c 6 × 4mm, pink. Pods (immature) c 5mm.

Bhutan: C–Tongsa district (Chendebi) and Mongar district (Mongar and Donga La), **N**–Upper Kulong Chu district (Lao). Shaded streamsides, 2300–2895m. June–July.

3. C. circaeoides Hook. f. & Thomson

Stems erect, 10–30cm, sparsely hairy. Lateral leaflets elliptic, c 1.5 × 0.5–1cm, acute, sometimes absent; terminal leaflet ovate, 2–4 × 1–2.5cm, acuminate, base rounded or cuneate, margins irregularly sinuate-toothed or lobed, glabrous; petioles 2–5cm. Racemes few-flowered. Sepals oblong, c 3 × 1mm. Petals c 4.5mm, white. Pods c 25 × 1.5mm.

Bhutan: S–Gaylegphug and Deothang districts, **C**–Punakha, Tongsa and Tashigang districts; **Sikkim.** Moist broad-leaved forests, 1500–2400m. April–May.

4. C. trifoliolata Hook. f. & Thomson

Similar to *C. circaeoides* but stems trailing or decumbent, 15–45cm; leaflets oblanceolate to broadly ovate or suborbicular, 0.7–2 × 0.2–1cm, obtuse or acute, often 3-lobed, base rounded, cuneate or attenuate, glabrous or minutely pilose on upper surface; petioles 2–6cm; sepals oblong, c 3 × 1mm; petals 6–8 × 2.5–3mm; pods 15–20 × 1mm.

Bhutan: S–Deothang district (Tsalari Chu), **C**–Thimphu district (Barshong and Dochu La) and Punakha district (E of Wangdu Phodrang); **Sikkim.** Streamsides in broad-leaved forests, 1675–3100m. April–May.

5. C. macrophylla Willdenow

Stem 45–150cm, finely pubescent at first. Leaves 5–25cm with 2–7 pairs of lateral segments or leaflets; segments elliptic, ovate or lanceolate, 1.5–10 × 0.7–2.5cm, obtuse or acuminate, base cuneate or rounded, margins coarsely serrate or subpinnatifid, sparsely pubescent; terminal leaflet similar. Flowers few to numerous on pedicels 1–2cm. Sepals oblong-elliptic, 5–8 × 1.5–3mm, glabrous or sparsely hirsute. Petals purplish, obovate or oblanceolate, 8–16 × 2–5.5mm. Pods 3–4.5 × 0.2–0.3cm.

Bhutan: C–Ha to Tashigang districts, **N**–Upper Mo Chu, Upper Bumthang Chu and Upper Kulong Chu districts; **Sikkim.** Damp shaded situations in forests, etc., 2440–4100m. May–July.

A common and variable species so far as height, number and division of leaflets and size of flowers are concerned, but displaying little correlation of characters.

6. C. griffithii Hook. f. & Thomson

Stems 20–45(–60)cm, glabrous, angular, grooved. Leaves 1.5–9cm; lateral leaflets 3–5 pairs, ovate, 3–10 × 2–7mm, obtuse, base rounded, margins entire or irregularly sinuate; lowermost pair auriculate against stem; terminal leaflet suborbicular, up to 1.3cm long and broad. Racemes few-flowered. Sepals oblong-elliptic, 3–3.5 × 1mm, obtuse. Petals white or purplish, obovate, 8–9 × 3–3.5mm. Pods (immature) c 1cm.

Bhutan: C–Thimphu to Bumthang districts, **N**–Upper Mo Chu to Upper Kulong Chu districts; **Sikkim.** Streamsides, 2000–3800m. May–July.

7. C. impatiens L.

Similar to *C. griffithii* but leaves with narrow sagittate 2–3mm auricles at base; leaflets pinnately-lobed or pinnatifid; racemes many-flowered; sepals 1.5mm; petals white, 3mm; pods c 2 × 0.1cm.

Bhutan: C–Ha district (near Ha) and Mongar district (near Zimgong); **Sikkim.** Fields and shady streamsides, 1475–2750m. April–May.

8. C. yunnanensis Franchet; *C. inayatii* Schulz, *C. sikkimensis* Hara

Stems 15–25cm. Leaves (including petiole) 5–9cm sparsely pubescent; lateral leaflets 2–3 pairs, ovate, 1–2.5 × 0.75–1.5cm, obtuse, crenately toothed; terminal leaflet larger, up to 5.5 × 2cm. Flowers few, white or pink. Sepals oblong-ovate, 2.5–3 × 1–1.5mm. Petals c 4 × 2mm. Pods 2.5–3cm.

Bhutan: C–Thimphu district (Dotena and Dochu La), **N**–Upper Mo Chu, Upper Bumthang Chu and Upper Kulong Chu districts; **Sikkim.** Moist shaded places, 2440–2700m. May.

9. C. flexuosa Withering; *C. scutata* Thunberg subsp. *flexuosa* (Withering) Hara, *C. hirsuta* L. var. *sylvatica* sensu F.B.I. non Link.

Annual or biennial. Stems 10–35cm, flexuose. Basal leaves few, scarcely forming a rosette; stem-leaves 4–10, 2.5–5cm; lateral leaflets 2–4 pairs, ovate-elliptic, 3–7 × 2–5mm, obtuse or subacute, entire or irregularly sinuate, glabrous; terminal leaflet obovate or suborbicular, 1–1.5 × 1cm, sometimes 3–4-lobed. Flowers few to numerous. Sepals oblong-elliptic c 2×0.7mm. Petals white, c 3 × 1mm. Stamens usually 6. Pods c 2 × 0.1cm, not or slightly overtopping unopened buds.

Bhutan: S–Samchi and Gaylegphug districts, **C**–Thimphu, Tongsa and Tashigang districts; **Sikkim.** Moist soil and streamsides, 500–2500m. January–June.

10. C. hirsuta L.

Similar to *C. flexuosa* but always annual; stems ± straight; basal leaves 2–7cm, numerous, forming a rosette, hairy on upper surface; stem-leaves 2–4, smaller than basal ones; sepals elliptic c 2 × 0.7mm; petals c 2.5 × 0.75mm, sometimes absent;

stamens usually 4, sometimes 6; pods 15–20 × 1.5mm, overtopping unopened buds.

Sikkim. Damp ground, 1500–2000m. March–October.

11. C. multijuga Franchet

Similar to *C. flexuosa* and *C. hirsuta* but stems erect, 5–30cm; leaves not rosetted 2–5cm; lateral leaflets 8–9 pairs, ovate, 3–4 × 1–1.5mm, acute, margins usually with 1 or 2 small teeth, terminal leaflet narrowly obovate or elliptic; sepals broadly elliptic, c 3 × 1.5mm, margins pale membranous; petals white or purplish, 4–6 × 2–3mm.

Bhutan: C – Sakden district (Gamri Chu, Takhtoo). Damp shady ground, 2000m. March.

The above description and record refers to the more slender var. **gracilis** Schulz.

12. C. elegantula Hook. f. & Thomson

Slender annual. Stems 10–15cm, ascending, not bulbiliferous. Leaves 3–4.5cm; leaflets 5–7 pairs, elliptic, 3–7 × 1–2mm, entire or with an angular lobe on lower margin. Racemes 2–9-flowered. Sepals narrowly ovate, c 1.5mm. Petals white or purplish, c 5mm. Pods 12–15 × 1mm.

Bhutan: S – Chukka district, **C** –Thimphu and Tongsa districts, **N** – Upper Mo Chu district. Wet rocks, etc., 1100–2800m. April–May.

13. C. loxostemonoides Schulz; *C. pratensis* sensu F.B.I. non L.

Similar to *C. elegantula* but more robust, stems up to 30cm, bearing several ovoid 2–3mm bulbils at ground level; leaflets elliptic up to 1.5 × 0.5cm, entire or 3-lobed; sepals 4–5 × 1.5–2mm; petals 7–13 × 3–5mm; pods 1.8–3.5 × 0.15cm.

Bhutan: locality unknown (73).

12. LOXOSTEMON Hook. f. & Thomson

Slender perennial herbs with thin bulbiliferous rhizomes. Stems simple, glabrous. Leaves pinnate or trifoliate, bearing simple hairs. Racemes few-flowered, ebracteate. Flowers similar to those of *Cardamine* but filaments longer, broadened and toothed above. Pods linear, cylindrical.

1. L. pulchellus Hook. f. & Thomson

Slender herb, 5–10cm tall, bearing a cluster of ovoid bulbils, each c 2 × 1mm, at stem base. Leaves trifoliate or pinnate with 2(–3) pairs of lateral leaflets, elliptic or oblanceolate, 5–10 × 2–4mm, acute or obtuse, base rounded or cuneate, margins entire, ciliate. Flowers (1–)2–4, ± corymbose. Sepals ovate, c 2.5 × 1.5mm. Petals obovate, c 4.5 × 3mm, pink. Pods (immature) c 5mm.

Bhutan: N – Upper Mangde Chu district (Mangde Chu); **Sikkim.** Rocky ground, 4100m. July.

13. PEGAEOPHYTON Hayek & Handel-Mazzetti

Small perennial herbs with thick or thin rootstocks, glabrous or with simple hairs. Leaves all basal forming a rosette, spathulate. Flowers solitary on slender scapes. Sepals ± spreading, rounded at apex, not pouched at base. Petals white or purplish, rounded, shortly beaked. Ovary narrowly ovoid. Valves of pods with faint midrib and veins, septum incomplete or lacking, indehiscent; style thick, short; stigma flat; seeds few, in 2 rows.

1. P. minutum Hara

Rootstock slender, c 1mm thick, branching. Leaves obovate, spathulate, ovate or linear, 1–4 × 1–1.8mm; petioles 3–20mm, glabrous. Scapes 4–25mm, minutely pubescent on one side. Sepals 1.2–2 × 0.8–1mm. Petals obovate-spathulate, 2–4.5 × 1–3mm, white or purplish. Pods ellipsoid, 8–15 × 1–1.5mm; style 1–2mm.

Bhutan: N – Upper Mo Chu (Lingshi), Upper Mangde Chu, Upper Bumthang Chu (Narim Thang) and Upper Kulong Chu (Me La) districts; **Sikkim:** Oma La to Migothang. Moist peat and scree, 3960–4900m. May–August.

2. P. scapiflorum (Hook. f. & Thomson) Marquand & Shaw; *Cochlearia scapiflora* Hook. f. & Thomson, *P. sinense* (Hemsley) Hayek & Handel-Mazzetti, *Braya sinensis* Hemsley.

Similar to *P. minutum* but larger and stouter with tap-root up to 10mm broad, branching above; leaves spathulate or oblong, 1–7 × 0.1–0.6cm, margins entire or with a few shallow teeth; scapes 1–5cm, glabrous; sepals 2–4 × 1–2.5mm; petals 3–6 × 2–4.5mm; pods ellipsoid or suborbicular, 7–10 × 4–8mm, slightly compressed.

Bhutan: C – Ha, Thimphu, Tongsa and Bumthang districts, **N** – Upper Mo Chu to Upper Kulong Chu districts; **Sikkim.** Wet screes and streamsides, 3960–4880m. May–August.

14. PHAEONYCHIUM Schulz

Perennial herbs with thick woody rootstocks. Leaves in a basal rosette, entire, bearing stellately-branched hairs. Flowers several on ebracteate scapes. Sepals erect or somewhat spreading, inner ones not pouched. Petals obovate, narrowing below into a claw. Pods linear-elliptic, compressed; seeds 3–6 in 1 row; style short, thick.

1. P. parryoides (Kurz) Schulz; *Cheiranthus parryoides* Kurz

Rootstocks up to 1.5cm thick, densely covered above with remains of dead leaves. Stems 5–15cm, ± pubescent with whitish hairs. Leaves elliptic, 1–7 × 0.5–2.5cm, acute, base attenuate into petiole 1.5–6cm, ± pubescent. Sepals elliptic, 3–4.5 × 1–1.5mm, obtuse, pubescent. Petals 7–10 × 3–4mm, white or mauve, shading to dark purple in the claw. Pods c 1.5 × 0.2cm, densely greyish pubescent or white-tomentose.

Bhutan: N – Upper Mo Chu district (Cheypechey). Dry cliff-faces, 3655m. June.

15. BARBAREA Brown

Biennial or perennial herbs with erect angular stems, glabrous or with sparse simple hairs. Lower leaves lyrate, pinnatisect; upper ones pinnatifid or nearly entire. Flowers in ebracteate racemes. Inner sepals not pouched at base. Petals yellow, distinctly longer than sepals. Pods linear, 4-angled, valves with prominent midribs; stigma slightly 2-lobed; seeds c 10, unwinged, in a single row.

1. B. intermedia Boreau; *B. vulgaris* Brown var. *sicula* sensu F.B.I. non Presl

Stems 20–45(–60)cm. Lower leaves 6–10cm, with 3–8 pairs of elliptic or obovate lateral segments; upper leaves similar with linear or oblanceolate segments. Sepals ovate-elliptic, c 2.5 × 1mm. Petals obovate, c 4 × 1.5mm, yellow. Pods 20–30 × 1.5mm; style not persistent.

Bhutan: C –Thimphu, Punakha and Tashigang districts, **N** – Upper Mo Chu, Upper Bumthang Chu and Upper Kulong Chu districts; **Sikkim.** Weed around buildings and in open grassy places, 2895–4265m. May–July.

2. B. elata Hook. f. & Thomson

Similar to *B. intermedia* but stems up to 90cm, lower leaves with 1–3 pairs of lateral lobes; upper leaves with 1–3 pairs of lobes, or unlobed and sinuately toothed; pods 15–18 × 3mm; style persistent, 2–3mm.

Bhutan: N – Upper Mo Chu district (Lingshi); **Sikkim.** On sandy ground, 3800m. May.

16. ARABIS L.

Biennial or perennial herbs with simple, medifixed, forked or stellate hairs, sometimes glabrous. Stems erect, often simple. Basal leaves in a rosette, stem leaves sessile, semiamplexicaul. Flowers in bracteate or ebracteate racemes. Sepals erect or spreading, inner pair pouched at base. Petals white or purplish. Pods linear, flat, valves veiny with weak midrib; style short, stigma scarcely 2-lobed; seeds numerous, compressed, ± winged.

1. Annuals or biennials with thin simple rootstocks ....**Species 1 & 2**
\+ Perennials with thick branching rootstocks ....**Species 3 & 4**

1. A. pterosperma Edgeworth; *A. alpina* sensu F.B.I. non L.

Biennial. Stems 20–35cm, appressed-pubescent with forked hairs. Basal leaves rosetted, spathulate, 4–6 × 1–1.5cm, obtuse or subacute, base attenuate, margins shallowly crenate-serrate, ciliate and ± pubescent with simple and forked hairs; stem leaves ovate. Racemes ±dense, ebracteate. Sepals elliptic-oblong, 3–3.5 × 1mm, white-margined. Petals oblanceolate, c 7 × 1.5mm, purplish. Pods erect at first, later spreading, up to 4.5 × 0.15cm, glabrous, on pedicels 8–10mm; seeds orbicular, compressed, c 1.25mm diameter, narrowly winged.

Bhutan: C – Ha and Thimphu districts, **N** – Upper Kulong Chu district; **Sikkim.** Alpine meadows and woodland, 3500–3960m. May–August.

2. A. amplexicaulis Edgeworth

Annual or biennial. Stems erect, 20–60cm. Basal leaves rosetted, oblanceolate or spathulate, 3–9 × 0.5–3cm, subacute or rounded, base attenuate, margins with a few small rounded teeth or sometimes sinuate, sparsely pubescent; stem leaves ovate or oblong, sessile, 2.5–7 × 0.7–3cm, acute or obtuse, base rounded and semiamplexicaul. Racemes lax, ebracteate; pedicels 5–10mm. Sepals ovate, 3–5 × 0.5–1mm. Petals 7–10 × 1.5–3mm, white. Pods 4.5–5 × 0.1cm; seeds unwinged.

Bhutan: C –Tongsa district (near Chendebi). Cool broad-leaved forest, 2550m. April–May.

3. A. axilliflora (Jafri) Hara; *Parryodes axilliflora* Jafri

Perennial; rootstock surrounded by fibrous leaf-remains. Stems 10–20cm, leafless or with 1–3 leaves. Basal leaves elliptic or spathulate, 1–3.5 × 0.6–1.2cm, obtuse or acute, base attenuate into petiole up to 5cm, margins entire or shallowly crenate, sparsely pubescent with simple and forked hairs; stem leaves oblanceolate or spathulate, 1–2.5cm, Racemes bracteate; bracts elliptic, 7–25 × 2–8mm. Sepals oblong-elliptic, c 8 × 3mm. Petals obovate, 8–13 × 4–6mm, purplish. Pods 2.5–3 × 0.15cm, glabrous, on pedicels 8–15mm.

Bhutan: C –Thimphu district (Paro), Punakha district (Tang Chu) and Bumthang district (Kemphu), **N** – Upper Mo Chu district (Naha) and Upper Bumthang Chu district (Pangotang). Rocks and cliff ledges, 3650–4570m. May–June.

4. A. venusta Hara

Similar to *A. axilliflora* but stems 3–10cm with 1–2 leaves; basal leaves ovate-spathulate, 5–20 × 4–10mm, with 1–2 serrate teeth on each side, petiolate, minutely pubescent at margins and nerves; sepals oblong, 2.2–3.2mm; petals spathulate, 5–6 × 2–3.4mm; pods 10–15 × 1.2mm; pedicels 2–9mm.

Bhutan: N – Upper Kulong Chu district (Shingbe, Me La). Cliffs and damp ground, 3960m. June.

17. **NASTURTIUM** Brown

Perennial marshland herbs, glabrous or with few simple hairs. Leaves pinnate. Flowers small, in ebracteate racemes. Inner sepals slightly pouched at base. Petals white, spathulate. Pods linear, ± cylindrical with prominent ridges caused by numerous ovoid seeds.

1. N. officinale Brown; *Rorippa nasturtium-aquaticum* (L.) Hayek. Eng: *Water-cress.*

Stems 10–30cm, erect or spreading, copiously rooting at lower nodes. Leaves 1–12cm; lateral leaflets elliptic, 3–20 × 2–10mm; terminal leaflet ovate-cordate, 5–20 × 5–15mm, acute or obtuse, margins entire or sinuate. Sepals oblong, c 3 × 1mm. Petals obovate, 5 × 2 mm, tapering below to a narrow claw. Pods 10–18 × 1.75–2mm, somewhat upwardly curved, spreading ± horizontally; pedicels 6–12mm.

Sikkim: Darjeeling. Streams and ditches. June.

Eaten as a salad plant.

18. **RORIPPA** Scopoli

Erect branching annuals or perennials, glabrous or with simple hairs. Leaves ± deeply lyrate-pinnatifid. Flowers in bracteate or ebracteate racemes. Inner sepals pouched at base. Petals yellow, ± as long as sepals, sometimes absent. Pods cylindrical or ellipsoid, somewhat oblique, valves with indistinct midrib; style short, stigma rounded, entire; seeds few to numerous in 1–2 rows.

1. R. palustris (L.) Besser; *R. islandica* auct. non (Oeder) Borbas, *Nasturtium palustre* (L.) DC.

Stems 20–30(–60)cm. Leaves 5–10cm; lateral segments oblong-elliptic, 3–5 pairs; terminal segments ovate, obtuse, margins crenately toothed, ±glabrous. Racemes ebracteate. Sepals c 2mm. Pods 5–6 × 2mm, ± as long as pedicels; seeds in 2 rows.

Bhutan: **S** – Chukka district, **C** – Ha, Thimphu and Punakha districts. Fields and ditches, 1400–3350m. June.

2. R. madagascariensis (DC.) Hara; *Nasturtium indicum* sensu F.B.I. p.p. non (L.) DC.

Similar to *R. palustris* but leaves sparsely hairy, deeply pinnatifid, auriculate-amplexicaul at base, margins often more acutely toothed; pods 8–15 × 1.5–2.5mm, slightly upcurved; pedicels 2–4mm.

Bhutan: **S** – Phuntsholing district (Torsa River), **C** – Thimphu district (Thimphu). Dry shingle and tracksides, 200–2400m. February–May.

3. R. benghalensis (DC.) Hara; *Nasturtium indicum* (L.) DC. var. *benghalense* (DC.) Hook. f. & Anderson, *R. dubia* (Persoon) Hara var. *benghalensis* (DC.) Mukerjee

Similar to *R. palustris* and especially to *R. madagascariensis* but racemes bracteate; bracts 5–20mm, entire or minutely toothed; pods 6–17 × 1–1.5mm on pedicels 3–7mm.

Bhutan: **S** – Phuntsholing and Gaylegphug districts; **Sikkim.** 270–600m. May.

4. R. indica (L.) Hiern; *Nasturtium indicum* sensu F.B.I. p.p. non (L.) DC., *N. montanum* Hook. f. & Anderson

Similar to *R. palustris* but glabrous; lower leaves with 1–2 pairs of lateral segments, upper ones entire; pods 12–25 × 1–1.5mm, on pedicels 4–8mm.

Sikkim: Kurseong (95).

5. R. dubia (Persoon) Hara; *Nasturtium indicum* sensu F.B.I. p.p. non (L.) DC.

Similar to *R. palustris* and *R. indica* but annual, more slender, glabrous; flowers often lacking petals; pods thin, straight, 12–25 × 0.6–0.8mm, on pedicels 4–8mm; seeds in 1 row.

Sikkim: Gangtok. 1650m. April–June.

19. LEPIDOSTEMON Hook. f. & Thomson

Small annuals bearing branched hairs. Stems erect, simple. Leaves oblanceolate, usually toothed. Flowers in ebracteate corymbose racemes. Sepals spreading, inner ones somewhat pouched at base. Petals yellow or purplish, obovate, emarginate, clawed at base. Filaments (at least of longer stamens) broadened into a tooth at the apex on both sides. Pods linear, covered in branched hairs; style short, stigma globose or slightly 2-lobed, broader than style; seeds compressed, 5–6 in 1 row.

1. L. pedunculosus Hook. f. & Thomson

Stems 2–10cm. Leaves oblanceolate, 5–15 × 2–3mm, entire or with 2–3 teeth. Sepals c 2.5 × 1.25mm. Petals 3.5–5 × 1.5–3mm, yellow or purplish. Pods (immature), up to 6 × 1mm; style c 0.5mm.

Bhutan: **C** – Ha district (Ya La) and Tongsa district (Kemphu); **Sikkim.** 4570m. June.

20. DONTOSTEMON Ledebour

Caespitose or subprostrate, glandular annuals; hairs simple. Leaves pinnatifid. Flowers in ebracteate racemes, similar to those of *Arabis* but filaments flattened below and those of longer stamens sometimes connate in pairs in lower half. Petals pink or white. Pods linear, valves bearing minute glandular warts; seeds in 1 row; style very short; stigma broad, rounded.

1. D. glandulosus (Karelin & Kirilow) Schulz; *Arabis glandulosa* Karelin & Kirilow

Stems up to 12cm. Leaves elliptic-lanceolate in outline, up to 3cm, with narrow, linear segments c 2–3mm broad. Sepals oblong, 2 × 0.5mm, bearing long simple hairs. Petals pink or white, 2.5–3 × 1mm. Pods c 15 × 0.75mm.

Bhutan: **N**–Upper Mo Chu district (between Chamsa and Yabu Thang); **Sikkim:** Chugya and Kalaeree. 3300–4880m. May.

21. PARRYA Brown

Perennials with thick branched rhizomes; hairs simple, glandular or eglandular. Leaves all basal, linear or spathulate, entire, distantly dentate or shallowly pinnatifid. Flowers purple or white, solitary or several in ebracteate scapose racemes. Sepals erect, inner pair somewhat pouched at base. Pods flattened, valves smooth, midrib and veins prominent; style short, stigma decurrent; seeds compressed, winged or wingless.

1. P. nudicaulis (L.) Boissier; *P. scapigera* (Adams) G. Don, *P. macrocarpa* sensu F.B.I. non Brown, *Neuroloma nudicaulis* (L.) DC. Fig. 33g.

Rootstock densely tufted and surrounded by leaf remains. Leaves oblanceolate or spathulate, 3.5–18 × 0.3–1cm, acute or subacute, base attenuate, margins sinuate or weakly pinnatifid, shortly and sparsely glandular-pubescent; petioles up to 6cm, sheathing at base. Racemes 5–15-flowered, on scapes 6–30cm, glandular-pubescent. Sepals oblong, 8–10 × 2–3.5mm. Petals purplish or white below, obovate, 8–10 × 5–8mm, emarginate, claw 10–13mm. Pods linear, 6–10 × 0.5–0.6cm, acute; seeds ovate, c 5 × 3mm, wing 1–3mm.

Bhutan: **C**–Thimphu district (Barshong), **N**–Upper Mo Chu district (Gyengothang); **Sikkim.** Rock crevices and boulders, 3950m. May.

2. P. platycarpa Hook. f. & Thomson

Similar to *P. nudicaulis* but leaves ovate-elliptic, 1.5–2.5 × 0.3–0.6cm, acute, margins entire, ciliate, otherwise glabrous, base attenuate to sheathing petiole; scapes c 7cm; pods ovate-elliptic, 2–3 × 1.2–1.5cm; seeds unwinged.

Sikkim: Lhonak. 4880–5100m.

22. ERYSIMUM L.

Annuals or perennials with appressed 2–3-branched, medifixed hairs. Stems leafy; leaves simple, attenuate at base, not auriculate. Flowers in corymbose racemes, ebracteate, or breacteate only at base. Sepals erect, the inner ones ± pouched at base. Petals clawed, yellow or orange. Pods linear, valves rounded or keeled; stigma capitate, ± bilobed; seeds numerous in 1 row.

1. Dwarf perennials, stems up to 12cm; pods 1.5 – 5cm. ....**Species 1 & 2**
+ Taller annuals or biennials, stems 25 – 30cm or more; pods 5 – 11cm **Species 3 – 5**

1. E. deflexum Hook. f. & Thomson

Stems decumbent, 7 – 12cm, ascending at apex; rootstock thick, woody. Leaves linear-spathulate, 10 – 30 × 1.5 – 3mm, obscurely toothed, appressed-pubescent, hairs 2-branched. Flowers c 1cm diameter, orange-yellow. Pods 3 – 5cm; seeds ellipsoid c 1.5 × 0.7mm, funicles c 1mm.

Sikkim: Kongra Lama and Dongkong. 3950 – 5500m.

2. E. funiculosum Hook. f. & Thomson

Stems erect, c 5cm. Leaves oblanceolate, 20 – 22 × 2 – 3mm, entire, bearing 2-branched hairs. Flowers as in *E. deflexum*. Pods 15 – 20 × 2 – 2.5mm, valves ± angular; seeds ellipsoid c 2 × 0.7mm, funicles 2 – 3mm.

Sikkim: Kongra Lama. 4265 – 4570m.

3. E. hieracifolium L. agg.; *E. bhutanicum* W. W. Smith. Med: *Gongthokpa.*

Stems up to 75cm. Leaves oblanceolate, 3 – 10 × 0.5 – 1.5cm, acute, base attenuate, bearing 3 – 5 small teeth on either side, sparsely pubescent mostly with 2-branched hairs (a few 3-branched ones may be present near margin). Sepals narrowly ovate or oblong, 5 – 8mm. Petals orange, claw 5 – 9mm, limb oblong or obovate, 5 – 7 × 2 – 5mm, rounded or emarginate. Pods 5 – 8cm.

Bhutan: C – Thimphu, Punakha and Bumthang districts, **N** – Upper Mo Chu and Upper Kulong Chu districts. River banks and hillsides, 3650 – 4100m. May – August.

A widespread and variable aggregate species; further research is necessary if it is to be subdivided in Asia as has been done in Europe.

4. E. pachycarpum Hook. f. & Thomson

Stems robust 45 – 60cm; hairs scattered, 2 – 3-branched. Leaves oblanceolate, 7 – 9 × 0.75 – 1.5cm, sinuate-dentate. Flowers orange-yellow, c 8mm diameter, on very short pedicels 3 – 6mm. Pods 5 – 7 × 0.25 – 0.3cm, valves obtusely keeled.
Bhutan: N – Upper Mo Chu district (Lingshi); **Sikkim**: Lachen Valley and Tungu. 3050 – 3950m.

5. E. longisiliquum Hook. f. & Thomson

Tall, erect annual. Stems robust, 30 – 50cm, acutely angled; hairs 2 – 3-branched. Leaves oblanceolate, obscurely toothed. Pods 10 – 11 × 0.15cm.

Bhutan: N – Upper Mo Chu district (Laya) and Upper Kulong Chu district (Me La); **Sikkim**: Tungu. 3050 – 3950m. May – June.

23. PYCNOPLINTHOPSIS Jafri

Perennial herbs with thick rhizomes. Leaves all basal, in a rosette, simple, glabrous or bearing branched hairs. Flowers solitary on short scapes. Outer sepals ± pouched at base. Petals spathulate, white or pink, longer than sepals. Pods linear, oblong or elliptic, compressed, often curved at apex; seeds 5–7 in 1–2 rows.

1. P. bhutanica Jafri; *P. minor* Jafri, *Pegaeophyton bhutanicum* Hara

Basal leaves spathulate or oblanceolate, 1.2–6 × 0.4–1.5cm, apex rounded or subacute with (5–)9(–15) teeth, base attenuate, usually sparsely hairy. Scapes 2–3cm, glabrous. Sepals oblong-elliptic, (2–)5–7 × (1.8–)2.5–3.5mm. Petals obovate, 10–15 × 5–8mm, white or purplish. Pods linear-oblong, up to 15 × 2–2.5mm; style 1–1.5mm.

Bhutan: C–Thimphu and Punakha districts, **N**–Upper Mo Chu to Upper Kulong Chu districts; **Sikkim**. Wet cliff faces and screes, 3200–4570m. May–July.

24. EUTREMA Brown

Perennials with leafy stems, glabrous or with simple hairs. Leaves simple, basal ones petiolate, stem leaves sessile or petiolate. Racemes ebracteate. Sepals somewhat spreading, rather broad, rounded, not saccate at base. Petals white or purple, obovate, shortly clawed at base. Ovary with 3–10 ovules. Pods short, oblong-lanceolate or elliptic, somewhat compressed, valves ± keeled at midrib, septum often incomplete; style short; stigma capitate, somewhat bilobed; seeds few, large, in 1 row.

1. Stems (15–)30–50cm .... **Species 1 & 2**
\+ Stems 5–12cm .... **3. E. heterophyllum**

1. E. deltoideum (Hook. f. & Thomson) Schulz; *Sisymbrium deltoideum* Hook. f. & Thomson

Leaves ovate-deltoid, 3–8 × 1.5–6cm, acute or acuminate, base cordate, margins entire or crenate near base, glabrous; petioles up to 12cm. Sepals broadly elliptic, 1.5–2 × 1mm. Petals 6–7 × 2.5–2.75mm, white or purplish. Pods 10–12 × 2–3mm, somewhat curved; seeds c 3.

Bhutan: C – Ha, Tongsa and Mongar districts, **N** – Upper Mo Chu and Upper Kuru Chu districts; **Sikkim.** Streamsides and swampy ground, 3650–4570m. June–August.

2. E. himalaicum Hook. f. & Thomson; ? *E. lanceifolium* (Franchet) Schulz

Similar to *E. deltoideum* but leaves ovate-lanceolate, 3–7 × 1.5–4cm, subacute, base rounded; lower ones shortly (c 3cm) petiolate; upper ones sessile, semiamplexicaul; sepals suborbicular, 2.5–3 × 1.5–2mm; petals white, c 4.5 × 2mm; pods elliptic, c 15 × 2mm; seeds 3–6.

Bhutan: **C** –Thimphu district, **N** – Upper Mo Chu, Upper Mangde Chu and Upper Kulong Chu districts. Meadows and streamsides, 3350 – 3960m. May – July.

3. E. heterophyllum (W. W. Smith) Hara; *Braya heterophylla* W. W. Smith

Basal leaves ovate, 7 – 10 × 3 – 6mm, acute, base rounded; petioles up to 15mm; upper leaves lanceolate, 8 – 15 × 3 – 5mm, acute, base rounded, amplexicaul, entire, glabrous. Flowers in short dense corymbose racemes. Sepals ovate-elliptic, c 2.5 × 1.5mm. Petals white, c 4 × 2mm. Pods narrowly ovoid, c 6 – 10 × 2mm, 4-seeded.

Bhutan: **N** – Upper Mo Chu (Lingshi), Upper Mangde Chu (Changkhab) and Upper Bumthang Chu (Waitang) districts. 4400 – 4730m. May – June.

25. MICROSISYMBRIUM Schulz

Slender branching annuals with simple and branched hairs. Leaves simple, toothed, petiolate or sessile. Flowers in bracteate racemes. Sepals suberect, not pouched. Petals narrowly spathulate, rounded. Filaments filiform. Ovary cylindrical, ovules 10 – 42. Pods linear on slender recurved pedicels, valves weakly nerved; style short, slender; stigma minute; seeds numerous in 1 row.

1. M. axillare (Hook. f. & Thomson) Schulz; *Sisymbrium axillare* Hook. f. & Thomson

Lower branches spreading, 6 – 10cm, pubescent with simple and branched hairs. Lower leaves ± rosulate, obovate or spathulate, 2.5 – 4 × 0.75 – 1.5cm, acute or obtuse, margins with up to 6 pairs of teeth. Racemes 6 – 10-flowered, bracts leaf-like, pedicels 2 – 6mm. Sepals oblong, c 2 – 3mm. Petals white, 3 – 5mm. Pods linear, 25 – 30 × 0.75mm, glabrous; style very short.

Bhutan: **C** – Punakha district (below Norbding); **Sikkim.** Roadsides in Evergreen oak forest, 2450 – 5000m. April

The above description and records refer to the typical var. **axillare.**

var. **dasycarpum** Schulz

More densely hairy; pods covered in forked hairs.

Bhutan: **C** –Thimphu district (Chapcha) and Tongsa district (Tongsa). Walls and roadsides, 2300m. April – May.

26. ARCYOSPERMA Schulz

Dwarf perennials with thick tap-roots. Basal leaves rosulate, glabrous or with simple hairs. Racemes ebracteate. Sepals somewhat spreading, not pouched at base. Petals obovate, emarginate at apex, narrowed and shortly clawed at base. Ovary cylindrical, ovules c 35. Pods linear, somewhat curved, valves thin, unveined; seeds few to numerous in 2 rows.

1. A. primulifolium (Thomson) Schulz; *Eutrema primulifolium* (Thomson) Hook. f. & Thomson

Stems 4–10cm, slender. Basal leaves oblanceolate, 4–12 × 1.5–3cm, acute or obtuse, base attenuate, margins sinuous or denticulate, sparsely pilose with simple hairs; stem-leaves oblong-elliptic, up to 3 × 1cm. Sepals ovate-oblong, c 3× 1mm. petals white or purplish, c 4–5 × 1mm. Pods 10–20 × 1.5–2mm.

Bhutan: C–Thimphu district (Drugye Dzong and Barshong). Screes and rocky ground, 3200–3400m. May.

27. TORULARIA (Cosson) Schulz

Small annual or perennial herbs with shortly stalked 2-branched and simple hairs. Leaves pinnatifid or toothed, rarely entire. Racemes bracteate. Sepals spreading, inner pair not pouched at base. Petals spathulate or obovate. Pods linear, cylindrical, ± constricted between seeds; style short; seeds c 12, in 1 row.

1. T. humilis (Meyer) Schulz; *Sisymbrium humile* Meyer. Fig. 33h.

Stems erect, 1–20cm. Leaves elliptic or oblanceolate, 1.5–4 × 0.2–0.7cm, acute or obtuse, base attenuate or shortly petiolate, margins with up to 4 rounded teeth on either side, greyish-pubescent. Sepals elliptic, 1.5–3 × 0.7–1.25mm, obtuse. Petals purplish, 3.5–6 × 1.25–3mm. Pods 1.7 × 0.15cm, appressed-pubescent.

Bhutan: C–Ha and Thimphu districts, **N**–Upper Mo Chu, Upper Mangde Chu, Upper Bumthang Chu and Upper Kulong Chu districts; **Sikkim.** Rocky slopes and screes, 3650–4200m. May–June.

28. BRAYA Sternberg & Hoppe

Dwarf caespitose perennials bearing 2-branched and simple hairs, rarely with only simple hairs. Stems erect, often scapose. Leaves simple, oblanceolate, entire or with a few teeth, basal ones usually in a rosette. Flowers in dense bracteate corymbose racemes becoming elongated in fruit. Sepals oblong, not pouched at base, often tinged with purple, margins hyaline. Petals obovate, white or purplish. Pods ovoid or subcylindric; style slender; seeds few, in 2 rows.

1. B. tibetica Hook. f. & Thomson

Stems 3–5cm, leafless or sometimes bearing a single leaf, hairs mostly forked. Leaves narrowly oblanceolate, 1–1.5 × 0.1–0.2cm, acute, base attenuate, margins entire or with 1–2 teeth on either side, hairs mostly forked. Inflorescence bracteate. Sepals c 2 × 1mm, sparsely pubescent with forked and simple hairs. Petals c 3 × 1mm, purplish. Pods c 4 × 1.5mm, attenuate at apex into a thin style, sparsely pubescent with forked hairs.

Bhutan: N – Upper Mo Chu district (Laycho and Yale La); **Chumbi**: Chomolhari. 4500 – 4880m. May – July.

The Sikkim record of *B. rosea* (Turczaninow) Bunge possibly refers to this species. The two are very similar but *B. rosea*, originally from the Altai Mountains, seems best distinguised by having an ebracteate inflorescence and pods rounded at apex and bearing a thick style.

2. B. oxycarpa Hook. f. & Thomson; *Aphragmus oxycarpus* (Hook. f. & Thomson) Jafri

Similar to *B. tibetica* but leaves always entire, glabrous, stems bearing minute, deflexed, simple hairs; flowers white or mauve; pods glabrous, c 10 × 2mm.

Bhutan: C – Thimphu district, **N** – Upper Mo Chu, Upper Bumthang Chu and Upper Kulong Chu districts; **Sikkim**. Screes and open peaty slopes, 4100 – 4600m. May – June.

3. B. forrestii W. W. Smith

Similar to *B. tibetica* but densely greyish-pubescent with simple hairs; leaves oblanceolate, 1 – 1.5 × 0.15 – 0.2cm, entire; flowers somewhat larger, sepals c 2.5 × 1.5mm, petals oblanceolate c 4 × 2mm, white to mauve.

Bhutan: C – Thimphu district (Gyengothang). In dry peaty turf, 3960m. May.

29. CHRYSOBRAYA Hara

Similar to *Braya* but leaves obovate with 3 – 5 coarse teeth; flowers apparently solitary on naked pedicels, later racemose by growth of inflorescence axis; petals bright yellow; pods linear-cylindric covered with branched hairs; seeds c 5 in 1 row.

1. C. glaricola Hara

Plants 1.5 – 4cm. Leaves oblanceolate, 7 – 20 × 1.5 – 4mm, subacute, base attenuate, margins entire or 3 – 5-toothed near apex. Pedicels 7 – 20mm. Sepals oblong, 2 – 2.5mm, persistent in fruit. Petals obovate, 3 – 6mm. Pods 7 – 11 × 1.5 – 2mm, densely stellate-pubescent.

Bhutan: C – Thimphu district, **N** – Upper Mo Chu, Upper Mangde Chu, Upper Bumthang Chu and Upper Kulong Chu districts. Rock crevices and screes, 3960 – 4730m. May – July.

30. ARABIDOPSIS Heynhold

Annuals or biennials bearing simple and branched hairs. Leaves oblong; basal ones petiolate; upper sessile, narrowed at base or auriculate. Racemes usually bracteate, at least at base. Sepals somewhat spreading, inner pair not pouched. Petals obovate or spathulate, rounded, pink, white or purple. Ovary terete, ovules

20–72. Pods linear, valves weakly nerved; style short, stigma capitate; seeds few to numerous in 1–2 rows.

1. Stem-leaves semiamplexicaul or auriculate at base ....**Species 1 & 2**
+ Stem-leaves narrowed at base, not auriculate...................**3. A. lasiocarpa**

1. A. himalaica (Edgeworth) Schulz; *Sisymbrium himalaicum* (Edgeworth) Hook. f. & Thomson

Stems ± erect, 10–40cm, ± hirsute, branching from base. Leaves oblong or spathulate, 0.7–5 × 0.3–1.5cm, acute or obtuse, margins shallowly pinnatifid; auricles of stem-leaves acute. Racemes bracteate to apex, bracts leaf-like. Sepals ovate-oblong, c 2.5 × 1mm. Petals pink, c 4 × 1.5mm. Pods 20–30 × 1mm, glabrous; style c 0.25mm.

Bhutan: C–Mongar district (Ghijamchu), **N**–Upper Mo Chu, Upper Bumthang Chu and Upper Kulong Chu districts; **Sikkim**. Rocky hillsides and open Fir forests, 3450–4500m. May–July.

2. A. mollissima (Meyer) Schulz; *Sisymbrium mollissimum* Meyer

Similar to *A. himalaica* but leaves entire or shallowly toothed; racemes bracteate only at base; flowers white or purplish.

Bhutan: C–Thimphu district, **N**–Upper Mo Chu district; **Chumbi.** 3400–4000m. May.

3. A. lasiocarpa (Hook. f. & Thomson) Schulz; *Sisymbrium lasiocarpum* Hook. f. & Thomson non F. Mueller, *S. bhutanicum* Balakrishnan

Greyish pubescent throughout. Stems erect, c 25cm, branched from base. Basal leaves in a rosette, spathulate, sinuate-pinnatifid, 1.5–2 × 0.4–0.6cm, obtuse, base attenuate; stem-leaves ovate or elliptic-oblong, entire or sinuate. Racemes bracteate at base. Sepals oblong, c 2 × 0.75mm. Petals white or purplish, c 3.5 × 1mm. Pods terete, 1.5–2 × 0.15cm, pubescent.

Bhutan: C–Ha and Thimphu districts. Roadsides and fields, 2450–2750m. April–May.

31. CHRISTOLEA Cambessedes

Rhizomatous perennial herbs bearing only simple hairs. Leaves obovate, toothed or lobed at apex. Racemes usually many-flowered, bracts linear. Sepals oblong, obtuse, inner pair scarcely pouched. Petals spathulate or obovate, purplish. Pods oblong, cylindric, pilose; seeds 12–15 in 1 row.

1. C. himalayensis (Cambessedes) Jafri; *Cheiranthus himalayensis* Cambessedes, *Ermania himalayensis* (Cambessedes) Schulz

Densely greyish pilose throughout. Stems ± erect, 3–10cm, leafy especially near base. Leaves obovate, 7–15 × 3–5mm, 3–5-toothed at apex, base cuneate. Sepals usually tinged dark purple, c 2.5 × 1mm. Petals purplish, c 4 × 1.75mm, distinctly veined. Ovary c 4mm at flowering time, increasing in size rapidly in fruit. Pods c 4cm; stigma stout, bilobed.

Sikkim: Lugnak La and between Donkia and Samdong. 4000–5000m. July.

32. DESCURAINIA Webb & Berthelot

Erect annuals, pubescent with branched and simple hairs. Leaves finely pinnatiscct. Racemes ebracteate. Sepals yellowish, not pouched, usually soon deciduous. Petals yellow, spathulate, shorter than sepals. Pods linear, cylindrical with very short thick style; seeds c 20 in 1 row.

1. D. sophia (L.) Prantl; *Sisymbrium sophia* L.

Stems 30–75cm, sparsely pubescent. Leaves bi- or tripinnatisect, elliptic in outline, 5–8(–12) × 2–5(–7)cm, ultimate segments linear or elliptic, 1–2mm broad, pubescent. Racemes elongate, ultimately 20cm or more long. Sepals oblong, c 2.5 × 0.5mm. Petals c 2mm. Pods 2–2.5 × 0.1cm, glabrous.

Sikkim: Chamgong. 4570m.

Family 64. MORINGACEAE

by A. J. C. Grierson & D. G. Long

Small trees with soft wood and resinous bark. Leaves alternate, 2–4-pinnate, base of petiolules and pinnae articulated and with stalked glands; leaflets opposite, entire; stipules knob-like, deciduous. Flowers in axillary panicles, bisexual, zygomorphic. Calyx a short tube with 5 reflexed petal-like lobes. Petals 5, free, unequal, upper erect, others reflexed. Stamens 5, epipetalous, alternating with 5 staminodes. Ovary superior, 1-celled, ovules 10–20 on 3 parietal placentae; style filiform with minute stigma. Capsule linear, triquetrous; seeds 3-winged, in cavities in spongy valves of capsule.

1. MORINGA Adanson

Description as for Moringaceae.

1. M. oleifera Lamarck; *M. pterygosperma* Gaertner. Eng: *Horse-radish Tree.* Fig. 33r.

Tree to 10m. Leaves 15–30 (–60)cm; leaflets ovate or obovate, 1–2 × 0.5–1.5cm, obtuse, base rounded. Panicles erect, 20–30cm; flowers white, fragrant.

Calyx-tube 3–4mm; sepals and petals spathulate, 1–1.4cm. Capsules pendent, 15–30 (–45)cm; seeds ovoid, c 1cm with 3 hyaline wings up to 7mm broad.

Bhutan: S – Phuntsholing district (Phuntsholing); **Sikkim.** Cultivated near habitations, 230m. January–February.

Native of NW India. Leaves and roots used medicinally and young fruit eaten as a vegetable (34); seeds yield 'Ben Oil' used by watchmakers (16); young roots make a good substitute for the true Horse-radish (16).

INDEX OF COMMON NAMES